Industry 4.0 from the MIS Perspective

Sevinç Gülseçen / Zerrin Ayvaz Reis /
Murat Gezer / Çiğdem Erol (eds.)

Industry 4.0
from the MIS Perspective

PETER LANG

Bibliographic Information published by the Deutsche Nationalbibliothek
The Deutsche Nationalbibliothek lists this publication in the Deutsche
Nationalbibliografie; detailed bibliographic data is available online at
http://dnb.d-nb.de.

Library of Congress Cataloging-in-Publication Data
A CIP catalog record for this book has been applied for at the Library of Congress.

Printed by CPI books GmbH, Leck

Cover illustration: iStock.com/Besjunior

ISBN 978-3-631-75768-0 (Print)
E-ISBN 978-3-631-75769-7 (E-PDF)
E-ISBN 978-3-631-75770-3 (EPUB)
E-ISBN 978-3-631-75771-0 (MOBI)
DOI 10.3726/b15120

© Peter Lang GmbH
Internationaler Verlag der Wissenschaften
Berlin 2019
All rights reserved.

Peter Lang – Berlin · Bern · Bruxelles · New York · Oxford · Warszawa · Wien

This publication has been peer reviewed.

www.peterlang.com

Preface

Information and communication technology has been particularly influential in new directions of society. There is a hypothesis that societies undergo three phases: archaic, socio-semiotic and techno-semiotic. In archaic phase, people and communities live still, nature-bound lives. The socio-semiotic phase was the avenue chosen by the western world in the emergence of modernity leading to industrialization and democratic institutions. In techno-semiotic phase, the society is under the influence of modern electronic and information technology, especially internet and the "extase" of communication, and this changed the previous structures of social organization radically. In this third phase, an end-to-end industrial transformation called Industry 4.0 sets new goals for manufacturing and impacts on business outcomes.

Information is not worth much if it does not serve a purpose. What the purpose of information for an organization is or how businesses use information to improve the company's operations is in the scope of Management Information Systems (MIS). As an integrative field, MIS focuses on both business processes and information technology. Waves of innovation spread the fundamental virtues of coherent information systems across all corporate functions and to all sizes of businesses in the 1970s, 1980s and the 1990s. Within companies, major functional areas developed their own MIS capabilities; often these were not yet connected: engineering, manufacturing and inventory systems developed side by side sometimes running on specialized hardware.

With some of its characteristic elements such as IoT (Internet of Things), digital twin simulation models, cyber-physical systems, advanced robots, big data analytics, cloud platform and virtual/augmented reality, Industry 4.0 originally was only used for manufacturing; it is de facto going further. Today, we clearly see how the several parties which were involved in Industry 4.0 themselves move it to smart transportation and logistics, smart buildings, smart healthcare and even smart cities.

This book aims to provide relevant theoretical frameworks and the latest empirical research findings in the area of MIS with the scope of Industry 4.0. It is written by and for professionals who want to improve the general understanding of the strategic role of Industry 4.0 in the distributed business environment and the necessity to protect and properly utilize the key elements of Industry 4.0 at different levels of organizations as well as in society.

The book has originated in Turkey and it has an exemplary organization. In 24 chapters, the authors examine the issues raised by the emerging Industry 4.0 technologies which are relevant for MIS and their application in SMEs, especially in Turkey.

The target audience of this book will be professionals in SMEs and researchers working in the field of information technology, MIS, informatics, computer science, information and communication sciences.

On behalf of the editors' board
Prof. Dr. Sevinç Gülseçen

Contents

List of Contributors

Şebnem Akal
Marmara University, İstanbul, Turkey, sebnemakal@marmara.edu.tr
Musab Talha Akpınar
takpinar@ybu.edu.tr
Mehmet Ali Akyol
Middle East Technical University, Ankara, Turkey, aliakyol@metu.edu.tr
Ourania Areta
oareta@ybu.edu.tr
Halil Arslan
Cumhuriyet University, Sivas, Turkey, harslan@cumhuriyet.edu.tr
İdil Atasu
Bogazici University, İstanbul, Turkey, idil.atasu@boun.edu.tr,
Doğan Aydın
Bahcesehir University, İstanbul, Turkey, dogan.aydin@vs.bau.edu.tr
Mehmet N. Aydin
Kadir Has University, Istanbul, Turkey, mehmet.aydin@khas.edu.tr
Bertan Badur
Boğaziçi University, İstanbul, Turkey, badur@boun.edu.tr
Ahmet Cihat Baktir
Bogazici University, Istanbul, Turkey, cihat.baktir@boun.edu.tr
Dr. Türksel Kaya Bensghir
Todaie, Ankara, Turkey, tbensghir@gmail.com
Aysun Bozanta
Boğaziçi University, İstanbul, Turkey, aysun.bozanta@boun.edu.tr
M. Hanefi Calp
Karadeniz Technical University, Trabzon, Turkey, hcalp25@hotmail.com
Gülşah Çifçi
Erciyes University, Kayseri, Turkey, gulsahhcifci@gmail.com
Cihan Çılgın
Dokuz Eylül University, İzmir, Turkey, cihan.cilgin@deu.edu.tr
Halil Cicibaş
Middle East Technical University, Ankara, Turkey, halil.cicibas@metu.edu.tr
Abide Coşkun Setirek
Boğaziçi University, İstanbul, Turkey, abide.coskunr@boun.edu.tr
Kadir Alpaslan Demir

Turkish Naval Research Center Command, Istanbul, Turkey, kadiralpaslandemir@gmail.com
Fatma Seray Demirkan
Boğaziçi University, İstanbul, Turkey, seray.demirkan@boun.edu.tr
Ebru Dilan
Kadir Has University, Istanbul, Turkey, ebru.dilan@khas.edu.tr
Ahmet Doğan
Osmaniye Korkut Ata University, Osmaniye, Turkey, ahmetdogan@osmaniye.edu.tr
Çağrı Doğu
Energy Holding A.S., Kavacik Meydani, Energy Plaza Kat: 8, 34805, Beykoz, Istanbul, Turkey
Yasar University, Dep. of Comp. Eng., Izmir, Turkey, cdogu@enerjeo.com
Gülay Ekren
Sinop University, Sinop, Turkey, gekren@sinop.edu.tr
M. Lemi Elyakan
Cumhuriyet University, Sivas, Turkey, mlelyakan@gmail.com
Tuncay Ercan
Yasar University, Dep. of Comp. Eng., Izmir, Turkey, tuncay.ercan@yasar.edu.tr
Alptekin Erkollar
Sakarya University, Sakarya, Turkey, erkollar@sakarya.edu.tr
P. Erhan Eren
Middle East Technical University, Ankara, Turkey, ereren@metu.edu.tr
Ebru Gökalp
METU, Ankara, Turkey, egokalp@metu.edu.tr
Mert Onuralp Gökalp
Middle East Technical University, Ankara, Turkey, gmert@metu.edu.tr
Sevinç Gülseçen
Istanbul University, İstanbul, Turkey, gulsecen@istanbul.edu.tr
Abdulkadir Hızıroğlu
abdulkadir.hiziroglu@asbu.edu.tr
Dilan Özcan Kalfa
dokalfa@ybu.edu.tr
Emre Karagöz
Dokuz Eylül University, İzmir, Turkey, emre.karagoz@deu.edu.tr
Umut Kaya
Kavram Vocational High School, İstanbul, Turkey, umut.kaya@kavram.edu.tr
Kerem Kayabay
Middle East Technical University, Ankara, Turkey, kayabay@metu.edu.tr

Ugur Keles
Istanbul University, İstanbul, Turkey, ugur.keles@ogr.iu.edu.tr
Kadir Keskin
Kavram Vocational High School, İstanbul, Turkey, kadir.keskin@kavram.edu.tr
Tuğba Koç
Sakarya University, Sakarya, Turkey, tcekici@sakarya.edu.tr
Altan Koçyiğit
Middle East Technical University, Ankara, Turkey, kocyigit@metu.edu.tr
Birgul Kutlu
Bogazici University, İstanbul, Turkey, birgul.kutlu@boun.edu.tr
Sona Mardikyan
Boğaziçi University, İstanbul, Turkey, mardikya@boun.edu.tr
Bilgin Metin
Bogazici University, Istanbul, Turkey, bilgin.metin@boun.edu.tr
Birgit Oberer
Sakarya University, Sakarya, Turkey, oberer@sakarya.edu.tr
Meltem Ozturan
Bogazici University, İstanbul, Turkey, ozturanm@boun.edu.tr,
Zümrüt Ecevit Satı
Istanbul University, İstanbul, Turkey, zsati@istanbul.edu.tr
Umut Şener
METU, Ankara, Turkey, sumut@metu.edu.tr
Vahap Tecim
Dokuz Eylül University, İzmir, Turkey, vahap.tecim@deu.edu.tr
Aslıhan Tüfekci
Gazi University, Ankara, Turkey, asli@gazi.edu.tr
Dr. Ufuk Türen
Turkish Military Academy, Ankara, Turkey, uturen2011@gmail.com
Ceyda Ünal
Dokuz Eylül University, İzmir, Turkey, ceyda.unal@deu.edu.tr
Atınç Yılmaz
Beykent University, İstanbul, Turkey, atincyilmaz@beykent.edu.tr
Dr. Yücel Yılmaz
Marmara University, İstanbul, Turkey, yucelyilmaz@marmara.edu.tr
Ahmet Gürkan Yüksek
Cumhuriyet University, Sivas, Turkey, agyuksek@cumhuriyet.edu.tr

İdil Atasu[*], Meltem Özturan and Birgül Kutlu

A Roadmap for Turkey's Industry 4.0 Transformation Based on Germany's Strategy

1. Introduction

Since the end of the 18th century, the world has gone through three distinct industrial revolutions triggered by technology. The first one was around 1784 following the introduction of water-/steam-powered mechanical manufacturing facilities. The second one was following the introduction of electrically powered mass production based on the division of labor in the 1870s. The third one started in the early 1970s and has since continued to the present, which involved the employment of electronics and information technology (IT) in order to achieve increased automation of manufacturing processes. The machines have since then taken over most of the manual labor and some of the brainwork (Kagermann, Helbig, Hellinger & Wahlster, 2013). Thus, whereas up until and including the second industrial revolution the production was labor intensive, with the third industrial revolution a shift has begun toward capital-intensive production. When we look at the previous industrial revolutions, the commonality among them is that they have been ignited by technology improvements that have appeared through mechanization, electricity and IT.

Nowadays, what is capturing the industrial world is the advancement of the fourth industrial revolution, or Industry 4.0. The term was first used in Germany at the Hannover Fair in 2011 (Vogel-Heuser & Hess, 2016). In North America, similar ideas are gathered around the term "Industrial Internet", which is used to refer to a broader concept than industrial production (Drath & Horch, 2014).

Manufacturing needs to shift from mass production to mass customization, and it needs to move toward lack of inventory and instead create "on demand" products close to the centers of demand. This would require optimizing the capital used, and the fourth industrial revolution is shaped around this idea (Dujin & Geissler, 2016).

[*] Corresponding author: Idil Atasu, idil.atasu@boun.edu.tr, Bogazici University, MIS Department.

The fourth industrial revolution has been occurring as a result of the incorporation of the Internet of Things (IoT) into the manufacturing environment (Kagermann, Helbig, Hellinger & Wahlster, 2013). The IoT is basically the connection of objects to the Internet instead of human beings. These objects exchange data through embedded sensors. The application of the IoT into the factory environment will be via cyber physical structures. A Cyber Physical Structure (CPS) is the interaction between humans and cyber physical systems. Since a CPS is made up of a physical and virtual, digital component, the interaction between the human and the CPS occurs by either manipulating the physical component directly or via a user interface where the virtual component is managed by humans (Gorecky, Schmit, Loskyll & Zuhlke, 2014). In a manufacturing environment, the CPS will comprise of smart machines, storage systems and other production facilities that would be capable of autonomously exchanging data, triggering actions in related objects and all of these smart objects define smart factories. Within this framework, smart objects or products know their own history, current status and all possible routes to reaching their target. They are also uniquely identifiable, which makes it possible to monitor their movements and states. Business processes within the factories are vertically linked to these manufacturing systems and horizontally connected to the business value networks that are managed starting from when an order is placed that continues until the end when the final product reaches the customer (Kagermann, Helbig, Hellinger & Wahlster, 2013). Thus, with the creation of the smart factories, there will begin a phase of cost-cutting in terms of capital, energy and personnel as well as more flexibility, reduced lead times and adaptation to customer requirements with small batch sizes (Heng, 2014).

Industry 4.0 will also address and is expected to solve some of the challenges of the world today, which include resource and energy efficiency, urban production and demographic change. That is because Industry 4.0 is expected to enable resource productivity and efficiency gains that will affect the entire value network (Kagermann, Helbig, Hellinger & Wahlster, 2013). Industry 4.0 overall is that term that describes the changes of the manufacturing and production industry landscape of the developed world (Brettel, Friederichsen, Keller & Rosenberg, 2014).

Unlike the previous industrial revolutions that were triggered from technological innovations, Industry 4.0 is a planned attempt for a revolution of production systems. Globalization has increased competition of companies producing in high-wage countries (Schuh, Klocke, Brecher & Schmitt, 2007). Cost advantages in production, especially that of labor cost, have caused relocation of production from high-wage countries to low-wage countries. Therefore, companies in low-wage countries are mainly focusing on mass production (economies of scale) and those in high-wage countries have to balance their production among mass production and producing customized products (economies of scope). In order to maintain a balance, the latter types of companies have to optimize their processes using capital-intensive tools and systems. Thus in the case of production economy there is a dilemma between economies of scale and scope. Since mass production requires standardized automation, it undermines flexibility that is attained through economies of scope. Another dilemma in production is the one between value orientation and planning orientation in the production planning process. The first comes with costly planning efforts and is beneficial in the long run while the latter focuses on maximizing value by efficient planning processes (Schuh et al., 2013). These two dilemmas make up the polylemma of production (Brecher et al., 2012). In Germany, The Cluster of Excellence "Integrative Production Technology for High-Wage Countries" of RWTH University focuses on resolving this polylemma through individualization, virtualization, hybridization and self-optimization. All of these four research areas have strong links to the topic of Industry 4.0 (Brettel, Friederichsen, Keller & Rosenberg, 2014). Germany has a special interest in the topic because according to German Chamber of Commerce and Industry, one in four German industrial companies have relocated to low-cost countries. However, relocation of production also results in relocation of R&D and services, and for high-wage countries, securing domestic production is a vital part of securing national wealth (Brecher et al., 2012).

German firms create one-third of EU's total value added (Heng, 2014) and Germany has significantly improved its return on capital employed over the last 15 years, despite a 9 % drop in employment. The value added of German industry has risen 80 % between 2000 and 2014 and the profits have risen 158 % during the period. The rate of use of production equipment rose from 85 % in 1998 to 95 % in 2014.

Thus, for Germany, a successful transformation of the manufacturing industry is of high importance as it contributes over 25 % of the Gross Domestic Product (GDP) and provides over seven million jobs as announced by Eurostat and the Federal Statistical Institute of Germany (as cited in Brettel, Friederichsen, Keller & Rosenberg, 2014, p. 37). Therefore, a strategic proposal for the implementation of Industry 4.0 was issued in 2013 after the collaboration of German government, industry and research institutions. In December 2013, Germany released the standardization roadmap, and German Industry 4.0 has become the national strategic plan to increase German global competitiveness by 2020 (Zhang, Peek, Pikas & Lee, 2016).

German Industry 4.0 uses certain strategies in its transformation to Industry 4.0 that can be listed as: attaining standardization and open standard reference systems, setting up models to manage the complex systems, providing a comprehensive broadband infrastructure, establishing security mechanisms, innovating the organization and design models, emphasizing training and continuous professional development and establishing sound rules and regulations to improve resource efficiency (Zhang, Peek, Pikas & Lee, 2016).

As a result of the transformation to Industry 4.0, German manufacturing sectors are expected to boost their productivity by 5–8 % or 90–150 billion Euros over the next ten years. Increased corporate demand in advanced technologies as well as demand for customized products is expected to add on additional revenues of 300 billion Euros and an increase employment by 6 %. In addition capital investment is expected to revert the cost advantage in production back to Germany by cutting the costs by 20 % (Numanoglu, Eynehan, Morkoc-Nikelay & Aksoy, 2016).

In comparison to the German economy, Turkish economy is the 17th largest economy in the world as of 2014. The 75 % urban population of Turkey supports its industrial manufacturing based economy (Tuncel, 2014). Until 2008, the EU's share in Turkey's exports was higher than 50 %. But after 2008, this share began declining, whereas the share of Middle East and North Africa began increasing. Given the demographic structures of EU with an aging population and Turkey with a young and active human capital stock, Turkey has been exporting labor-intensive products and importing capital-intensive products from the EU (Alpaslan, 2012).

In order to assess the manufacturing productivity in Turkey, manufacturing value added can be observed. Manufacturing value added is the net output of a sector after adding up all outputs and subtracting intermediate inputs. It is calculated without making deductions for depreciation of fabricated assets or depletion and degradation of natural resources. As of 2013, Turkey ranks 16th in current USDs in terms of manufacturing value added.

The growth in manufacturing value added between 2000 and 2013 in 100 millions of USDs is seen in Fig. 1. As can be seen in Fig. 1, there is a huge gap in terms of volume with respect to the forerunners of the world and Turkey in terms of manufacturing value added.

In terms of its position in the global value chain, according to the research done by OECD (Organisation for Economic Co-operation and Development), as a result of globalization, like it did in many countries, Turkey's integration into global value chains has increased, with the foreign content of exports going from 10 % in 1995 to 25.7 % in 2011. The foreign content, which is the imported inputs of the industries, has risen significantly in all sectors between 1995 and 2011, mostly in basic metals, motor vehicles and fabricated metals. Around 54.1 % of Turkey's value added exports are exports of final products, indicating that the integration into Global Value Chains is more orientated toward downstream activities such as assembly (OECD, 2015).

Fig. 1: Growth in Manufacturing Value Added in 100 Million USDs. Source: http://data. worldbank.org/indicator/NV.IND.MANF.CD

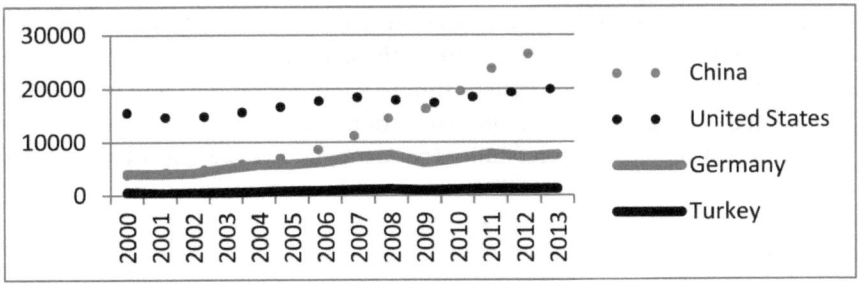

Markusen in his 2002 study states that multinational firms have kept skilled labor-intensive activities such as R&D and management in the headquarters and moved production of the final goods to countries where labor is cheaper (Markusen, 2002). Leveraging on labor-intensive products usually produces low added value. Thus, there are two structural obstacles for competitiveness in the present situation. Turkey depends highly on imports in order to export the final good. And these final goods are produced in Turkey by multinationals as a result of the labor-intensive cost advantage of Turkey. This cost-advantage, however, produces low added value, which results in less profits. Thus, as a result, despite the rising global demand in high-tech products, the share of such products among the exported manufacturing goods is 4 % in Turkey. In addition to these the high-skilled labor is limited in Turkey, and the labor resources in the industrial sectors are lately shifting toward service sectors (Numanoglu, Eynehan, Morkoc-Nikelay & Aksoy, 2016).

In 2011, the Ministry of Science, Industry and Technology (MSIT) has expressed the necessity of increasing the competitiveness and efficiency of Turkish industry with higher share in high-tech products that have high added value. In order to achieve this, three basic strategic objectives have been set. The first is increasing the weight of mid- and high-tech sectors in production and exports, the second is the transition to high added value products in low-tech sectors and the third is increasing the weight of companies that can continuously improve their skills (Tuncel, 2014).

If Germany gains a cost advantage via Industry 4.0 implementation, then the cost advantage of low-cost and labor-intensive countries will diminish substantially. As a result, Turkey will lose competitiveness and lose its global market share, which will cause unemployment. If Turkey does not implement Industry 4.0, it will fall into a cycle of low value added production and low investment. The main sectors where Industry 4.0 could be utilized in Turkey are automotive, white goods, textile, chemical, food and beverage and machinery since they contribute the most to the economy. The relevant Industry 4.0 modifications that could be done within these sectors are in the fields of flow of information and materials, integration with suppliers, simulation of the product and production process in the design phase, flexible production and smart product and production lines that increase predictability (Numanoglu, Eynehan, Morkoc-Nikelay & Aksoy, 2016).

2. Methods

In order to assess the gap between Germany and Turkey, certain characteristics of both countries that are essential in a leap of the manufacturing industry will be compared. These characteristics are R&D funding levels, innovation possibilities, the state of information technologies and education.

2.1 R&D Funding levels

As of 2013 Turkey ranks 34th in terms of R&D expenditure as percentage of GDP among the world countries. The percentage of GDP expenditure for Germany is 2.83 % of the GDP whereas for Turkey it is 0.94 %. As collected by World Bank, expenditures for R&D are current and capital expenditures (both public and private) on creative work undertaken systematically to increase knowledge, including knowledge of humanity, culture and society, and the use of knowledge for new applications. The comparison of R&D expenditures with other countries can be seen in Fig. 2.

Fig. 2: R&D as a Percentage of GDP. Source: http://data.worldbank.org/indicator/ GB.XPD.RSDV.GD.ZS

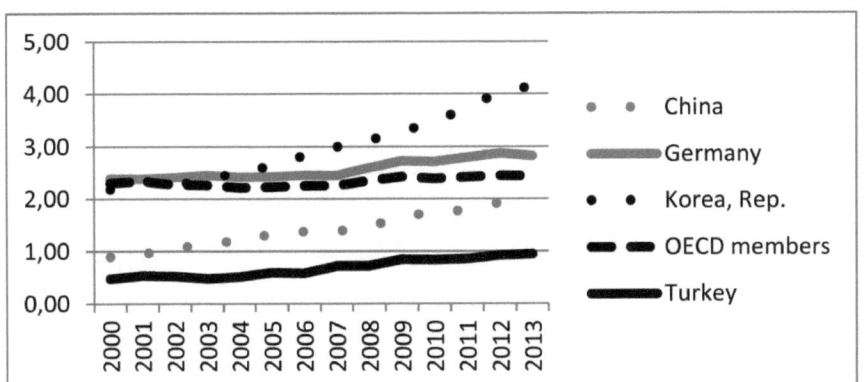

Even though R&D expenditures as a percentage of GDP have increased over the 2000–2013 period for Turkey, they are still below the OECD average and that of Germany as seen in Fig. 2. Business R&D Expenditure as a percentage of GDP is also below the OECD averages and Germany for Turkey. The percentage of Business R&D expenditure was close to 2 % for Germany in 2013 whereas for Turkey it was around 0.5 %.

The research quality of the increased R&D endeavor was analyzed in a research performance and evaluation event in Ankara on March 12, 2013. In terms of research output, Turkey ranks 18th among the top 20 countries and in that being so Turkey can be deemed as competitive. In terms of numbers of articles, it is on the lower end; however, in terms of growth as of 2011, it was number four. In terms of citations, around half of Turkish papers were cited between 2006 and 2010. Turkish research has also become increasingly interdisciplinary; however, international scientific collaboration is not yet sufficient. It has been concluded in the event that in order to attract international research funds, Turkey should keep its steady rate of growth in R&D and increase its international collaboration (Basal & Keskin, 2013).

Structural composition of business R&D when compared to Germany is insufficient. As of 2009, Turkey's business R&D is geared more toward medium to low-tech manufacturing rather than high-tech manufacturing.

Fig. 3: Structural composition of business R&D (% of business R&D expenditure (BERD) 2009 Data Source: https://stats.oecd.org.

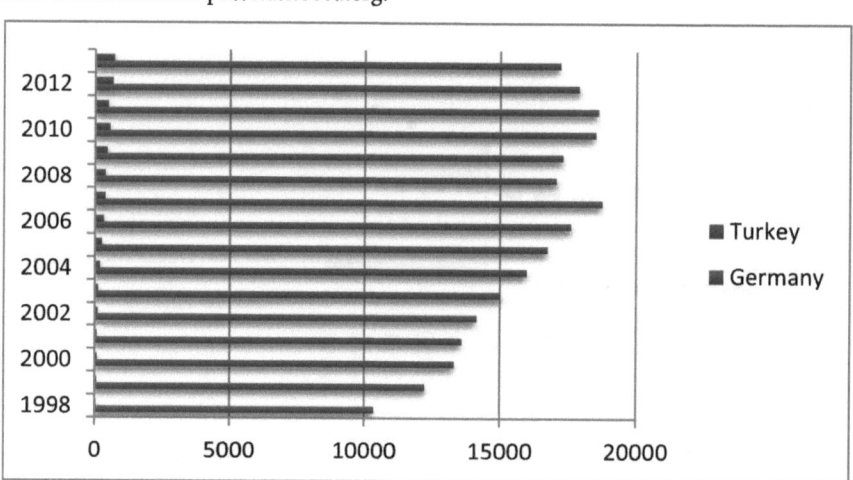

As seen in Fig. 3, as of 2009, Turkey's business R&D is geared more toward medium to low-tech manufacturing rather than high-tech manufacturing.

2.2 Lack of Embedded Innovation

Pyka, Kustepeli and Hartmann in their 2016 study compare the total number of patents and technological advantages of Turkey, Germany and the EU between 1981 and 2010. They have found out that Turkey has been catching up in several technology fields, such as "pharmaceuticals", "thermal processes and apparatuses" and "civil engineering". However, in absolute terms, the number of patents of Turkey is significantly behind Germany and the EU. In terms of diversification and knowledge breadth of the patent portfolio, Turkey applied for 14 patents in 9 technology fields between 1981 and 1985, while applying for 1753 patents in 35 technology fields between 2006 and 2010. However, this number is still weak compared to Germany applying for 59,825 patents in all available 35 technology fields between 1981 and 1985 and applying for 161,238 patents in all of the available technology fields between 2006 and 2010. More importantly the applicants in Turkey's patent network consist only of international companies and there are no Turkish companies in the top ten list of applicants. This emphasizes Turkish technological dependency on especially German and US enterprises (Pyka, Kustepeli & Hartmann, 2016). The evolution of the number of patent applications for Turkey and Germany can be observed in Fig. 4.

Fig. 4: Patent Applications of Turkey compared with Germany (2007-2009) Data Source: https://stats.oecd.org/viewhtml.aspx?datasetcode=PATS_IPC&lang=en#

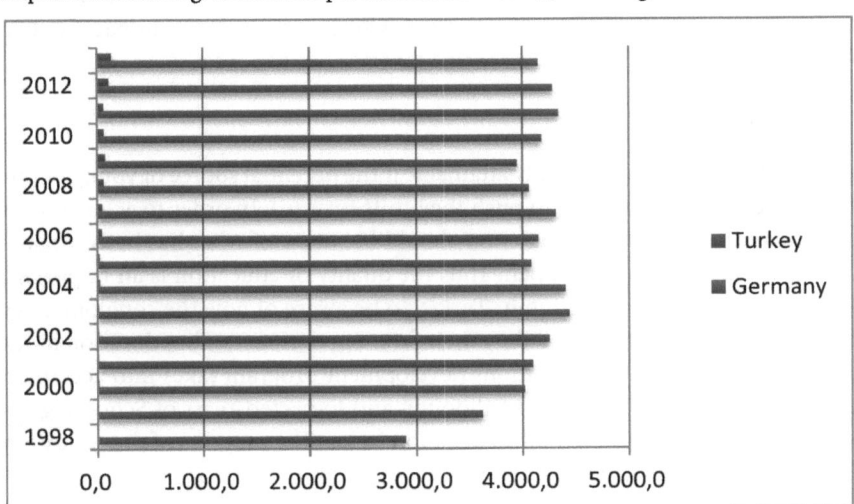

2.3 Information Technology Gap

The Network Readiness index that is evaluated by the World Economic Forum assesses countries over four categories in terms of readiness to technological transition. The first category is the Environment sub-index that evaluates the political and regulatory environment as well as the business and innovation environment. This category assesses in sum the overall environment for technology use and creation. The second category is the Network Readiness sub-index that assesses the infrastructure, affordability and skills in terms of the ICT infrastructure. The third category assesses the usage in terms of government, individual and business usage. It in a way measures the adaptation. The last and the fourth category is the impact sub-index that assesses economic and social impacts. The overall score determines the Network Readiness of a country and in terms of the overall ranking, Turkey has ranked 48th with a score of 4.4 whereas Germany has ranked 15th with a score of 5.6. With respect to the sub-index of the first category of environment, Turkey has ranked 49th with a score of 4.2 whereas Germany's score is 5.2. With respect to the second category of the sub-index, which is readiness, Turkey has ranked 40th with a score of 5.5 with Germany ranking at 13th with a score of 6.1. With respect to the third category of usage, Turkey has ranked 59th with a score of 4 and Germany ranking at 14th with a score of 5.6, and lastly with respect to the fourth category of impact Turkey has ranked 58th with a score of 3.8 with Germany ranking at 15th with a score of 5.3. This shows that Turkey lags far beyond Germany with respect to IT and more so in the third and fourth categories which are usage and impact (Baller, Dutta & Lanvin, 2016).

2.4 Education

A high-skilled labor force is only attainable if the potential source of young population is capable and willing to push the limits forward. In order to achieve such a momentum, the young population should be aware of what has been, what is and what can potentially happen. Thus the quality of education is an important factor for the youth population in Turkey to be aware of how the conglomerates of the world are possibly scheduling their future investments. The education index as calculated by the United Nations development program takes into account a calculation using Mean Years of Schooling and Expected Years of Schooling. As of 2013, Turkey ranks 92nd in the education index. The progress of the education index over the years is depicted in Fig. 5.

Fig. 5: Human Development Report – Education Index. Source: http://hdr.undp.org/en/content/education-index

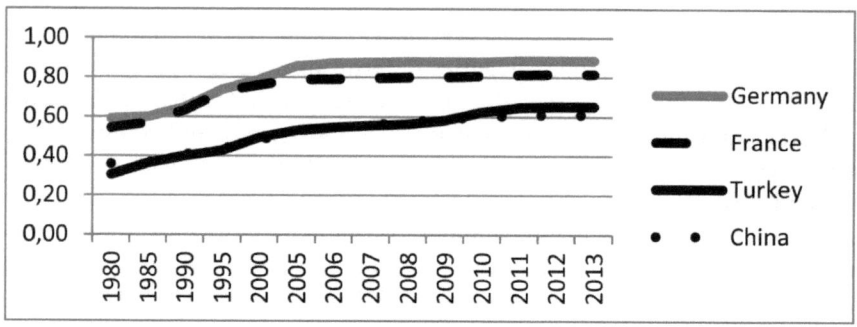

3. Findings

As a result of the comparison of statistics between Germany and Turkey, it is found that Turkey needs to transform its resources, such as labor force, and invest in production technology in order to transfer to Industry 4.0, otherwise it will lose its competitive advantage very fast as the demand for low-skilled labor force decreases. In a survey conducted by TUSIAD (Turkish Industry & Business Association) with Turkish industry executives, it is agreed that government and non-governmental organizations should work in conjunction to adapt the appropriate infrastructure. If Industry 4.0 production is adopted, Turkey is expected to experience 3 % of yearly additional growth and 150–200 billion USD increase in revenues (Numanoglu, Eynehan, Morkoc-Nikelay & Aksoy, 2016).

With the adoption of Industry 4.0, a high-skilled labor force that can manage the interactions within the smart factories will be needed. In order to bring up such a labor-force, education needs to be upgraded as well as be geared toward a more scientific nature that should encourage innovation and improvement. According to the survey by data analysis portal TURKSTAT (2016), employment of young people was 34.2 % in 2015 with an increase from 2014. 18.5 % of these youths were employed in agricultural sector, 31.4 % in the industrial sector and 50.1 % were employed in the service sector (TURKSTAT, 2016). Among the policy recommendations for Turkey in the OECD 2017 Reform agenda, reducing the wide quality gaps persisting among schools, school types and universities, by granting them more autonomy and resources per student is one item. Further developing pre-school education and strengthening vocational education with the co-operation of the business sector is another recommendation (OECD, 2016). School curricula, training and university programs should strengthen entrepreneurial approaches and increase the IT-related skills and innovation (Numanoglu, Eynehan, Morkoc-Nikelay & Aksoy, 2016).

As explained in the previous sections, current expenditure or investment in R&D is very insufficient compared with technology powerhouses like Germany. Due to its logistical position and current manufacturing sector, potential for enlarging the Turkish market is big. R&D activities therefore need to be encouraged to transform Turkish manufacturing industry and Turkey should not be dependent on developed countries for technological inputs totally and instead be in continuous collaboration with the world leaders. Taking advantage of different manufacturing technologies should be encouraged, and if this is done at the beginning, economic losses can be minimized.

In addition, Turkish producers will have to invest in the upcoming ten years around 10–15 billion TRY into Industry 4.0 production technologies; this is equivalent to 1–1.5 % of manufacturers' yearly revenues. This sort of long-term investment is usually not preferred in Turkey; producers are usually willing to take the return on investment in two years' time, and however, this sort of long-term investment is essential in the long-term survival of Turkish manufacturing industry (Numanoglu, Eynehan, Morkoc-Nikelay & Aksoy, 2016).

Finally, in order to standardize the transformation to Industry 4.0, Germany has founded an Industry 4.0 Steering Committee. The rationale behind the Steering Group was that the development and implementation of a host of new concepts and technologies could only be possible if they are backed by consensus-based standards. Standards would be necessary to create security and confidence among manufacturers and users. The Steering Group thus identifies the concrete needs for standardization, coordinates their implementation and advances their development (DIN/DKE, 2016). Such a committee to develop standards could also be founded in Turkey, especially since Industry 4.0 is a new concept and an encouraging and motivating foundation is necessary to lead and gain confidence of the industry. Plus standards will expedite the actual transition to novel production technologies.

4. Discussion and Conclusions

The analysis and comparison of the current technological positions of Germany and Turkey reveal that in terms of human capital and technological capital there are wide gaps between the two. The adoption of Industry 4.0 is essential for Turkey in terms of leveraging its logistical advantage, and increasing its trade volume in the upcoming years. Furthermore this increase should better be in high-value added products, which are more effective in enlarging the economy. If Turkey invests in increasing its human and technological capital rigorously, it can use its advantages efficiently.

5. Acknowledgment

This study was supported by the Bogazici University Research Fund, Grant Number 17N03P2.

References

Alpaslan, İ. B. (2012). Turkey's External Economic Relations: Does the MENA Region Steal the EU's Role? In B. Akcay & B. Yilmaz (Eds), *Turkey's Accession to the European Union: Political and Economic Challenges* (pp. 211–226). Rowman & Littlefield: USA.

Baller, S., Dutta, S., & Lanvin, B. (2016). World Economic Forum Global Information Technology Report 2016. Retrieved October 5, 2018, from http://www3.weforum.org/docs/GITR2016/WEF_GITR_Full_Report.pdf

Basal, T., & Keskin, G. (2013). Turkey's Scientific Research Output Is Booming – But What about the Quality? (Rep.). Retrieved from: https://www.elsevier.com/connect/turkeys-scientific-research-output-is-booming-but-what-about-the-quality. Accessed on May 10, 2017.

Brecher, C., Jeschke, S., Schuh, G., Aghassi, S., Arnoscht, J., Bauhoff, F., & Orilski, S. (2012). *Integrative Production Technology for High-Wage Countries.* Springer, Berlin Heidelberg, pp. 17–76.

Brettel, M., Friederichsen, N., Keller, M., & Rosenberg, M. (2014). How Virtualization, Decentralization and Network Building Change the Manufacturing Landscape: An Industry 4.0 Perspective. *International Journal of Mechanical, Industrial Science and Engineering,* 8(1), 37–44.

DIN/DKE (2016). German Standardization Roadmap. Retrieved October 5, 2018, from https://www.din.de/blob/65354/57218767bd6da1927b181b9f2a0d5b39/roadmap-i4-0-e-data.pdf

Drath, R., & Horch, A. (2014). Industrie 4.0: Hit or hype? [Industry Forum]. *IEEE Industrial Electronics Magazine,* 8(2), 56–58.

Dujin, A., & Geissler, C. (Eds.). (2016). The Industry 4.0 Transition Quantified: How the fourth industrial revolution is reshuffling the economic, social and industrial model. RB Publications. Retrieved October 5, 2018, from https://www.rolandberger.com/publications/publication_pdf/roland_berger_industry_40_20160609.pdf

Gorecky, D., Schmitt, M., Loskyll, M., & Zuhlke, D. (2014). Human-Machine-Interaction in the Industry 4.0 Era. Paper presented at "*12th IEEE International Conference on Industrial Informatics (INDIN, Porto Alegre, Brazil, 23–27 July,* pp. 289–294. Piscataway, NJ, USA: IEEE.

Heng, S. (2014). Industry 4.0: Upgrading of Germany's Industrial Capabilities on the Horizon (April 23, 2014). Available at SSRN: https://ssrn.com/abstract=2656608

Kagermann, H., Helbig, J., Hellinger, A., & Wahlster, W. (2013). *Recommendations for implementing the strategic initiative INDUSTRIE 4.0: Securing the future of German manufacturing industry; final report of the Industrie 4.0 Working Group.* Forschungsunion.

Markusen, J.R. (2002). Multinational Firms and the Theory of International Trade. Retrieved May 10, 2017, from https://mpra.ub.uni-muenchen.de/8380/1/.

Numanoglu, N., Eynehan, M.E., Morkoc-Nikelay, G., & Aksoy, E. (Eds.). (2016). Industry 4.0 in Turkey as an Imperative for Global Competitiveness, an Emerging Market Perspective. Retrieved October 5, 2018, from https://tusiad.org/tr/tum/item/download/7847_49fa245cf57d64611148ac42c0de4a7e

OECD. (2015). Trade in Value Added: Turkey. Retrieved May 10, 2017, from https://www.oecd.org/sti/ind/tiva/CN_2015_Turkey.pdf.

OECD. (2016). Reform Agenda for 2016: Turkey [Press Release]. Retrieved May 10, 2017, from https://www.oecd.org/eco/growth/Going-for-Growth-Turkey-2017.pdf.

Pyka, A., Kustepeli, Y., & Hartmann, D. (Eds.). (2016). *International Innovation Networks and Knowledge Migration: The German–Turkish Nexus*. Routledge: London and New York.

Schuh, G., Brettel, M., Wesch-Potente, C., Potente, T., Rosenberg, M., Keller, M., & Schmitz, T. (2013). The Potentials of "Effectuation" for the Resolution of the "Polylemma of Production". International Journal of Business and Management Studies, 5(2), 75–84.

Schuh, G., Klocke, F., Brecher, C., & Schmitt, R. (2007). *Excellence in Production*. Apprimus Wissenschaftsverlag, Aachen.

Tuncel, O. (2014). US Government of State: 2014 Investment Climate Statement. Retrieved May 10, 2017, from https://www.state.gov/documents/organiza tion/229296.pdf.

TURKSTAT. (2016). Youth in Statistics, 2015 [Press Release]. Retrieved May 10, 2017, from http://www.turkstat.gov.tr/PreHaberBultenleri.do?id=21517.

Vogel-Heuser, B., & Hess, D. (2016). Guest Editorial Industry 4.0-Prerequisites and Visions. *IEEE Transactions on Automation Science and Engineering*, 13(2), 411–413.

Zhang, X., Peek, W. A., Pikas, B., & Lee, T. (2016). The Transformation and Upgrading of the Chinese Manufacturing Industry: Based on German Industry 4.0. *The Journal of Applied Business and Economics*, 18(5), 97.

Ahmet Gürkan Yüksek[*], Halil Arslan, Gülşah Çifçi
and M. Lemi Elyakan

Development of a Central Controlled Automation Project on the IoT Platform

1. Introduction

The widely accepted definition for the Internet of Things (IoT) is "The Internet of Objects is described by van Kranenburg, 2008 as "a dynamic global network infrastructure that is based on standard and interoperable communication protocols, where physical and virtual objects have physical attributes and virtual personalities, use intelligent interfaces, integrate seamlessly into the information network". The IoT is a technology that allows physical objects and devices to communicate with each other through different communication technologies, sensors/actuators and RFID (Radio Frequency Identification) tags on the Internet. Today, many industrial field IoT technologies are being studied intensively (Li, Hou, Liu & Liu, 2012). In this context, the fact that objects can communicate independently over the Internet, the most common and powerful network, has become a key factor in many automation ideas.

The idea of merging computers, networks and sensors to control and monitor has become a much more popular concept in the last decade. Important factors such as the spread of IP-based networks, the miniaturization of computers, the rise of cloud computing technology trends and the developments in data analysis have triggered this popularity. Along with all these developments, sensor networks using RFID tags and advances also in near-field communication (NFC) technologies have constituted new visions for researchers (Whitmore, Agarwal & Xu, 2015). Based on the idea of the IoT, it includes that objects used in everyday life can be equipped with the ability to monitor, perceive and manage. As long as this vision reaches its goals, the objects that are subject to it will understand, perceive and interact with people and will have more complex, more intelligent networking abilities. Here are two basic elements: the devices themselves and the server-side architectures supporting those devices stand out. Another important

[*]   Corresponding author: Ahmet Gurhan Yuksek, agyuksek@cumhuriyet.edu.tr, Cumhuriyet University Computer Engineering, Sivas Turkey.

factor supporting them is that in most cases there may be a low-power gateway that can collect between devices and the wider Internet, for example, collecting, event handling, bridging (Fig. 1).

Fig. 1: Two Basic Connection Modes. https://theexuberantindian.wordpress.com/2017/04/16/iot-architecture-reference-model/

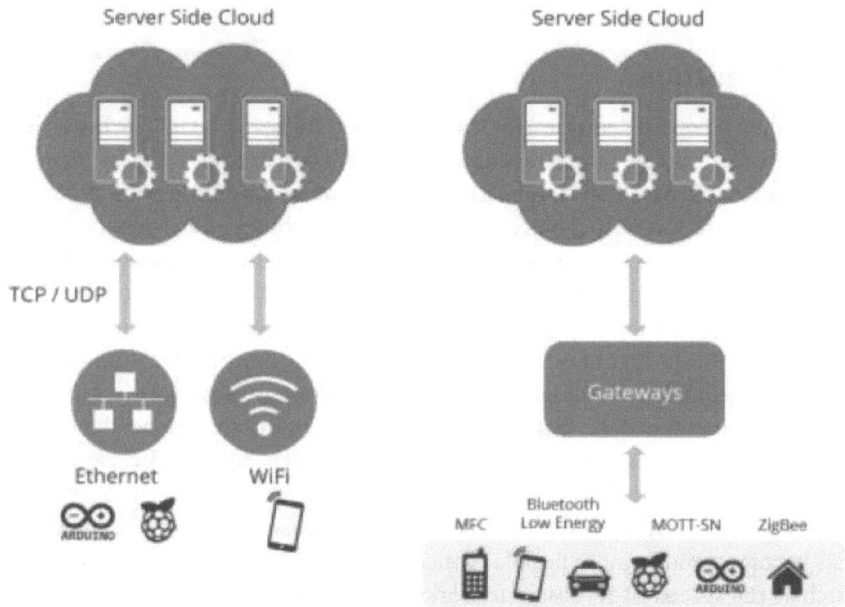

Although IoT models, which are simply proposed, appear to be constructed on the communication of RFID and other radio signal-based edge technologies, in reality a more inclusive model is needed that has the potential to interact with the physical world and create interfaces. Solutions in the IoT architecture should be in a way that will provide communication level and service level interoperability on different platforms. IoT architects and standards are being developed with this important feature in mind (Alexander Salinas et al., 2017; Fremantle, 2014; Marginean & Tran, 2016). Determining such standards motivates the development of a common understanding on IoT technologies first. Second, businesses which want to create their own appropriate IoT solutions should be supported by a "Reference Architecture" that defines basic building blocks and design options to overcome clashing requirements related to functionality, performance, installation and security. Interfaces are standardized, best practices and information usage are provided in terms of functionality. As is easily understood, the key factor for ensuring communication between objects is standardization. A well-known fact for application developers is that it is easy to claim such standards in a research environment, but it is difficult to achieve in real-world applications. The reference architectures developed here due to the fact that it defines the guidelines that can be used while the implementation or development of an IoT system is a common standard for different developers. The network topology of IoT systems consists of a central controller that connects to the Internet and cloud services or simple nodes that collect and send data to a limited number of networks. Nodes and gateways should be designed to minimize power consumption, provide reliable and robust network connections and expand wireless connectivity as much as possible. At the heart of the IoT systems is a processor unit or microcontroller (MCU) that runs software stacks that process data and it is connected to a wireless device for link (Fig. 2).

Fig. 2: Wireless Sensor Node Architecture.

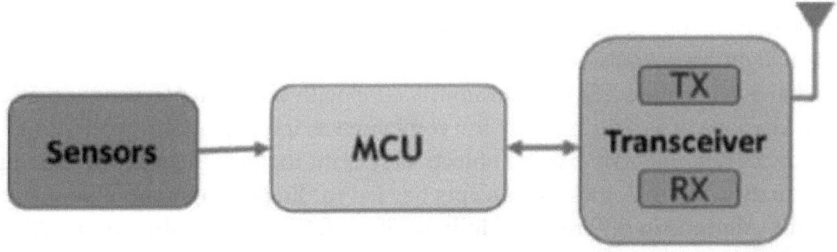

The requirements for both the MCU and the wireless device are specific to the last implementation and system requirements. Advanced IoT sensor nodes reinforce sensor functions and it uses an 8-bit MCU or 32-bit device to run a small stack of radio-frequency (RF) protocols. These devices generally operate with battery and are connected to heavier operations and into gateway at which the data transfer takes place. Sensor nodes typically transmit a small amount of data and often have to work with IoT system is a common standard batteries that are capable of lasting for several years. The devices should be connected portable and reliable and it must be able to operate in a variety of environmental conditions, independent of RF interference or physical hindrances. Since these devices are part of networks, it should take into consideration setup of networks for rallying of sensor data and displaying of data. The combination of the appropriate MCU and wireless or RF connection as well as development development tools and software stack for application development are critical for successful designs for these devices.

Foreseen study was designed and realized to provide a summary of the basic requirements, structure and development steps of an IoT project. As an exemplary application, an automated door control system to centrally colligate three different platforms (mobile android, web, embedded systems) on an IoT project has been selected and developed. In the method part, general lines of the project, block diagrams and the technologies used are summarized, while in the findings and conclusions section, project outputs and future suggestions are presented.

2. Method

One of the numerous advances in Internet technologies is the communication of things, and in this context, it is possible to allow the objects in different settlements to be managed through the defined mechanisms. Here, the concept expressed as object is the surrounding to the people in everyday life and are assets that can be accessed with various hardware mechanisms (refrigerator, car, motor, lighting systems, etc.). These devices are called as devices in the hardware mechanisms that provide data communication with each other (sensor, actuator, tag). The idea of providing daily life convenience, improving the quality of life and the development of existing objects as systems that will incorporate existing information technologies as concepts has led to the development of very successful solutions in different disciplines (Shrouf, Ordieres & Miragliotta, 2014; Theodoridis, Mylonas & Chatzigiannakis, 2013; Wang, 2016; Xu, He & Li, 2014). System is known as it will occur in a combination of many subsystems and all of these structures are merged structures together in such a way that they will perform the overall task of the system in harmony so as to form a target solution. An IoT system is the theory which is built upon the interoperability of many different subsystems. These systems are well defined, planned, designed, implemented and tested and purified from their faults and it is necessary to improve its existence throughout the life span. In short, it is necessary to plan the architectures of the system development life cycle of such projects to cover the whole structure. As mentioned, this work is intended to be an example guide for examining, cascading and similar project developments of an IoT project's system development steps.

2.1 System Identification and Design

The developed system is designed as an IoT project to manage the door systems, which are planned to be managed centrally, with mobile devices in different places. The structure summarized by the operation of the system is visualized in Fig. 3. The overall system consists of four different subsystems (modules):

- As web-based registration, management and reporting systems and management portal have all done using PHP + MySQL technology, JavaScript and HTML5 + CCSS + JavaScript frameworks.
- In a mobile (Android) based 2echnol, Detecting and controlling the existing door in the current location (Bluetooh or WiFi echnologies), using the Android Studio + Java programming language to transmit the door-related information to the central management system over the internet.
- The transmission module uses the MQTT (Message Queuing Telemetry Transportation) protocol, where communication and routing between objects is performed. A broadcast/subscriber is an extremely simple and lightweight messaging protocol designed for restricted devices and low bandwidth, high latency or unreliable networks. While design principles reduce network bandwidth and device resource requirements to a minimum, they are also trying to provide some degree of reliability in terms of reliability and delivery. These principles also make it possible to idealize the protocol in the context of the "machine to machine" or "IoT" making it emerged for mobile applications where bandwidth and battery power are high (FAQ—Frequently Asked Questions, MQTT, 2017).
- The hardware and software of the SBC (Small Board Computer Arduino – Raspberry Pi 3 – NodeMCU), which is responsible for the management of sensors (sensors), actuators and RFID tags, is combined and managed in one module.
- As clearly understood from Fig. 3, the system is based on the management of gate systems in different locations from a single center. The goal here is not to manage a single door, but to manage systems that have doors together. The gates are grouped in the relational database tables created in the MySQL centralized database server by grouping them according to the campus and the infrastructure in these campuses (building, etc.). With the improved web-based application, the tables in which these systems are kept are managed. The structure of the general database is designed to keep tables for the properties of all assets in the system and to report all mobility of these assets for future use. The system consists of a combination of subsystems with four different technological structures. Such schemes can lead to

incompatibilities and problems that cannot be calculated before the system is designed. It was also necessary to bring together expert knowledge in different areas. Taking all these situations into consideration, Agile Software Development Methodology was chosen as the project development method ("What is Agile Software Development?", 2015; Yüksek, A. G., Arslan H., Kaynar O., Çifçi G., 2016).

Fig. 3: Summary Architecture of System Operation.

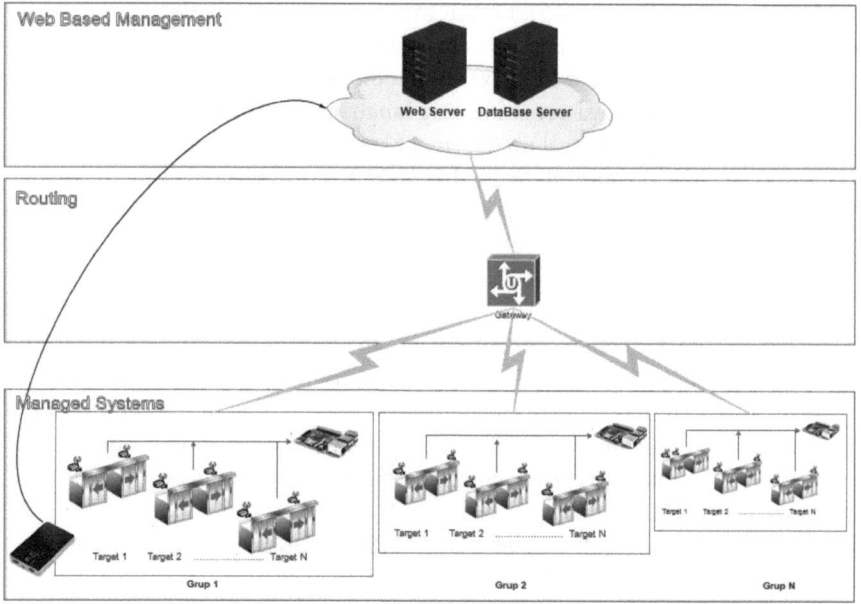

Fig. 4 shows the data flow diagram of the system. When personnel arrive in front of any door defined in the system (if the detection program is present and active as a service on the mobile device), unique door ID information continuously broadcast by the door is detected. The detected information and the MAC information of the portable device are sent to the control software over the Internet.

Fig. 4: Data Flow Diagram of the System (General System).

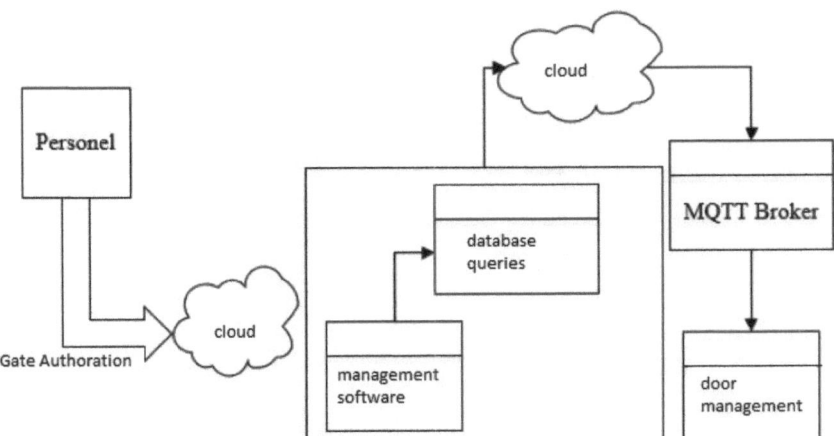

Here, apart from the MAC information of the device, information to be specially defined for the device to be produced by the developed mobile software may be preferred. However, for security reasons and the most decisive credentials of the device, the use of a MAC address is more preferable. The betting on the doors can be chosen so that this ID information can be broadcast to devices used in different technologies such as beacon ("Beacons") and NodeMCU ("NodeMcu–An Open-Source Firmware Based on ESP8266 wifi-soc.", 2017) using Bluetooth technology. But the software to be developed as a mobile sensor should be developed with the ability to detect these two different technologies. The module that extracts the system from a web-based inquiry project and turns it into an IoT project is the part that determines the door ID through the mobile device in the system. With this module it is the part that manages the opening of the doors by transferring the result of the identity inquiry which is returned from the database to the objects in the doors. The objects on the doors first produce inputs for mobile systems by broadcasting the continuous gate ID based on the structure of the overall system. The same objects or additional object support as a result of the developed structure; information coming from the Internet is becoming the manager of the door mechanism as shown in Fig. 5.

Fig. 5: General Software and Design Architecture.

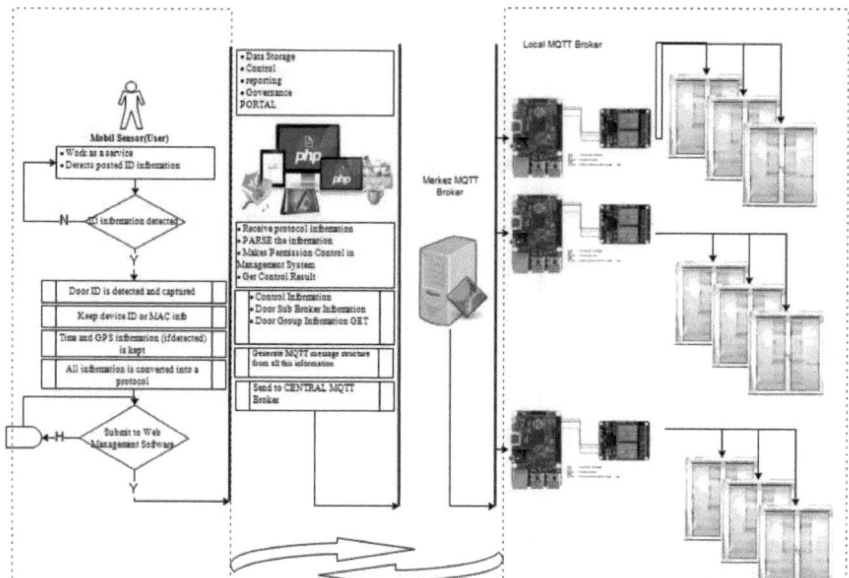

2.2 Communication of Things

We can say that our objects in the system as a result of a general overview are the doors and the personnel to use this door. The doors are equipped with the necessary hardware structures (sensors, actuators, RFID tags) and the personnel are included in the system with the mobile device they use. Here, the mobile devices and gates are actually communicating. There are many different ways of communicating with these objects. In this study, the MQTT protocol, one of the widely used network protocols in the IoT field, is dealt with. MQTT has been developed as a machine-to-machine (M2M) message transport protocol (http://mqtt.org/). The telemetry expression here includes remotely wired or wireless management of systems. It is primarily designed as a message transport or routing protocol for clients who subscribe to a publisher itself (publish/subscribe).

It has been optimized with the aim of connecting devices and events around the world via corporate servers. MQTT sensors, actuators, software technology of mobile devices are designed to create a structure that will make it easier to connect to each other (5 Things to Know about MQTT, 2014).The basic characteristics of MQTT can be summarized as follows: It is a lightweight messaging, queuing and transport protocol, has an asynchronous communication model, provides low network bandwidth for applications, works publish/subscribe (pub/sub) in the model. The data producer (publisher) and the data consumer (subscriber) are separated from each other, low complexity is a protocol aimed at low power and low footprint applications (Vaidya, 2016). The publisher/subscriber model (pub/sub) is an alternative to the client-server model that communicates directly with the endpoint of the client. In this model, there is a mechanism (broker) that communicates the publisher's messages to all subscribed clients. The publisher and the subscriber are not aware of each other's existence. H Here, known both by the publisher and the subscriber, filter all incoming messages and there is a third component called a broker that distributes them accordingly as seen in Fig. 6. The main element in the publisher/ subscriber concept is the parsing of the publisher and receiver, which can be further divided into different dimensions.

Fig. 6: MQTT Message Management. https://blog.knoldus.com/iot-what-is-mqtt-how-it-is-lightweight/ Accessed on Sep 10, 2018

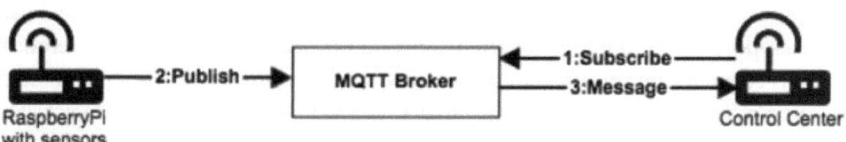

IoT from a single low-end device covers a single low-end device of real-time embedded technologies (SCB) and broad industry areas based on cross-platform of cloud systems. All these systems can have different platform structures. Platform independence is the most basic feature of the structures that will allow the communication of objects (Derhamy, Eliasson, Delsing & Priller, 2015).

The proposed study is based on MQTT broker architecture. The "Central Broker" shown in Fig. 5 resolves the "TOPIC" structure sent to it. The request within the hierarchy specifies which subscriber is "Local Broker" and this broker sends the door information and the door action determined by a new TOPIC structure (e.g door open or non-opening information with x ID number). This information is received by the device subscribing to the broker and the necessary action is taken via the actuator to handle the door this way keeping all of these transactions as logs in the central database which is necessary for future analysis and reports.

3. Conclusion

The most obvious disadvantages that can be faced when automation systems are built on the management of physical devices is the compatibility of hardware systems which are extremely effective on system performance and output, and manageability with software systems. Participating in IoT technologies very quickly in today's world and taking part in many industrial projects has led developers of software and hardware to improve the performance and compatible products in this area. It is possible to reach the products which will produce the most optimum solution from the wide range according to the properties, physical structure and conditions and the cost targets of the projects. For this reason the needs and conditions of the system must be well defined and the projects must be well managed. Only well-designed IoT projects with analytics, requirements, constraints, algorithms, models and their architectures in combination can achieve the targeted success. Since it is an entire system project, it is necessary that the sub-systems are planned and managed to reach the solution of the real problem.

With the help of a web-based automation system, the logic of IoT project development has been summarized through this project. It is defined on the basis of an appointment like logic of the users and on using services to be offered in the direction of identification in mobile devices (granting or denying permission to open a door). In spite of the fact that it is a very comprehensive project, it is focused on the concepts that outweigh the IoT direction in summary. In this study, the development process of a system that will be accommodated in many different platforms and management has been examined. Because these structures have been continuously evolving and due to the sudden negative situations that occur, the compatibility of the selected agile software development method has been observed. It also provides the expected criteria for system output, making it a good template for similar projects. However, since the system is an Internet-based project, all the resources and management are Internet centric, and the problems in the Internet directly affect the project. Internet interruptions, speed problem, etc., are the usual problems. However, since the hardware components that make up the system are available to work locally, the system is designed to work in line with plans B (by integrating some crucial features though not all features) into the system. For example, when there is no Internet to open the door, the business logic is planned and implemented so that the system can continue to operate even if the system has low characteristics, using the information in the temporary tables maintained locally.

References

5 Things to Know about MQTT—The Protocol for Internet of Things (5 Things to Know IBM Redbooks Blog) [CT915]. Rahul Gupta, Retrieved from https://www.ibm.com/developerworks/community/blogs/5things/entry/5_things_to_know_about_mqtt_the_protocol_for_internet_of_things?lang=en Accessed on Sep 10, 2018.

Alexander Salinas, M. B., Mathieu Boussard, NicolaBui, FrancoisCarrez, ChristineJarda, JourikDe Loof, (2013) ssne. ,"Internet of Things – Architecture IoT-A Deliverable D1.5 – Final architectural reference model for the IoT v3.0", Retrieved from: https://www.researchgate.net/publication/272814818_Internet_of_Things_-_Architecture_IoT-A_Deliverable_D15_-_Final_architectural_reference_model_for_the_IoT_v30 Accessed on May 3, 2017.

Beacons. (2017). Retrieved from https://developers.google.com/beacons/. Accessed on May 6, 2017.

Char, K. (2015). *Internet of Things System Design with Integrated Wireless MCUs*. Retrieved from https://www.silabs.com/documents/public/white-papers/Internet-of-Things-System-Design-with-Integrated-Wireless-MCUs.pdf Accessed on May 10, 2017.

Derhamy, H., Eliasson, J., Delsing, J., & Priller, P. (2015). A Survey of Commercial Frameworks for the Internet of Things. *2015 IEEE 20th Conference on Emerging Technologies Factory Automation (ETFA)* (pp. 1–8). https://doi.org/10.1109/ETFA.2015.7301661

FAQ—Frequently Asked Questions, MQTT. (2017). Retrieved from http://mqtt.org/faq. Accessed on May 6, 2017.

Fremantle, P. (2014). A Reference Architecture for the Internet of Things. *WSO2 White Paper*. https://www.researchgate.net/profile/Paul_Fremantle/publication/308647314_A_Reference_Architecture_for_the_Internet_of_Things/links/57ea00b708aef8bfcc963153.pdf. Accessed on May 3, 2017, September 10 2018.

Hannelore Marginean, D. K., Turan-SiTran. (2016, Jan 29). A Reference Architecture for the Internet of Things. https://www.infoq.com/articles/internet-of-things-reference-architecture. Accessed on May 3, 2017.

Li, Y., Hou, M., Liu, H., & Liu, Y. (2012). Towards a Theoretical Framework of Strategic Decision, Supporting Capability and Information Sharing under the Context of Internet of Things. *Information Technology and Management*, 13(4), 205–216. https://doi.org/10.1007/s10799-012-0121-1

MQTT Essentials Part 5: MQTT Topics & Best Practices. (2015, February 9). http://www.hivemq.com/blog/mqtt-essentials-part-5-mqtt-topics-best-practices. Accessed on May 10, 2017.

NodeMcu—An Open-Source Firmware Based on ESP8266 wifi-soc. (y.y.). Tarihinde 06 Mayıs 2017, adresinden erişildi http://nodemcu.com/index_en.html

Shrouf, F., Ordieres, J., & Miragliotta, G. (2014). Smart Factories in Industry 4.0: A Review of the Concept and of Energy Management Approached in Production Based on the Internet of Things Paradigm. *2014 IEEE International Conference on Industrial Engineering and Engineering Management* (pp. 697–701). https://doi.org/10.1109/IEEM.2014.7058728

Theodoridis, E., Mylonas, G., & Chatzigiannakis, I. (2013). Developing an IoT Smart City framework. In *IISA 2013* (pp. 1–6). DOI: 10.1109/IISA.2013.6623710, IEEE Explorer Digital Library.

TongKe, F. (2013). Smart Agriculture Based on Cloud Computing and IoT. *Journal of Convergence Information Technology*, 8(2), 210–216. https://doi.org/10.4156/jcit.vol8.issue2.26

Vaidya, M. (2016). Effective Processing of MQTT Protocol in Internet of Things (IoT) for Smart System. http://www.academia.edu/25290027/EFFECTIVE_PROCESSING_OF_MQTT_PROTOCOL_IN_INTERNET_OF_THINGS_IoT_FOR_SMART_SYSTEM. Accessed on September 7 2018

van Kranenburg, R. (2008). *The Internet of Things : A Critique of Ambient Technology and the All-Seeing network of RFID*. Amsterdam: Institute of Network Cultures. Amsterdam, September 2008. ISBN/EAN 978-90-78146-06-3 http://eartexte.ca/21011/, http://www.networkcultures.org/_uploads/notebook2_theinternetofthings.pdf, Accessed on Sep 10, 2018.

Wang, D. L. (2016). The Internet of Things the Design and Implementation of Smart Home Control System. In *2016 International Conference on Robots Intelligent System (ICRIS)* (pp. 449–452). https://doi.org/10.1109/ICRIS.2016.95

What Is Agile Software Development? (2015, June 29). , Retrieved from https://www.agilealliance.org/agile101/. Accessed on May 6, 2017.

Whitmore, A., Agarwal, A., Xu, L. D. (2015). The Internet of Things: A Survey of Topics and Trends. *Information Systems Frontiers*, 17(2), 261–274. https://doi.org/10.1007/s10796-014-9489-2

Xu, L. D., He, W., & Li, S. (2014). Internet of Things in Industries—A Survey. *IEEE Transactions on Industrial Informatics*, 10(4), 2233–2243. https://doi.org/10.1109/TII.2014.2300753

Yüksek, A.G., Arslan, H., Kaynar, O., Çifçi, G. (2016). *Smart Technology & Smart Management* (pp. 71–80), İzmir: Gülermat Matbaacılık.

Fatma Seray Demirkan*, Sona Mardikyan and Bertan Badur

A Scientometrics Study on Internet of Things (IoT)

1. Introduction

Antoine de Saint-Exupéry said, "The meaning of things lies not in the things themselves, but in our attitude towards them" ((De) Saint-Exupery, 1948). Likewise, for the Internet of Things (IoT) paradigm, the meaning does not lie in the objects but in the communication and connectivity between them. A simplified definition of the paradigm is denoted as follows: "The basic idea of this concept is the pervasive presence around us of a variety of things or objects – such as Radio-Frequency IDentification (RFID) tags, sensors, actuators, mobile phones, etc. – which, through unique addressing schemes, are able to interact with each other and cooperate with their neighbors to reach common goals" (Morabito, Giusto, Iera & Atzori, 2010).

The popularity of the IoT emanates from the development of the enabling technologies. RFID tags and the wireless sensor networks are foundations but there are others which are denoted in the various sources, such as identification and communication technologies, middleware, service-oriented architecture, data and signal processing, power and energy storage technologies, network technology and more (Atzori, Iera & Morabito, 2010; Bandyopadhyay & Sen, 2011; Vermesan, Guillemin & Friess, 2009; Xu, He & Li, 2014.

Big data is a prominent concept which has an intertwined relationship with IoT. The prevalent usage of IoT generates a huge amount of data both in quantity and diversity, which helps the development of big data. Meanwhile, big data analytics assists the developments of new business models and research areas of IoT (Chen, Mao & Liu, 2014).

IoT is dramatically gaining ground in the modern world not only because of its ubiquitous transformation effect on daily life such as healthcare systems and learning environments but also because it disrupts the business processes such as logistics and manufacturing. The application of this novel paradigm, therefore, becomes prominent. Main application areas are discussed in literature in different domains such as transportation and logistics, healthcare, smart environments,

---

* Corresponding author: Fatma Seray Demirkan, f.seraydemirkan@yahoo.com, İstanbul, Turkey.

personal and social domain (Atzori, Iera & Morabito, 2010) as well as personal and home, enterprise, utilities and mobile domains (Gubbi, Buyya, Marusic & Palaniswami, 2013). Also, there are some researches which give industry-specific applications (Bandyopadhyay & Sen, 2011; Xu, He & Li, 2014). Detailed application areas can also be found in the IoT strategic roadmap of 2009 of European Commission (Vermesan, Guillemin & Friess, 2009).

Future research directions and open challenges are widely discussed in the literature. Main concerns entail the lack of standards, security and privacy of personal information. Further discussions can be found on these papers (Al-Fuqaha, Guizani, Mohammadi, Aledhari & Ayyash, 2015; Bandyopadhyay & Sen, 2011; Gubbi, Buyya, Marusic & Palaniswami, 2013; Miorandi, Sicari, De Pellegrini, & Chlamtac, 2012); Stankovic, 2014).

The bilateral presence of this emerging issue in the everyday life of the people, as well as the business environment, makes this topic intriguing for distinct milieus. Therefore, this cutting-edge paradigm is not only researched by electronics, computer engineering and other technical areas but also business environments intent to keep themselves up to date on latest developments. Management Information Systems can play an important role to bind different perspectives of the issue.

The development of web and social networking and easy access to large databases pave the way to the development in the network sciences. Besides, it increases the importance of scientometrics in the evaluation of research entities. The very first definition of scientometrics is stated as "the quantitative methods of the research on the development of science as an informational process" (Nalimov & Mulchenko, 1971).

Even though there are some other study areas in scientometrics, the most important notion is citation. Citation creates linkage among authors, ideas, journals and institution in time, which then can be analyzed quantitatively as networks (Mingers & Leydesdorff, 2015). Therefore, citation data has started to be used for the modeling of network dynamics (Leydesdorff & Milojević, 2015).

Scientometrics puts an emphasis on communication in the sciences. According to Mingers and Leydesdorff (2015), its main themes can be specified as "measuring research quality and impact, understanding the process of citations, mapping scientific fields and the use of indicators in research policy and management."

This study aims to investigate the literature of IoT in terms of identifying the growth, trends and patterns to reveal the hidden relationship between disciplines by social network analysis based on the Web of Science bibliometric data.

This study covers 1333 articles and 6047 citing articles which are refined by categories such as "computer science and information systems", "management" and "business" and also by the period from 2000 to 2017 in Web of Science database.

This chapter entails four sections. The methods section explains the data collection process and the tools which are used to create the results of section findings such as InCites and VOSviewer. The findings section displays the analysis of author-given keywords, countries, authors, journals and organizations. The final section will be discussions and conclusions.

2. Methods

2.1 Data Collection

Web of Science online database is used as the main source of data. Primarily, the search inquiry is created as "Internet of Things" or "IoT" and both of the concepts are chosen as topics, the time span is selected as 2000 to 2017 regarding the emergence of the concept and no other parameters are used in this step. At the second step, search results are refined based on the Web of Science categories and the document types. To create more solid and relevant relationship with Management Information Systems, three categories have been chosen for refinement which are "computer science and information systems", "management" and "business". In addition to the category, "article" is chosen as the document type to refine the analysis.

Web of Science database has a dynamic structure, therefore, the number of results found changes over time. The data is retrieved on April 23, 2017. It includes 1333 articles from Web of Science core collection. To better understand interdisciplinary collaboration and map the relatedness of different disciplines, cited references of articles are also included in the data. Ultimately, 6047 records of citing articles without self-citation have been obtained from Web of Science database for the network analysis.

2.2 Tools Used

InCites is defined on the website of Thomson Reuters as "a customized, web-based research evaluation tool that allows you to analyze institutional productivity and benchmark your output against peers worldwide" (Thomson Reuters InCites, 2017). With an internal functionality of Web of Science, the search results dataset can be saved to the InCites. Soon after, robust visualization tools assist to create a range of descriptive analysis including bar graphs, treemap, trend graph and radar chart on entities such as people, organization, region, research areas and journals.

VOSviewer is a software tool which is used to produce and analyze bibliometric networks. VOS stands for visualization of similarities and the program uses a special unified mapping and clustering technique which was also developed by the researchers Nees Jan van Eck and Ludo Waltman at Leiden University in the Netherlands. This tool is freely available at http://www.vosviewer.com for research purposes and has a focus on the graphic representation of bibliographic networks. It was first introduced in the paper (van Eck & Waltman, 2010) and the technical details can be found in other articles and book chapters (Measuring Scholarly Impact, 2014; Waltman, van Eck & Noyons, 2010).

3. Findings

Descriptive analysis of the current literature is presented to give insight into the evolution of IoT, to reveal top contributing countries, authors, journals and organizations of the field. Additionally, the network analysis is conducted to discover co-occurrence of keywords and co-authorship of countries as well as organizations.

Fig. 1: Published Items Each Year. (Figures are generated by ourselves using InCites and VOSViewer).

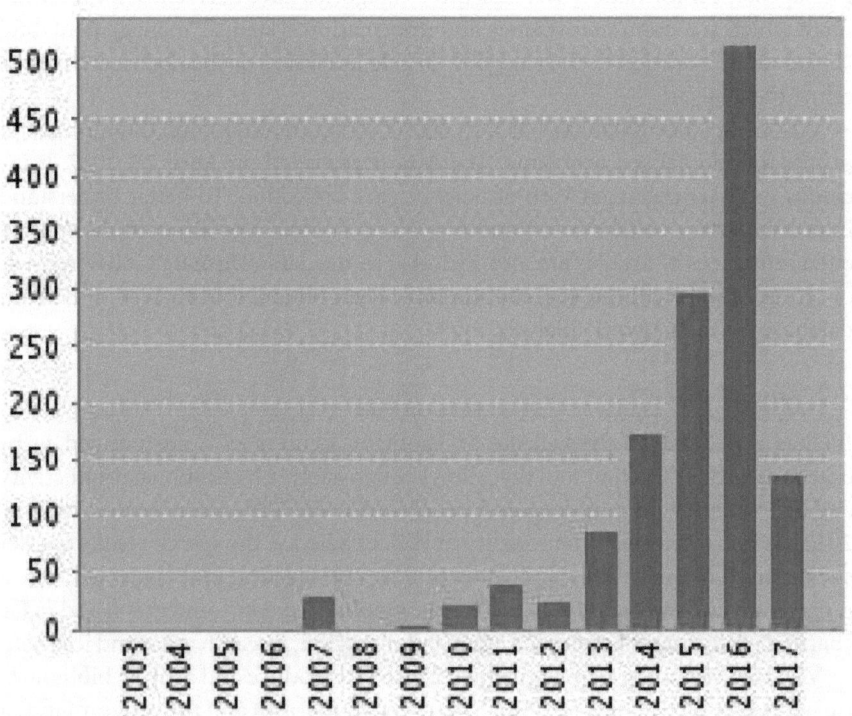

Fig. 1 shows that there is a growing trend on the IoT literature. From 2014 on-wards, the number of published items has increased exponentially.

Fig. 2: Citations Each Year.

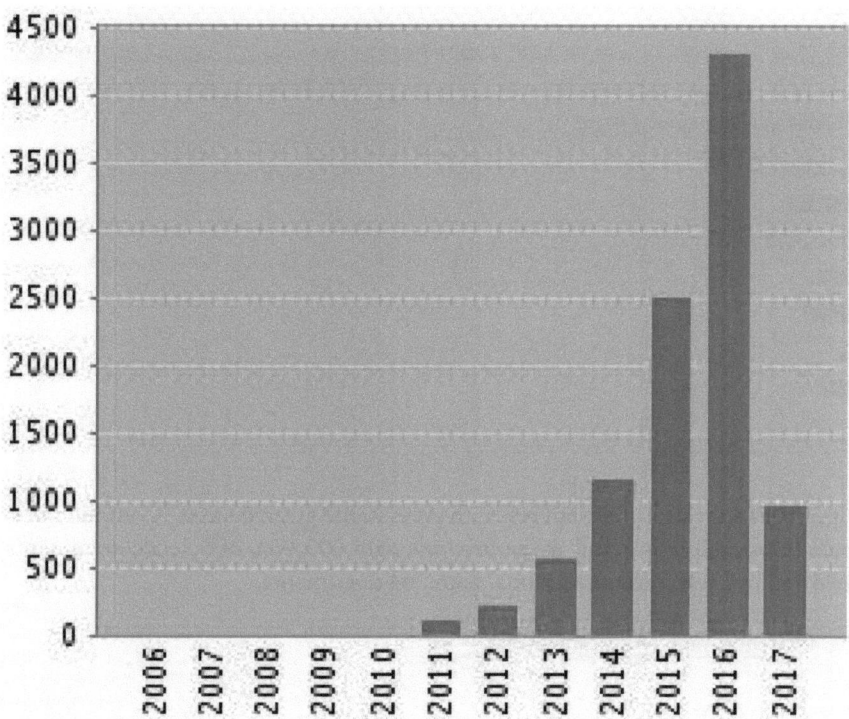

Fig. 2 shows the same patterns with Fig. 1. Citation counts have increased dramatically especially over the last three years.

Fig. 3: Top Ten Research Areas

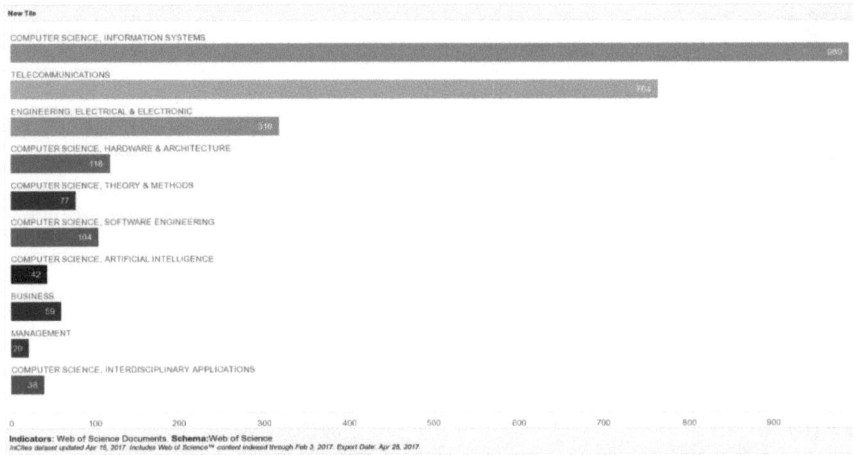

Fig. 3 shows the most relevant research areas in the given dataset. "Computer Science Information Systems" is the first one with 999 Web of Science documents followed by "Telecommunications" with 764 documents.

Fig. 4: Trend Graph of Research Areas.

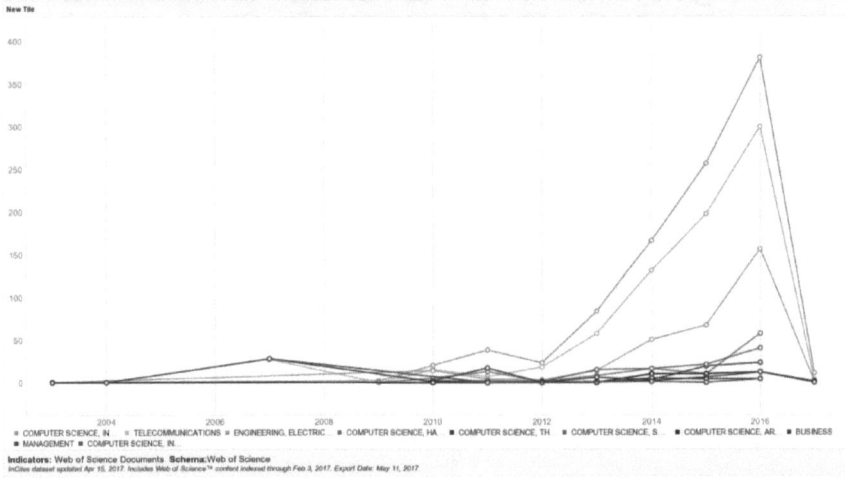

Fig. 4 illustrates the trend graph of top ten research areas based on their number of Web of Science documents. Almost all of the research areas have increased their production on the IoT literature for the last years but "Computer Science Information Systems" and "Telecommunications" outscore all the others.

3.1 Analysis of Author Keywords

Fig. 5 is the co-occurrence network of author keywords. The minimum occurrence of a keyword is selected as 10. One hundred and sixty two keywords meet the threshold. Undoubtedly, IoT is the most frequent one but the other words which are profusely used with it are RFID, Cloud Computing, Big Data, 5g, security and wireless sensor networks. This figure helps us to gain insights into IoT-related issues, enabling technologies and future directions.

Fig. 5: Co-occurrence of Author Keywords.

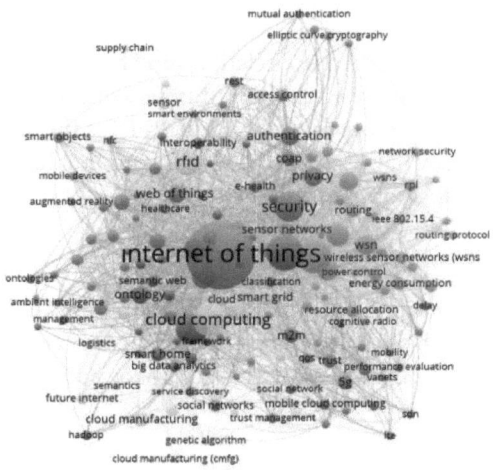

Fig. 6: Co-occurrence of Author Keywords Overlay.

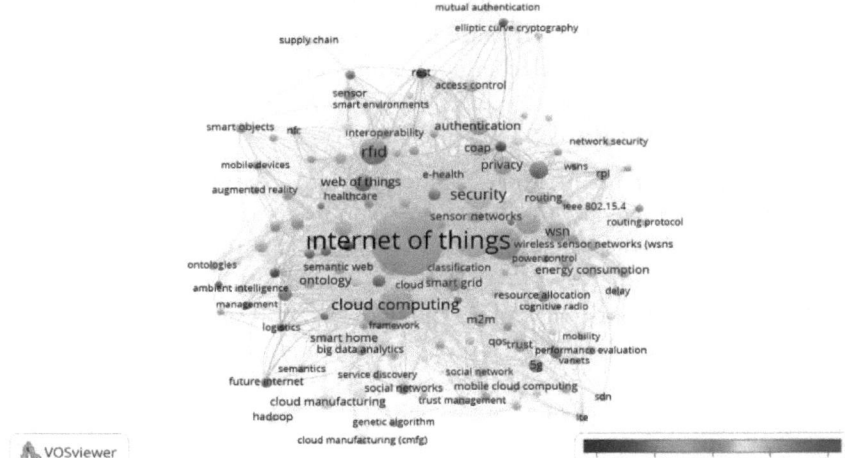

Fig. 6 shows how the co-occurrence relationship changes over time. There is an increasing trend in the past few years and this network shows which terms have been associated with IoT more recently. While RFID is the oldest concept, cloud computing and security have appeared later. It seems that big data and 5g are the most recent keywords associated with IoT.

3.2 Analysis of Countries

Fig. 7 illustrates the ten countries (according to first author's country) which make more contribution to the IoT literature. China is the first and the United States is the second most productive country. They are followed by South Korea and the United Kingdom.

Fig. 7: Top Ten Contributing Countries.

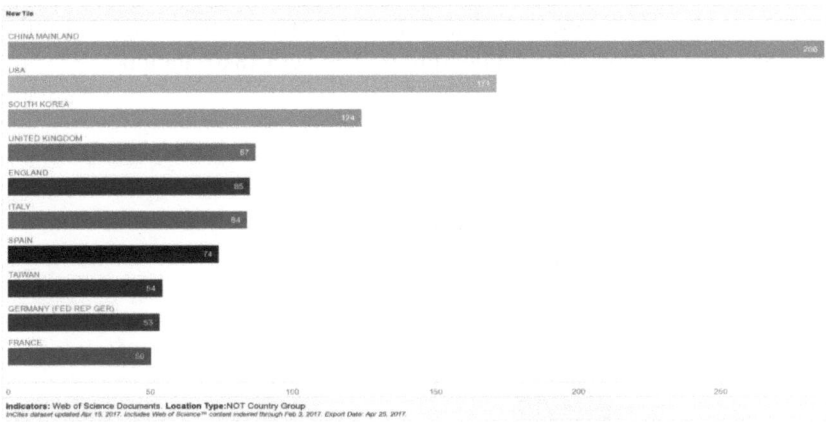

Fig. 8: Co-authorship of Countries.

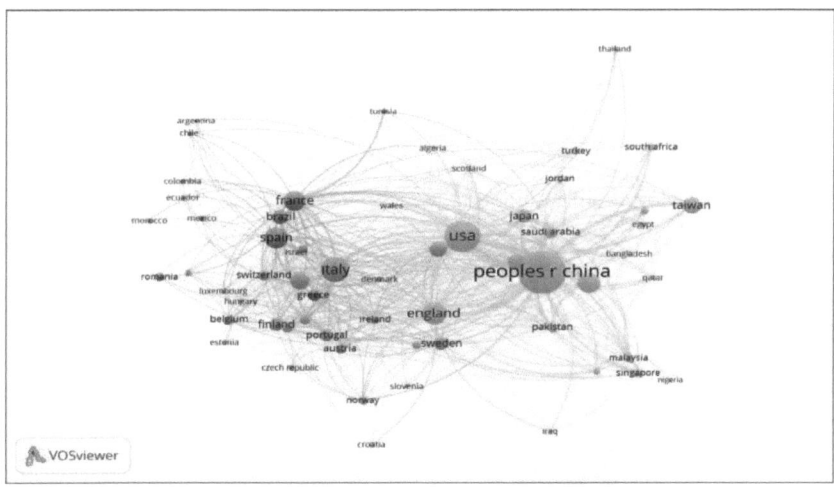

Fig. 8 displays the co-authorship relationship of countries. To create this figure, a minimum number of documents of a country and a minimum number of citations of a country are selected as five for clarification. Sixty-four countries meet this threshold. Each node named with a country and the size of the nodes represents the number of documents written. The number of the links and their strengths between two countries shows the level of collaboration between these countries. It is apparent that there is a strong collaboration between China and the United States.

3.3 Analysis of Authors

Fig. 9 shows the top ten most-cited authors. Luigi Atzori is the most-cited author with 1887 total citation followed by Antonio Iera with 1840 citation.

Fig. 9: Top Ten Authors Based on the Total Citation.

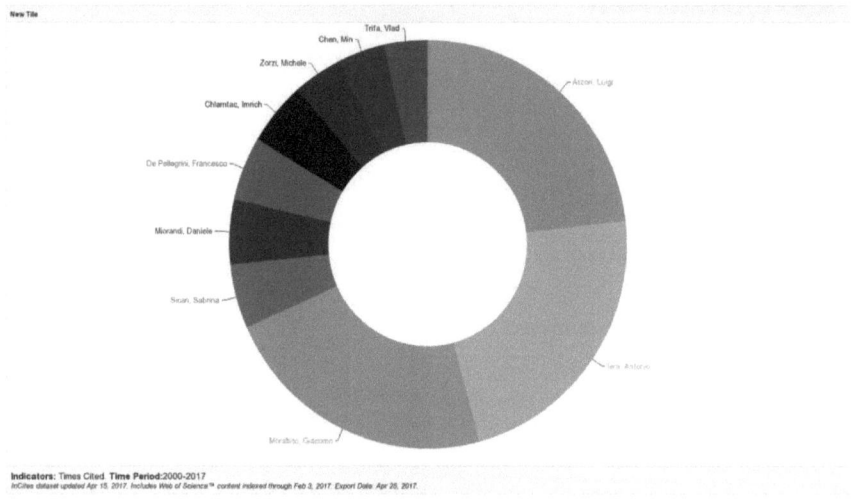

Fig. 10: Top Ten Most Prolific Authors.

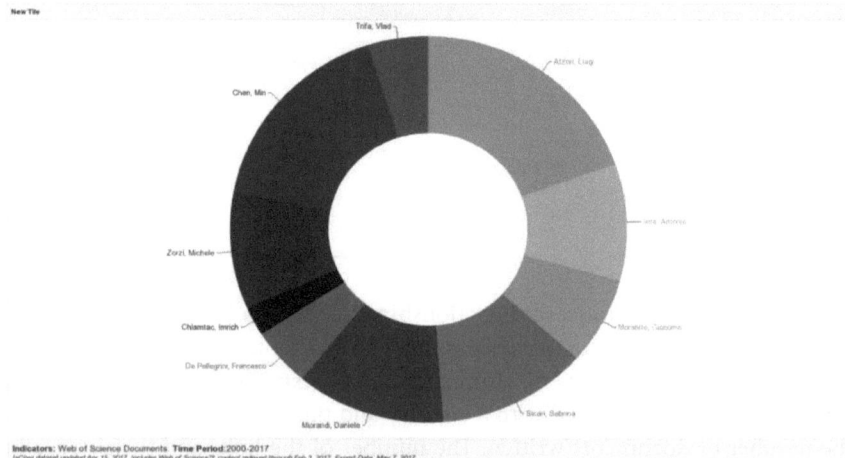

Fig. 10 shows the most prolific author as Luigi Atzori with eight documents who is followed by Min Chen with seven documents.

3.4 Analysis of Journals

Fig. 11 shows the most preferred ten journals according to the indicator of Web of Science Documents which publish articles related to IoT. *International Journal of Distributed Sensor Networks* has the highest number of publications with 132 documents in IoT field.

Fig. 11: Top Journals in the Field of IoT Preferred for Publications.

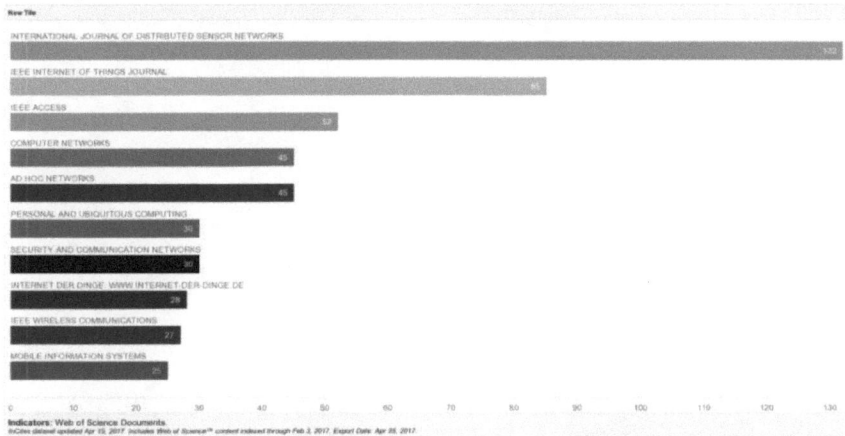

Fig. 12: ost-Cited Journals.

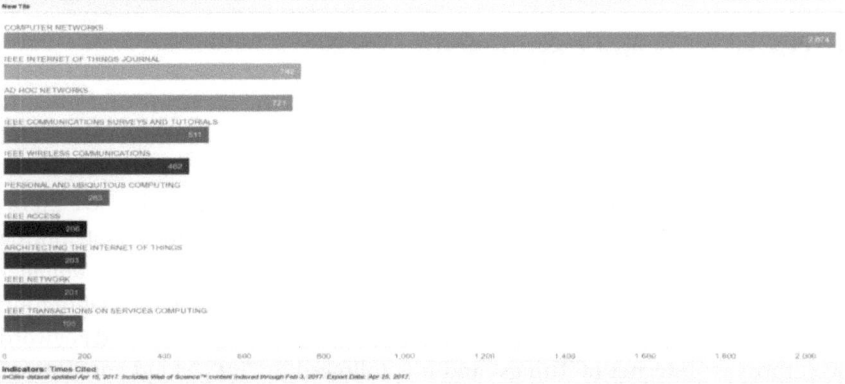

Fig. 12 describes top ten most-cited journals. *Computer Networks* is the most-cited one with 2074 citations followed by *IEEE Internet of Things Journal* with 742 citations.

3.5 Analysis of Organizations

Fig. 13 illustrates that most of the collaboration between organizations are densely concentrated around Chinese universities.

Fig. 13: Co-authorship of Organizations.

4. Discussions and Conclusions

The aim of the study is to understand the nature of the literature and identify core areas and key concepts to provide a roadmap for the future research and researchers in the IoT field.

Analysis of the data shows that there is a significant growth in the IoT literature over the past few years especially after 2014. China is the most contributing country. The most prolific and the most-cited author is the Luigi Atzori. *Computer Networks* is the most-cited journal with 2074 citation. The strongest collaboration links are between China and the United States. The most frequently used keyword by authors is "Internet of Things" and it is followed by "iot", "cloud computing", "security", "wireless sensor networks", "big data" and "Rfid".

References

Al-Fuqaha, A., Guizani, M., Mohammadi, M., Aledhari, M., & Ayyash, M. (2015). Internet of Things: A Survey on Enabling Technologies, Protocols, and Applications. *IEEE Communication Surveys and Tutorials*, 17, 2347–2376.

Atzori, L., Iera, A., & Morabito, G. (2010). The Internet of Things: A Survey. *Computer Networks*, 54(15), 2787–2805.

Bandyopadhyay, D., & Sen, J. (2011). Internet of Things: Applications and Challenges in Technology and Standardization. *Wireless Personal Communications*, 58(1), 49–69.

Chen, M., Mao, S., & Liu, Y. (2014). Big Data: A Survey. *Mobile Networks and Applications*, 19(2), 171–209.

(De) Saint-Exupery, A. (1948). Citadelle. *Paris: Gallimard.*

Gubbi, J., Buyya, R., Marusic, S., & Palaniswami, M. (2013). Internet of Things (IoT): A Vision, Architectural Elements, and Future Directions. *Future Generation Computer Systems*, 29(7), 1645–1660.

Leydesdorff, L., & Milojević, S. (2015). Scientometrics. *In J. D. Wright, & M. Lynch (Ed.), International Encyclopedia of the Social & Behavioral Sciences (Second Edition) (pp. 322–327). Elsevier.*

Ding, Y., Rousseau, R., & Wolfram, D. (2014). *Measuring Scholarly Impact.* DE: Springer Verlag.

Mingers, J., & Leydesdorff, L. (2015). A Review of Theory and Practice in Scientometrics. *European Journal of Operational Research*, 246(1), 1–19.

Miorandi, D., Sicari, S., De Pellegrini, F., & Chlamtac, I. (2012). Internet of Things: Vision, Applications and Research Challenges. *Ad Hoc Networks*, 10(7), 1497–1516.

Morabito, G., Giusto, D., Iera, A., & Atzori, L. (2010). *The Internet of Things.* DE: Springer Verlag.

Nalimov, V.V., & Mulchenko, Z.M. (1971). Measurement of Science: Study of the Development of Science as an Information Process. U.S. Air Force Systems Command, Foreign Technology Division.

Stankovic, J. A. (2014). Research Directions for the Internet of Things. *IEEE Internet of Things Journal*, 1(1), 3–9.

Thomson Reuters InCites. (2017). Clarivate Analytics- Research Analytics: Retrieved from Clarivate Analytics- Research Analytics: https://clarivate.com/products/incites/

van Eck, N. J., & Waltman, L. (2010). Software Survey: VOSviewer, A Computer Program for Bibliometric Mapping. *Scientometrics*, 84(2), 523–538.

Vermesan, O., Guillemin, P., & Friess, P. (2009). *Internet of Things Strategic Research Roadmap.* Brussels: The Cluster of European Research Projects.

Waltman, L., van Eck, N. J., & Noyons, E. C. (2010). A Unified Approach to Mapping and Clustering of Bibliometric Networks. *Journal of Informetrics, 4*(4), 629–635.

Xu, L. D., He, W., & Li, S. (2014). Internet of Things in Industries: A Survey. *IEEE Transactions on Industrial Informatics,* 10(4), 2233–2243.

Çağrı Doğu and Tuncay Ercan*

Secure Management Model for Scada Systems

1. Introduction

It is a well-known fact that the use of the networking technologies increases communication, sharing of information and mutual interaction among the different systems. The Internet of Things (IoT) is a new technological concept that the intelligent devices (different sensors, application outputs, production data in factories) somehow communicate with each other and form an intelligent and autonomous network based on end-to-end digitalization concept. Together with the developments in technology, many sensor devices can be integrated into the internet environment through WSN (Wireless Sensor Networks) systems (Akyıldız et al., 2002). Real-time continuous stream of data coming from the sensing devices that are used in the environment and from the data terminals in the production systems of a factory are transformed to be used by information systems (storage, database, application services) provided by service providers (like cloud computing services) on the Internet. This kind of Industrial IoT (IIoT) information flow will result in changes that can positively affect our daily life, business life and industrial production systems (Lojka, Bundzel & Zolotová, 2016).

The technological advances based on scientific discoveries in different fields, called the Industrial Revolution, back from the 18th century up to today had three important stages with remarkable features according to the industrial systems used in different periods. Starting with mechanical production systems using steam power, the process has left serial production with the help of electric power since the beginning of the 20th century, and production has become fully automatic with digital revolution, electronic systems and information technologies started in the 1970s. Rapidly growing industrial automation, coupled with internet technology, is moving toward the era of intelligent production, which is called the Industry 4.0 or 4th Revolution (Brettel, Friederichsen, Keller & Rosenberg, 2014).

---

*   Corresponding author: Doç.Dr.Ahmet Tuncay Ercan, Yaşar University Computer Engineering Dept., Bornova, İzmir, https://tercan.yasar.edu.tr/, tuncay.ercan@yasar.edu.tr, Tel: 0090 232 570 8241.

The integrated implementation of IoT and industrial automation systems is called IIoT (Lin et al., 2015). Intelligent production machines with IoT capability automatically communicate with each other over the network to control production and minimize operator contribution in the areas such that:

- Mechanical and electrical failures can be anticipated to reduce downtime due to failure.
- Rapid detection of raw material deficiency.
- Factory managers can receive production and malfunction information from any part of the world in real time.
- Every detail in Supply Chain Data Management can be shared with distribution channels and customers.

SCADA is an abbreviation for "Supervisory Control and Data Acquisition". It is the general name of a system including computers, communication equipment, sensors or other devices. In a SCADA system, continuous operational data coming from Programmable Logic Controller (PLC) systems in factories, distributed control systems and sensors within the production systems are being stored and evaluated real-time and transferred to end-users as reports or immediate warning for any negative event in the overall system. Information is displayed in a central Human Machine Interface (HMI) with graphics and can also be remotely accessed and monitored by Remote Terminal Unit (RTU). SCADA can be classified energy (electric, natural gas, thermal water, water, oil, etc.) or process control systems (industrial processes in a factory automation).

Although SCADA is generally a system used for industrial control purposes, this control process is executed by means of the tools like RTU, PLC and software planning systems like ERP (Enterprise Resource Planning). In this regard, it is not only a matter of computer engineering, but also an interdisciplinary area that includes different engineering fields such as electrical-electronics, machinery, computer, software, industry, and mechatronics. Although cloud computing is recently becoming a new environment for SCADA applications, its on-demand network access to a shared pool of configurable computing resources including networks, servers, storage, applications and services can be rapidly provisioned by service provider interactions. New cloud services like scalable data management and big data analytics allow new decision-making processes in SCADA systems (Church et al., 2015).

In this work, we mainly focus on SCADA systems by giving necessary architectural information and explaining the features of a working SCADA system in thermal energy power generation. The remaining part of the work is organized as follows. The next section explains the methodology followed in the conceptual framework of industrial SCADA systems. It explains the general SCADA architecture and ORC SCADA architecture we experienced. Then we present our management methodology and security perspectives in this system. We will explain our findings as advantages and disadvantages by comparing generic system features. We will conclude in the last section.

2. Methods

We all know that computer systems and networking technologies pass through rapid changes and this makes Information and Communication Technologies (ICT) one of the most developing sectors. There is no doubt that ICT is highly related with critical information systems in different industrial sectors. Therefore, SCADA became more important for many application areas (particularly in the energy sector) and is required to be used by the National Organizations of Energy Market Regulatory. SCADA has a subgroup of industrial control system (ICS) and another subgroup for distributed control system (DCS) in geographically distributed locations. ICSs can be used in different industries (electric, water, oil, gas, etc.). DCSs are supervisory and regulatory control systems and generally used to control production systems within a factory.

2.1 General SCADA Architecture

SCADA concept was first introduced in the mid-20th century and based on several production floors, industrial facilities and personnel to manually control and monitor remote sites for pushing alert buttons and making urgent analogue calls with the people in charge. The term "SCADA" emerged in the early 1970s, and for decades the rise of microprocessors and PLCs has increased the ability to monitor and control the automation processes of enterprises more than ever. The latest developments in technology have enabled automated SCADA systems according to the company with maximum efficiency at low cost. SCADA has some auditing and data collection systems running behind the scenes in almost every plant or in any workplace setting up a network (Hayden, 2014).

Advancements in Intelligent Instrumentation and RTUs/PLCs have made the process-control solutions to be easily managed and operated by a SCADA system. SCADA is an industrial computer-based control system employed to gather and analyze the real-time data to keep track, monitor and control industrial equipment in different types of industries. PLC is an automation device used in the control of processes such as control of machines or production departments in factories. Unlike normal computers, the PLC has many inputs and outputs (I/O). PLC plays a big role in the foreground of factors such as producing more and better-quality products in a short time, producing with very low error rates. General architecture of SCADA system is given in Fig. 1.

Fig. 1: General Architecture of SCADA. . Reprinted from Types of Network Architecture, In Dentrodelasala, n.d., Retrieved April 28, 2017, from http://dentrodelasala.com/wp-content/uploads/2018/09/types-of-network-architecture-on-architecture-intended-for-scada-network-architecture-on-architecture-with-scada-system.png.

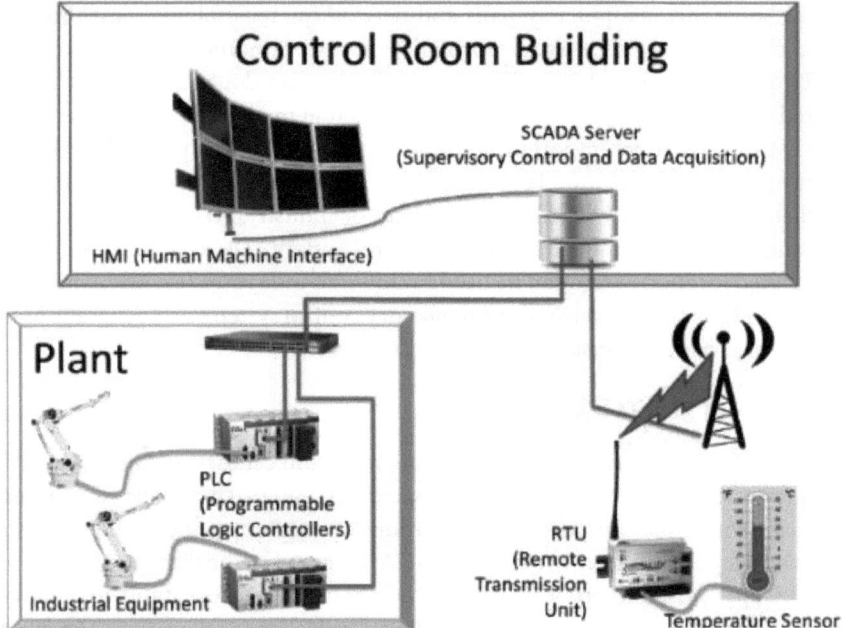

SCADA systems are basically software that can be used to monitor a wide area of facilities from a single center with devices such as computers, mobile phones or tablets. It can be used from a single device and can be controlled and monitored with multiple computers and portable devices via network connections. Together with the communication standards like RS-232, RS-422 and RS-485, SCADA uses the real-time encapsulated PROFINET protocol with TCP/IP. After all industrial devices communicate with PROFINET, industrial networks can be easily deployed and controlled in every network layer. Each command executed by operators is traceable with packet-by-packet. A physical industrial LAN schematic is given in Fig. 2 (PI, 2015).

Fig. 2: A Physical Industrial LAN Schematic. Reprinted from PROFINET network qualification by Peter Thomas, In PI-PROFINET PITC's OCTOBER 2015, n.d., Retrieved May 15, 2017, from https://image.slidesharecdn.com/profinetnetworkqualification-peter thomas-oct2015-151012075302-lva1-app6891/95/profinet-network-qualification-peter-thomas-oct-2015-3-638.jpg?cb=1445260571

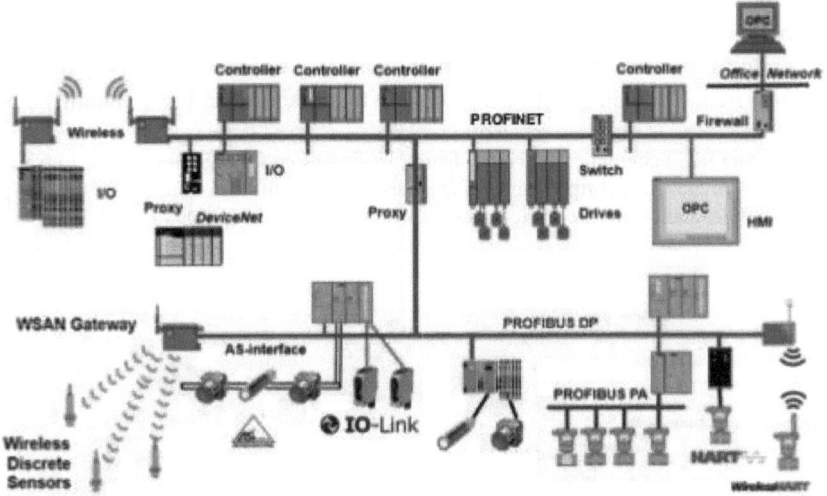

Main PLC, auxiliary PLCs, electronic protection and locking systems, motor control units are the main control units. The main part keeps the entire system under control and has the following characteristics:

- Multiple master stations should be able to talk at the same time through separate communication channels.
- RS-232, RS-485 physical communication layer, copper and fiber optic physical environment should be able to communicate with multiple protocols.
- Must have the ability to be easily expanded, configured and maintained.
- It should be able to carry out its own tests; the faults that occur should have a structure that stimulates both itself and the SCADA center.
- Redundancy is the most important aspect of continuing the production. When replacing a faulty module, there must be a hardware structure that does not require cutting of the energy.

2.2 ORC SCADA Architecture in Use

Organic Ranking Cycle (ORC) is currently used in geothermal energy utilities in order to generate electrical power (Ozden & Paul, 2011).Operators control the energy generation in ORC using HMIs. In ORC utilities, there are more than 1000 sensors in a typical site. These sensors collect data from all production and re-injection wells, brine transfer pumps, pentane levels, volume tank, turbines, generator and much more things which work in a utility. Operators manage all of these with HMIs in SCADA UI. ORC system is shown in Fig. 3 (Singh, 2009).

Fig. 3: An ORC Schematic. Reprinted from , In Turboden ORC Systems, n.d., Retrieved February 23, 2017, from https://image.slidesharecdn.com/introduction-140115161937-ph papp01/95/organic-rankine-cycle-macro-power-plant-11-638.jpg?cb=1389803359

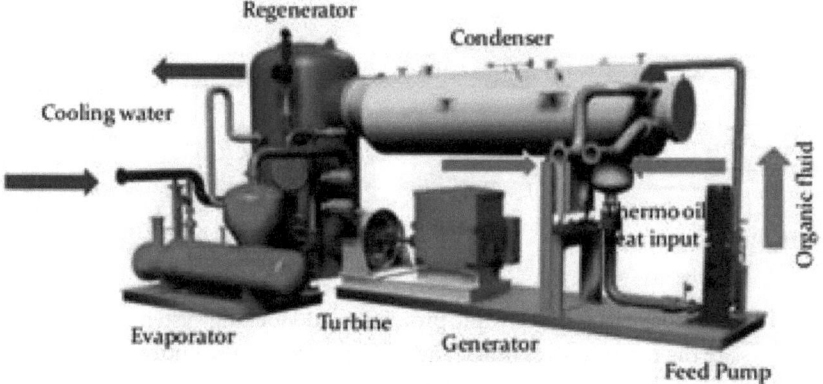

These types of industrial equipment come with PLCs in the SCADA system with an RTU capability. PLCs go to a network environment and communicate with HMI through special protocols. Moving all these devices to a network environment requires a series of safety precautions, although facilitating our work for more automation. Automation firms are not so sensitive for scheduled updates in control devices and even in the current operating system (Knapp & Board, 2011).

2.3 Management of SCADA

Main management issues in the facilities operated by the SCADA system are maintenance activities, efficient management of workforce planning and ensuring security measures. They can be listed as follows:

- Physical check for cleanup at scheduled periods.
- Electronically check for input voltages and communication in monitors.
- Software updates published by vendors.
- Backing up industrial PCs and PLC files in scheduled times to encrypted storage device.

2.4 Security of SCADA

Nowadays, cybersecurity threats like APTs (Stuxnet, Night Dragon, etc.) are a new phenomenon. As a result, there are many old systems that may be vulnerable to cyber-attacks, because cybersecurity was not a simple idea at the time of initial design and installation.

Complete solutions are given in Fig. 4.

Fig. 4: Complete Industrial Cybersecurity Solutions (Honeywell). Reprinted from Complete Industrial Cyber Security Solutions In Honeywell Process Solutions, n.d., Retrieved April 02, 2017, from https://image.slidesharecdn.com/honeywellindustrialcybersecurity-150427025213-conversion-gate01/95/honeywell-industrial-cyber-security-2-638.jpg?cb=1435649249

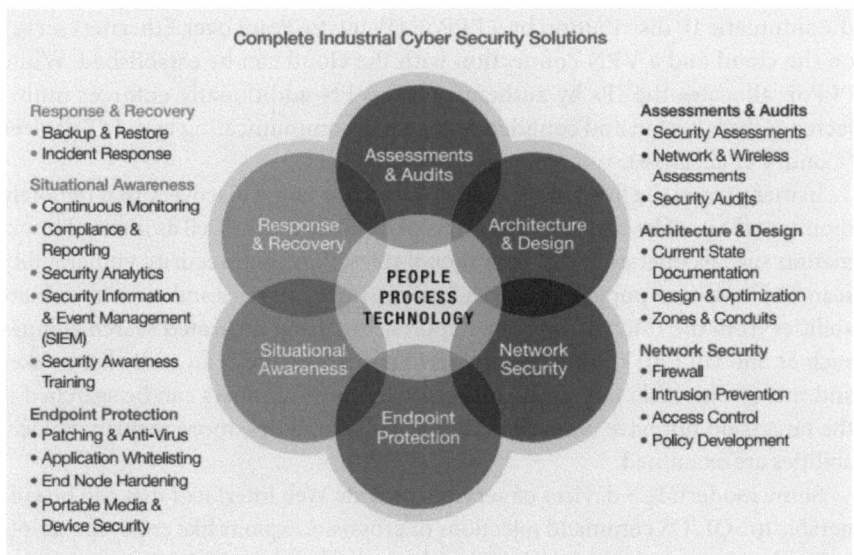

Power generation facilities, metropolitan traffic control systems, water treatment systems and factories are all at risk. Exploits freely available on the Internet make the ICS of leading vendors easy targets for attackers. They require a rugged and reliable security gateway solution to detect threats and control access to critical components in industrial network. The security gateway must detect industrial packets (MODBUS, BACnet, CIP, DNP3, IEC-60870-5-104, ICCP, MMS, OPC, PROFINET, Step7, etc.) and learn commands of SCADA. Then, it can define thresholds for industrial components of separated operator PCs. If infected PC sends a wrong command, the firewall can stop the communication of PLC from industrial network. It must install NGAV (next-generation anti-virus) and zero-day malicious detector software to industrial PCs (HMIs) to protect from risks. OS updates are also mandatory to fix the vulnerabilities in the OS. These systems must warn operators and information technology (IT) admins with e-mails or intranet messaging by warning and error information messages (Kobara, 2016).

Since every IoT device has an internet connectivity like 6LowPAN going to a direct server, this will not be an efficient choice for security (Hui, Culler & Chakrabarti, 2009). In our case, we applied SaaS cloud computing service that automatically provides load balancing, Dynamic DNS, VPN and hash mecha-nism features. These settings will enable both an instant data stream and an anonymous connection to IoT devices. Cloud server provider may take over the automatic IP distribution by a PPPoE (Point-to-Point over Ethernet) server on the cloud and a VPN connection with the cloud can be established. While PPPoE allocates the IPs by authentication, VPN additionally enforces multi-factor authentication and confidentiality when communicating with IoT devices (Condry et al., 2016).

In many cases, the first step in a cyber-attack is a target discovery that remotely monitors the profiles and configurations of destinations, as well as internal infor-mation such as operators and operational roles. Ports and security vulnerability scanners have been popular to search for open ports, services and security vulner-abilities from the Internet, but other approaches using dedicated search engines such as Shodan (2017) have become serious, because they can easily list weaker and more vulnerable targets. IP addresses and port numbers can be searched if the targets do not have publicly disclosed vulnerabilities; more security vulner-abilities are examined.

Some modern ICS devices or services provide Web interfaces that can be vul-nerable to SQL/OS command injections or cross-site exploits like cross-site script-ing and cross-situational fraud. It can also provide inappropriate remote access control mechanisms like default IDs, passwords for authentication and access control schemes. Additional jumping mechanisms that can be written manually cope with the loss of passwords. Another security measure for Internet discovery is to put devices and servers behind a firewall. ICS/SCADA honeypots are useful for understanding their discovery activities. They imitate the behavior of common industrial control protocols and monitor activities related to them. These honey pots can be created using CONPOT. Telescoping devices have IoTPOT, which is common in some IoT devices (Kobara, 2016; Pa et al., 2015).

Risk factors of any infrastructure should be identified under the name of the risk management framework of the organizations and short-, medium- and long-term security measures are planned. These measures can be examined by analyzing universal rules, cyber resources, preliminary risk analysis, threat and preparation levels and cyber threat tools (ISACA Journal, 2014; Stouffer, Pillitteri, Abrams and Hahn, 2015).

3. Findings

The occurrence of any breakdown in the power generation plant should be intervened quickly. The SCADA system we experienced in the plant uses SaaS in cloud computing. Security and management information can be displayed in accordance with the user's requests. Thousands of sensors connected to the ICS and DCS infrastructure ensure real-time data simulations within the system. Many data can also be collected from RTUs. In the selection of SCADA systems for energy sector, including more than one plant, the management capability includes several operating zones together with maintenance, cost and separate installation criteria. That SCADA applications should also be compatible with external applications in the company and support Turkish language as well is an important criterion in system administration. Thus, hidden additional costs that will arise after installation are reduced in advance (Moness & Moustafa, 2016).

Current technologies will have security weaknesses unless being constantly updated. In order to protect against these security weaknesses, it is necessary to educate employees. In some production sites, unfortunately, SCADA has not been produced and no automation has been considered (U.S. General Accountability Office, 2011).

ICS is an indispensable capability in the management of plants. Turkish Electricity Transmission Company (TEIAS) continuously communicates with the RTU to read the instant 154KV output values in the plant and measure power quality. In order to read the output quality, TUBITAK (Scientific and Technological Research Council of Turkey) also receives all data from its servers via the NTP protocol discussed by Alcaraz and Zeadally (2013).

This remote access through the internet connection should be symmetric. The DoS prevention system on the Internet service provider should be activated with certainty. Thus, the speed of symmetrical internet will always be constant. The seventh OSI layer firewall should be installed to read SCADA packets between HMI and sensors. All VLANs should be separated from each other and careful switching rules should be defined on the firewall. Hazardous packages should be blocked by opening IPS/IDS. All user devices must be easily identified with the help of the active directory server to be installed inside. It may be necessary to make group policy settings to restrict the people who will use SCADA. Using an MFA will be a good security measure if it is going to be accessed from outside using VPN. It is not necessary to forget the human factor in SCADA systems. Operators may have inappropriate security clearances or abuse the system. For this reason, training and supervision is very important.

Authors state that the faults in the system should be examined in two main categories, internal and external. These faults cause a definite stance in production systems. Therefore, the system must have a redundant structure (Alcaraz and Zeadally, 2013).

4. Discussion and Conclusions

SCADA is a very important business tool to produce efficient and better-quality products at minimum cost in order to reduce dependency on human power, to provide life and property security and to use resources efficiently. The SCADA systems should be planned, installed and managed in critical industrial systems. Together with many application areas, energy sector is one of the compulsory areas that need to be SCADA system required by the National Organizations of Energy Market Regulatory. This organization collects energy production data in all energy utilities and manage market control, purchase and sale prices in the energy sector.

The system also checks interruption records in the energy plants and updates daily pricing in the energy market. If necessary, the whole system is managed from the TEIAS center to remotely monitor possible energy interruptions, open cutters and maneuver for continuity of supply remote monitoring system (Gungor & Lambert, 2006; Official Gazette of Turkey, 2012).

Every IP-enabled device from the Internet is actually connected to each other. Therefore, it is necessary to maintain and develop the credibility of a critical infrastructure in the country. With industry 4.0, it is inevitable to protect these networks in the country and provide a continuous working scheme.

References

Akyıldız, L. F., Sankarasubramaniam, Y., Su, W., & Cayırcı, E. (2002). Wireless Sensor Networks: A Survey. *Journal of Computer Networks*, 38, 393–422.

Alcaraz, C., Zeadally, S. (2013). Critical Control System Protection in the 21st Century. *IEEE Computer*, 46(4), 74–83.

Brettel, M., Friederichsen, N., Keller, M., & Rosenberg, M. (2014). How Virtualization, Decentralization, and Network Building Change the Manufacturing Landscape: An Industry 4.0 Perspective. *Journal of Mechanical, Aerospace, Industrial, Mechatronic and Manufacturing Engineering*, 8, 37–44.

Church, P., Mueller, H., Ryan, C., Gogouvitis, S. V., Goscinski, A., Haitof, H., & Tari, Z. (2015). Moving SCADA Systems to IaaS Clouds. *2015 IEEE International Conference on Smart City/SocialCom/SustainCom Together with DATACOM 2015 and SC2 2015*. Chengdu, China

Condry, M. W., & Nelson. B. C. (May 2016). Using Smart Edge IoT Devices for Safer, Rapid Response with Industry IoT Control Operations. *Proceedings of the IEEE*, 104(5).P1–9

Gungor, V. C., & Lambert, F. C. (2006). *A Survey on Communication Networks for Electric System Automation*, NEETRAC Publications, Elsevier.

Hayden, E. (2014). *An Abbreviated History of Automation & Industrial Controls Systems and Cybersecurity*, A Sans Analyst Whitepaper.

Hui, J., Culler, D., & Chakrabarti, S. (January 2009). *6LoWPAN: Incorporating IEEE 802.15.4 Into the IP Architecture*, Internet Protocol for Smart Objects (IPSO) Alliance.

ISACA Journal, Volume 1, (2014). *"SCADA Cybersecurity Framework"*. Retrieved 1 May 2017 from http://www.isacajournal-digital.org/isacajournal/2014_volume_1?pg=20#pg2

Kobara, K. (April 2016). Cyber Physical Security for Industrial Control Systems and IoT. *IEICE Transactions on Information and Systems*, E99–D, 4.

Knapp, E., Broad, J. (2011). *Industrial Network Security*. Syngress, Elsevier.

K. Lin et al., Human localization based on inertial sensors and fingerprint in industrial internet of things, Computer Networks (2015), Accessed 9 May 2017. Retrieved from http://dx.doi.org/10.1016/j.comnet.2015.11.012

Lojka, T., Bundzel, M., & Zolotová, I. (2016). Service-Oriented Architecture and Cloud Manufacturing. *ACTA Polytechnica Hungarica*, 13, 6.

Moness, M., & Moustafa, M. (2016). A Survey of Cyber-Physical Advances and Challenges of Wind Energy Conversion Systems: Prospects for Internet of Energy. *IEEE Internet of Things Journal*, 3, 2.

Ozden, H., & Paul, D. (2011). *Organik Rankin Çevrim Teknolojisiyle Düşük Sıcaklıktaki Kaynaktan Faydalanılarak Elektrik Üretimi. Örnek Çalışma: Sarayköy Jeotermal Santrali*.X. Ulusal Tesisat Muhendisligi Kongresi – 13–16 April 2011, Izmir, TURKEY.

Pa, Y. M. P., Suzuki, S., Yoshioka, K., & Matsumoto, T. (2015). IoTPOT: Analysing the Rise of IoT Compromises. *9TH USENIX Workshop on Offensive Technologies (WOOT 15)*, August, 2015 Washington, DC.

PI. (2015). PROFINET—The Solution Platform for Process Automation. *PI White Papers*.

SinghSh. A. (2009). *Organic Rankine Cycle Power Plant for Renewable Energy Resources*, Maulana Azad National Institute of Technology, India, 462051

Shodan. (2017). Search Engine, Retrieved 11 April 2017 from https://www.shodan.io/

Stouffer. K., Pillitteri. V., Abrams. M., & Hahn. A. (May 2015). *Guide to Industrial Control Systems (ICS) Security*. NIST Special Publication, 800–82.

Official Gazette of Turkey. (2012). *Regulation on Service Quality in Electricity Distribution and Retail Sale Date and Number: 21/12/2012 – 28504 43th Article*, Energy Market Regulatory Authority.

U.S. Government Accountability Office. (2011). *Critical Infrastructure Protection*. Report to Congressional Requesters, GAO-12–92, Washington, DC.

Türksel Kaya Bensghir*, Ufuk Türen and Yücel Yılmaz

How a Workforce for Industry 4.0 Era? Labor 4.0

1. Introduction

Throughout history, mankind has experienced four main revolutions; one regarding agriculture and the other three regarding manufacturing industry. Revolution is defined as rapid radical change, renovation or restructuring of institutions which have significant impact on societies or fundamental change in a certain domain (Kuhn, 1996: 1–19).

Agricultural revolution, which is thought to be actualized in 10,000 BC and symbolize the transformation from hunting and gathering way of living to a settled one, is characterized as the era in which mankind has developed systems on agriculture and raising livestock and the concept of capital accumulation has emerged (Eğilmez, 2017). It is accepted that this era continued until the first industrial revolution (1760–1840) (Industry 1.0). Although some basic technical tools have been used in agriculture, livestock raising and manufacturing, such as weaving looms, the origin of the energy exerted in any kind of manufacturing system has remained as human or animal power in the period between the agricultural and the first industrial revolution.

In the 18th century, in line with the steam power being invented and supplying the required energy for manufacturing systems and transportation, unprecedentedly huge power generating machines and railways carrying raw materials and goods to unbelievably remote places have been put into use. Second industry revolution (Industry 2.0) has emerged with the feasible use of electricity in manufacturing systems such as controlling and powering. This capability caused the invention of assembly lines, by this means, conceptualization and proliferation of mass production (Acemoğlu & Robinson, 2013: 173–199).

In this process, besides railway systems, engines for land transportation and mobile systems has been manufactured, and thus the delivery of raw materials and goods to much more remote places, which cannot be reached by railroads, has been materialized. The third industrial revolution (Industry 3.0) has emerged

---

* Corresponding author: Türksel Kaya Bensghir, tbensghir@gmail.com, Hacı Bayram Veli Üniversitesi İİBF, Ankara.

thanks to the invention and proliferation of receptive systems, microprocessors, computers and networks, programmable automation systems and robot production lines. During this era, the human error oriented system failures have been tried to be eradicated by significantly diminishing the share of human labor in production and services. With this approach, Internet and smart telecommunication devices have become commonplace in industries and social life while the human contribution has been minimized in manufacturing and services (Eğilmez, 2017).

Since the first industrial revolution, manufacturing has been regarded as the main driving force for economic growth and increasing standards of living in countries. In the past, advanced technology, especially automation, was accepted as the principle factor for increasing the productivity in manufacturing. Besides, globalization and the impacts of globalization on society have caused some unprecedented challenges in manufacturing industries (Mital & Pennathur, 2004).

Instead of solely focusing on automation and mass production, henceforth the human factor, in terms of both labor and customer, has been acknowledged as the central element in manufacturing systems. This perspective becomes more of an issue in terms of productivity growth and competitive edge in globalized markets. The vision of "dark factory" (robotic systems do not require light to work by), which emerged as a result of automation efforts for decades, has brought forth some problems such as trend of disappearance of flexibility in production, loss of capability and creativity in workforce and employment contraction. In line with the improvements regarding Industry 4.0, the human factor that has the capability of flexibility, adaptability, creativity, reasoning and decision making, which can hardly be replaced by autonomous systems, is foreseen as a main and central role player in future manufacturing systems (Frey & Osborne, 2017).

During the age of Industry 4.0, factories, in which machines, products, tools, workers and customers are connected with smart automation systems, information technologies using highly developed sensor and control systems have emerged. Abovementioned interconnected subsystems will process all together, exchange data and information and try to maximize the value created in each process through the value chain (Jasperneite, 2014).

In the near future, it is foreseen that these systems may have extremely diverse and complicated impacts, and thus, especially the human factor's role and contribution to value chain could change significantly (Bonekamp & Sure, 2015). Labor, which has been in a struggle for value and existence since the first industrial revolution, has the state of gaining its pre-industry revolution age central role

(Menawat, 2016). These new approaches consider the users of the high-tech systems as invaluable partners for new ideas and innovations, and increasingly urge to utilize their knowledge and experience in realizing technological developments and predicting the market performance of new products (Franke & Piller, 2004; Fredberg & Piller, 2011). Along with the Industry 4.0, in the process of labor's seizing the central and the highest value-added position, the attributes which should be owned by the labor and the education which is necessary for them are considered important (Nelles, Kuz, Mertens & Schlick, 2016).

The main aim of this research is to explore the characteristics and attributes which will be required for the labor of future production systems. The issues emphasized in this chapter may shine a light on the effort of universities, which are the foundations for vocational and professional education and training, to focus on the most necessary domains in developing proper labor for next-generation production systems.

2. Expected Changes in Organizations in Industry 4.0 Era

Along with the Industry 4.0 era the expected changes which will embrace the organizations can be defined as environmental, technological, organizational and labor dimensions.

2.1 Environmental Changes

Surge in demand in local/global markets and new technologies evoke various changes in production systems (Wiendahl et al., 2007). In conjunction with the increase in demand for products which are highly customized and produced according to individuals and having relatively short life cycle, automatic mass production systems are not considered economically sustainable. Flexibility, agility and productivity of production systems are thought to be the most important aspects for capability to compete in global markets. In this regard, customer oriented, smaller batch sized, even manually assembled, products begin to gain importance and earn recognition (Magruk, 2016). As a natural result of variation in products, the complexity of production systems escalate. Use of innovative information systems gains importance in order to make this complex, multi-component co-operation, coordination and synchronization requiring structure functional (Kleindienst, Wolf, Ramsauer & Pammer, 2016). The factory of the future necessitates production organization wired with Internet of Things (IoT), which is able to produce goods that are customized according to individual, situation and requirement with less resources and labor, and people who can incorporate more intellectual capacity.

2.2 Technological Changes

The impact of technology on the future production systems should be evaluated in line with two different perspectives. The first one deals with the production technologies and product support systems owned by producer firms in value creation processes which are expected by their customer. The second point of view, influenced by the fact that technology shifts the demand, claims that the firms that have been producing worthless products and putting them in the market for many years have to provide their products and production systems with smart and holistic approaches. Product diversity materialized by the developments in technology, at least for the firms in developed economies, is defined as one of the most important strategic objectives of the future. The revolutions realized before the fourth industrial revolution (first: mechanization, second: electrification, third: automation) assure the required infrastructure and the able technologies and approaches for their sequential successors. In this sense, digital communication, embedded technologies, human-computer-interaction, sensors and triggering technologies, standardization and norms and software and systems technologies are regarded as the most significant enabling processors for Industry 4.0 (Moorhouse, 2002).

Today, in the industrial environment, robots have gradually become commonplace. They are designed for the tasks which are dirty, dangerous, requiring high physical power and speed, and shortly not suitable for human, and to operate flawlessly, accurately and indefatigably. Meanwhile, humans have been progressively employed in jobs which can barely be accomplished without a skillful human hand and cognition. For instance, in automobile industry while the robots in assembly line mostly lift, locate and install the heavy parts, humans deal with the tasks like connecting cables to pertinent electronic devices or setting up the machines in accordance with diverse production schemes. Similarly, today, in smart phone manufacturing, minute parts are generally assembled on the printed circuit cards manually. Nevertheless, some researchers predict that this situation will change in line with the robots being smarter and adept in the coming years. On the other hand, some other researchers support the view predicting that robots will not be able to replace the position of humans in the future industries but cooperate and collaborate with them (Gilchrist, 2016: 11). This view displays a parallel stance with Licklider's (1960) vision of human computer symbiosis. This vision has foreseen that humans and digital computers will develop a very close partnership and very harmonious symbiosis by completing each other's inadequate sides.

In the era of Industry 4.0, the understanding of humans and computers being the entities completing each other begins to find much more favor nowadays.

Inasmuch as humans have the cognitive skills and tactual sense which is extremely necessary for their sensitive and vulnerable capability to move and locate miniscule objects. On the other hand, robots become much more rapid, enduring, reliable and efficient in the tasks which are extremely weary and routine for humans. The trouble faced here is that the robots in manufacturing have not been designed to operate with humans shoulder to shoulder. For safety reasons, recently, these systems are being programmed to halt their operations by means of various sensors that are perceived as unsafe when approaching humans (Gilchrist, 2016: 11). These robotic manufacture lines, which have been used since the 1970s with regards to dark factories vision, can be anticipated to improve to provide much more coordinated and harmonized human and robot interaction in the future (Gilchrist, 2016: 12).

2.3 Organizational Changes

Industry 4.0 is expected to bring significant evolutions in production processes and a consequence in jobs and tasks. This situation necessitates changes in desired employee capabilities and qualities. Less complex and routine tasks such as machine loading and unloading are considered to continue to be executed through automation in this era. However, in this process, humans will spend much of their working hours performing complicated and indirect tasks in cooperation and collaboration with machines (Siemens, 2013). The principle task for labor will be supervising and regulating the automated complex processes and the applications executed and directed by machines (Frey & Osborne, 2017). In this way, dealing with information and vast amount of data, and communication with machines will become more of an issue as one of the most important aspects of the future jobs and tasks (Gehrke et al., 2015). It is foreseen that measuring employee performance will be much more objectively materialized through increasing digitalization and the use control systems. Organizations will be subject to support employee education and training in order to increase organizational performance. This trend will be a desired aspect by both organizations and labor unions. The future organization will be formed by means of human–machine cooperation, and organizational and environmental changes can be sustainably managed in proportion as the effort spent to improve both sides namely human and machine (Bonekamp & Sure, 2015).

In organizations' value chain, some evolutions have been expected in terms of relations with suppliers, dealers and customers. In line with inter-organizational information systems, which operate integrated with production processes, some tasks that were executed by knowledge workers previously will be taken over by information systems. Consequently, it is reported that more flawless and more just-in-time procurement processes will be materialized. The real-time customer demands will be available all along with the processes, and an integrated production process, which can be tracked from the beginning to end by all shareholders, will be created (Brynjolfsson & Hitt, 2000).

It is also foreseen that in the future organization close cooperation and interaction between humans and machines will increase significantly. Besides, it is predicted that job enrichment perceived by employees will rise; high technology information systems will encourage the trend of decentralization in high vital organizational tasks such as decision making and planning; thus, much more process integration and cross-functionality perspective will be required. Eventually, the hierarchical layers in organizational structure and the need for central management interventions are anticipated to diminish even more (Bonekamp & Sure, 2015).

2.4 Changes in Labor

In line with Industry 4.0, the labor is expected to change. Three basic aspects are prominent here. First, the age average of labor has been increasing progressively. Second, labor has been diversifying in terms of ethnicity in workplaces (Ziefle & Jakobs, 2010). Third, the demand for labor has been evolving. These changes have highlighted the importance of the need of sustainable organizational support and providing professional training and education for labor (Kleindienst, Wolf, Ramsauer & Pammer, 2016).The evolution in work methods, roles and tasks has inevitably made the required employee skills change. In this respect, "systems thinking" skills have been gaining importance as most fundamental quality of employees hence Industry 4.0 environment will subsume more complex systems and technologies such as automation, interconnectedness and interoperability between sub-systems (Deuse, Weisner, Hengstebck & Busch, 2015). This perspective basically focuses on the skills such as recognizing the systems and sub-systems of the systems, understanding the functions and interactions between systems and sub-systems and consequently seeing the big picture of the systems and predicting their behavior (Kleindienst, Wolf, Ramsauer & Pammer, 2016). Employee qualities, which are predicted as necessary in the future manufacturing environment, can be seen in Table 1 as grouped in terms of "absolutely compulsory", "compulsory" and "reason for preference" (Gehrke et al., 2015).

When the qualities categorized and presented in Table 1 are scrutinized, it is seen that required qualities for Industry 4.0 era are dealt as two groups, namely, personal and technical. Using sociotechnical systems approach, it is considered that labor having the qualities required by Industry 4.0 era should play a principle role in keeping pace with technological improvements and in developing and steering them. To this end, interdisciplinary and multicultural work places are thought to become widespread and as a consequence industries are predicted to need individuals who are well educated and experienced in social sciences. Moreover, the technological knowledge and skills acquired by individuals in the universities are thought to be up-to-date in order to establish enough mastering and awareness in the workplaces.

It is also considered important that the labor having qualities necessitated by Industry 4.0 environment should be technology friendly and skilled in thinking analytically, acquiring and using information and knowledge and learning continuously. Notwithstanding, these qualities are mainly provided by higher education institutions, in early stages, namely, primary and secondary education, and the efforts should be devoted to individuals' acquisition of such qualities. Being successful in the information age necessitates the opportunities of informatics infrastructure and other technological aspect provided to individuals, acquisition of knowledge regarding these technologies and use of them correctly and the overall government support for the process. In other words, whilst the labor is wired with the necessary skills and knowledge about up-to-date technologies, labor's capability to analyze and use these information and knowledge should be improved in order to enhance and bring production systems to upper levels.

Tab. 1: Necessary Qualities for Future Labor. Source: Adapted from Gehrke et al., 2015.

	Absolutely Compulsory	Compulsory	Reason for Preference
Technical Qualities	Information technology knowledge and capabilities	Knowledge management	Capability to program computers and coding
	Processing and analyzing data and information	Interdisciplinary/generic knowledge on technologies and organizations	Expert knowledge on technologies
	Statistics knowledge	Awareness on information technology and data security	Awareness on ergonomics
	Understanding organizations and processes	Expert knowledge on manufacturing activities and processes	Knowledge on legal matters
	Communication capability with modern interfaces		

	Absolutely Compulsory	**Compulsory**	**Reason for Preference**
Personal Qualities	Capability to manage oneself and time	Relying on new technologies	
	Being adapted and ability to change	Continuous improvement and lifelong learning	
	Team work skills		
	Communication skills		
	"Systems thinking" skills		

When the issue is evaluated from another angle, in today's circumstances in which produced and shared information and knowledge rapidly outdate, the world becomes a global village and competition conditions harden. It can be easily seen that the human who can execute the task that cannot be done by robots will be needed more than ever. In this sense, division of labor between humans and robots will be in question, and here, either side will assume the task they can best do. As mentioned above, robots can undertake the dangerous and routine tasks with good quality, low cost and much faster than people. Additionally, there are tasks which cannot be or probably will never be executed by robots. For instance, social, historical and cultural analyses have been made merely by people. Here, people do not refer to ordinary people but subject matter experts. These subject matter experts using their knowledge and experiences have the capability to analyze current settings and also intuit the future situations. In this frame, it can be evaluated that the importance of social sciences will ascend and while highlighting pure and applied sciences the social sciences should not be neglected in industries and educational systems.

3. Discussion

In general, studies on predicting the technology of the future can give accurate results and it is possible to make accurate estimations about what kind of technologies may emerge in the future through examining futurist science and through people's predictions and the trends of progress and development of existing technologies (Bonekamp & Sure, 2015). Simultaneously with Industry 4.0, it is predicted that the production process will increase by about 30 % and the efficiency by about 25 % by means of the established digital communication and interaction between machines, pieces/entities and people, and consequently, this situation will raise the mass customization to new horizons.

Thanks to transformation from production from single, automated cells to fully integrated factories that allows all the elements to communicate with each other and increase flexibility, speed, productivity and quality (for instance in Germany), it is evaluated that the productivity in total production costs in the next ten years will be around 5–8 % and will increase gross domestic product by 1 % each year, bring in new investments of € 250 billion and lead to a 6 % increase in the total number of jobs with 390,000 new jobs in the manufacturing sector. Besides, in terms of occupations that will become important in the future, again in Germany, it is reported that the need for mechanical engineers will increase by more than 10 %, while the acquisition of routine and simple tasks into automated systems will result in a shrinkage of less skilled jobs; in parallel to this, the growing need for software, digital communications and data analysis tasks will require much more specialists in the areas of computer, software and information systems (Rüßmann et al., 2015).

According to the results of a research carried out in England, which is regarded as the center of the first and second industrial revolution, 59 % of factory owners engaged in the manufacturing sector acknowledge that Industry 4.0 will cause radical changes in their sectors, however only 8 % understand the concept correctly. In the same survey it is reported that, due to Industry 4.0, 51 % of the respondents believe that productivity will increase, 47 % believe that it can provide better data for analysis of production processes, 46 % believe that it will increase competition, and 44 % believed that lower production costs can be reached. It is claimed that within the efforts to raise awareness and adapt the strategy of national innovation to Industry 4.0, the UK government is trying to design 2050's modular factory through establishing new research centers; under 2025 initiative China continues to work with new innovation research centers and aims to reduce production costs and production life cycle by 30 % using smart manufacturing approaches until 2025 (BDO UK, 2016).

It is observed that in the Netherlands within the concept called "intelligent industry" government focuses the efforts of increasing awareness, accepting the inevitability of this revolution and making the necessary investments and transformations in time, encouraging the private sector to reduce costs and promote customer orientation, and supporting local products to take the place of those products produced in China. In the United States, within the scope of the National Industrial Strategy adopted in 2011 by allocating large resources to R&D activities toward prioritized domains such as visualization, informatics and digital production, both the state and private sector accelerate plans and projects to bring more technology-oriented businesses into the market, thus accelerating the increase of its global trade competition (BDO UK, 2016). While preparations of Industry 4.0

of technology leading countries in the world are continuing at full speed, it seems that it is imperative for the follower countries to take the necessary precautions not to miss this transformation which has already started.

The impact of Industry 4.0 which defines the future manufacturing systems is not considered as limited to cyber-physical systems and local industrial processes. This approach, from supplier to dealer, affects the whole value chain. One of the most important issues faced by the early implementers of this evolution having such a broad and radical impact is described as "lack of qualified labor". It also underlined the necessity of measures to be taken by educational and training sector for skills and capacity acquisition toward the needs of Industry 4.0 revolution. Especially, revision of the current curricula of higher education institutions in order to satisfy these needs is thought to be essential. Besides, it is considered significant that institutions or companies operating in the software and technology development sector should update their skill sets according to current improvements by examining and comprehending the details and delicacies of industrial control systems. Moreover, in the context of the tasks to be done by governments, it is emphasized that they constitute the most significant power to increase industrial production and private sector should be encouraged toward Industry 4.0 revolution (Gilchrist, 2016: 204).

In this context, recommended subjects to add to the curricula of higher education institutions in the preparation of the human power that Industry 4.0 would need, primarily toward designing and developing new production systems within the context of Industry 4.0 can be listed as the norms and standards of information technologies, new production logistics and production infrastructures, work safety and security for new technologies, information ergonomics, human–machine interaction and modeling of technological systems through information systems. For the role of augmented operator which is necessary for the labor to work in production systems in the era of Industry 4.0, acquisition of competencies, such as audit and control of manufacturing systems using virtual production systems, influencing production targets, use information technology based support systems, remote maintenance and control of production lines, are claimed as vital in the scope of the curricula. In order to meet the human power need of this age, it is thought that the efforts of higher education institutions will not be enough and the need for universities and companies to work in cooperation on various issues is seen as a major priority. In this respect, deepened compressed and relevant university programs supported by industry experience, not only engineering but also technician skills providing curricula, gaining interdisciplinary skills for the growing need for project type team structures, new approaches to the development of job-oriented knowledge and skills and the development of electronic learning techniques and environments to support the acquisition of these skills are advised (Rüßmann et al., 2015). Besides, it is also envisaged that, in accordance with the changing demographics of increasingly aging and more heterogeneous human power, it is necessary to develop new learning and acquisition approaches (Würslin, 2013).

4. Conclusion

It is evaluated that pacing with this revolution which has started and accelerated will be a decisive factor for the sustainability and competitiveness of the nations in macro-scale and the companies in micro-scale. To this end, it is thought that companies should generate their migration plans consisting of adequate mental and investment associated preparations which are necessary for transition to Industry 4.0 era. A good grasp on these migration plans having both technological and social dimensions is crucial for the completeness and flawlessness of the transition process. The troubles which may be caused by resistance to change and/ or underinvestment to labor should be considered and dealt with.

The importance of family in rising individuals who can be easily attuned to Industry 4.0 era is clear. Through understanding the necessities of modern times, parents should be in charge of guiding children from the early ages toward the professions which will be favorite and desirable in the future.

References

Acemoğlu, D. & Robinson, A. J. (2013). *Ulusların Çöküşü: Güç, Zenginlik ve Yoksulluğun Kökenleri,* Çev: Faruk Rasim Velioğlu, Doğan Kitap, 7.Baskı, İstanbul.

BDO UK. (2016). The BDO Industry 4.0 Survey: 2016. https://www.bdo.co.uk/en-gb/insights/industries/manufacturing/industry-4-0-report Accessed on May 10, 2017

Bonekamp, L., & Sure, M. (2015). Consequences of Industry 4.0 on Human Labour and Work Organisation. *Journal of Business and Media Psychology*, 6, 33–40.

Brynjolfsson, E., & Hitt, L. M. (2000). Beyond Computation: Information Technology, Organizational Transformation and Business Performance. *The Journal of Economic Perspectives*, 14(4), 23–48.

Deuse, J., Weisner, K., Hengstebck, A., & Busch, F. (2015). Gestaltung von Produktionssystemen im Kontext von Industrie 4.0. In Botthoff A., Harmann E. A (Ed.), *Zukunft der Arbeit in Industrie 4.0*. Springer Vieweg, Berlin, Heidelberg.

Eğilmez, M. (2017). Endüstri 4.0, Kendime Yazılar. http://www.mahfiegilmez. com/2017/05/endustri-40.html?m=1. Accessed on May 10, 2017.

Franke, N., & Piller, F. (2004). Value Creation by Toolkits for User Innovation and Design: The Case of the Watch Market. *Journal of Product Innovation Management*, 21, 401–415. doi: 10.1111/j.0737-6782.2004.00094.x

Fredberg, T., & Piller, F. T. (2011). The Paradox of Tie Strength in Customer Relationships for Innovation: A Longitudinal Case Study in the Sports Industry. *R&D Management*, 41, 470–484. doi: 10.1111/j.1467-9310.2011.00659.x

Frey, C. B., & Osborne, M. A. (2017). The Future of Employment: How Susceptible Are Jobs to Computerisation? *Technological Forecasting and Social Change*, 114, 254–280.

Gehrke, L., Kühn, A. T., Rule, D., Moore, P., Bellmann, C., Siemes, S., … & Standley, M. (2015). A Discussion of Qualifications and Skills in the Factory of the Future: A German and American Perspective. *VDI/ASME Industry 4.0 White Paper*, 1–28. Hannover Messe 2015, 13–17 April 2015, Hanover, Lower Saxony, Germany, https://www.researchgate.net/publication/279201790_A_Discussion_of_Qualifications_and_Skills_in_the_Factory_of_the_Future_A_German_and_American_Perspective/citations, Accessed on May 10, 2017.

Gilchrist, A. (2016). *Industry 4.0: The Industrial Internet of Things*. Apress, New York.

Jasperneite, J. (2014). Towards the Smart Factory–Status and Open Issues. In *9th IEEE International Conference on Emerging Technologies and Factory Automation (ETFA)* 19th IEEE International Conference on Emerging Technologies and Factory Automation, 16.- 19 September 2014, Barcelona.

Kaya Bensghir, T. (1996). *Bilgi Teknolojileri ve Örgütsel Etkileri*, TODAİE, Yayın No. 274.

Kleindienst, M., Wolf, M., Ramsauer, C., & Pammer, V. (2016). What Workers in Industry 4.0 Need and What ICT Can Give—An Analysis. *Human Computer Interaction Perspectives on Industry 4.0 at the 16th International Conference on Knowledge Technologies and Datadriven Business (i-KNOW 2016)*. October 19, 2016, Graz.

Kleindienst, M., Wolf, M., Ramsauer, C., Winter, E., & Zierler, C. *Book of Proceedings of the 6th International Ergonomics Conference: ERGONOMICS 2016*, June 18, 2016. In C. E. S. (ed.). *Focus on Synergy*, Croatian Ergonomics Society, Zadar, Cratia, pp.179–188.

Kuhn, T. S. (1996). *The Structure of Scientific Revolutions*. 3rd Edition. The University of Chicago Press, Chicago.

Licklider, J. C. R. (1960). Man-Computer Symbiosis. *IRE (Institute of Radio Engineers) Transactions of Human Factors in Electronics HFE (Human Factors Engineering), 1(1), 4–11.*

Magruk, A. (2016). Uncertainty in the Sphere of the Industry 4.0-Potential Areas to Research. *Business, Management and Education*, 14(2), 275.

Menawat, A. (2016). Industry 4.0 – What It Is and What to Expect. http://menawat.com/industry-4-0-what-it-is-and-what-to-expect/. Accessed on May 8, 2017.

Mital, A., & Pennathur, A. (2004). Advanced Technologies and Humans in Manufacturing Workplaces: An Interdependent Relationship. *International Journal of Industrial Ergonomics*, 33(4), 295–313.

Moorhouse, D. J. (2002). Detailed Definitions and Guidance for Application of Technology Readiness Levels. *Journal of Aircraft*, 39(1), 190–192.

Nelles, J., Kuz, S., Mertens, A., & Schlick, C. M. (2016). Human-Centered Design of Assistance Systems for Production Planning and Control: The Role of the Human in Industry 4.0. In *Industrial Technology (ICIT), 2016 IEEE International Conference* (pp. 2099–2104). IEEE. 14–17 March 2016. Taipei, Taiwan.

Rüßmann, M., Lorenz, M., Gerbert, P., Waldner, M., Justus, J., Engel, P., & Harnisch, M. (2015). Industry 4.0: The Future of Productivity and Growth in Manufacturing Industries. *Boston Consulting Group*, 14. http://www.inovasyon.org/pdf/bcg.perspectives_Industry.4.0_2015.pdf Accessed on May 13, 2017

Siemens, A. G. (2013). Competencies for the Future of Manufacturing. *Siemens Industry Journal*, 2/2013, 11–25, https://www.siemens.com/content/dam/internet/siemens-com/customer-magazine/old-mam-assets/print-archiv/2/industry-journal/industry-journal-2-2013-en.pdf, Accessed on May 14, 2017.

Wiendahl, H. P., ElMaraghy, H. A., Nyhuis, P., Zäh, M. F., Wiendahl, H. H., Duffie, N., & Brieke, M. (2007). Changeable Manufacturing-Classification, Design and Operation. *CIRP Annals-Manufacturing Technology*, 56(2), 783–809. doi: 10.1016/j.cirp.2007.10.003.

Würslin, R. (2013). Integration of Industry 4.0 in Education Programs of German Universities of Applied Science. https://www.imove-germany.de/cps/rde/xbcr/imove_projekt_de/d_Education-Forum-2013_Session2_Wuerslin.pdf. Accessed on May 15, 2017.

Ziefle, M., & Jakobs, E.-M. (2010). New Challenges in Human Computer Interaction: Strategic Directions and Interdisciplinary Trends. *4th International Conference on Competitive Manufacturing Technologies*. University of Stellenbosch, South Africa, pp. 389–398. http://www.comm.rwth-aachen.de/files/coma_10_ziefle_jakobs_2.pdf Accessed on May 2, 2017.

Emre Karagöz* and Vahap Tecim

Design of Intelligent Direction Systems via Multi-Criteria Decision Making

1. Introduction

Many of the evacuation situations over the past 100 years have failed due to the lack of an evacuation model. Many of the deaths that occurred were not caused by fire, explosion, poison gas or other external hazards, but rather by the crowd itself (Helbing et al., 2013). For this reason, it is crucial for the crowd to be guided in the right way. Generally, simulation-based pedestrian models can be divided into microscopic and macroscopic models according to crowd dynamics (Yang and Dong, 2015). Chi Xie (2008) described evacuation planning models as optimize-based and simulation-based models. The optimize-based model refers to the functional form of a network flow in the sense of designing the problem to investigate the best evacuation model. The simulation-based model is based on the evaluation of defined plans (Xie, 2008). Schultz (2014) points out that human movements in the dynamic models of pedestrians divide into different behavioral directions. In 2011, 120,000 full-time employees and 50,000 visitor movements were followed to explain group dynamics (Schultz, 2014). Okaya and Takahashi (2015) examined the Great East Japan Earthquake and the terrorist attack on the World Trade Center on September 11, 2001. People in both events showed the same type of behavior in the same type of panic. Okaya and Takahashi categorized these behaviors in three parts. These are evacuation moments, tasks completed after evacuation and emergency evacuation (Okaya and Takahashi, 2015). Krausz and Bauckhage (2011) have prepared the optical flow calculation method for early recognition of critical crowd situations.

Santos and Aguirre (2004) talked about flow-based models, cellular automata and agent-based models. They recommended simulation programs such as Simulex for success on agent-based models (Santos and Aguirre, 2004). Kachroo and the others said that their evacuation model is consistent in function, architecture, design and impact. They also described evacuation process in three parts and these are before evacuation, during evacuation and after evacuation (Kachroo

* Corresponding author: Emre Karagöz, emre.karagoz@deu.edu.tr, Dokuz Eylül Üniversitesi İ.İ.B.F İzmir/Turkey.

et al., 2008). Lim and the others (2016) examined multi-commodity network flow optimization for alternative ways. Their model allows a real-time evacuation model for the users (Lim et al., 2016). Zhao and Winter (2015) said that deciding which way is right is very important for successful evacuation. They emphasized that mobile technologies like smartphones and tablets can bring new solutions to this problem (Zhao and Winter, 2015). Huovilainen and Oy (2010) proposed the real-time guidance software which is called Dynamic Discharge Glare System in increasing the safety of different evacuation models such as tunnels or metro stations. The system shows dangerous departure routes automatically. This system can be integrated with other security systems such as fire, chemical or biological hazard detectors (Walton & Budnick, 1997).

This study focused on optimizing and strengthening the integration of evacuation systems using new communication technologies. In this study, the effective and healthy usability of Industry 4.0, which creates the areas of use that minimize the human element, is detirmend by making use of the decision making techniques based on people's intuition.

2. Method

In daily life, many decision-making problems are encountered for various purposes. Wrong decisions can cause inevitable consequences. Decision-making process consists of the following steps (Yaralıoğlu, 2010).

- Identifying the problem.
- Collecting information about the problem.
- Classifying, analyzing and interpreting the information.
- Discovering the decision options.
- Determining the best option.
- Decision making about the choice.
- Implementing the choice.
- Evaluating the choice.

In some cases, decision-making process is very difficult. Because the decision maker is not fully informed about the options to be decided. Multi-criteria decision-making methods can help to solve such problems. There are some multi-criteria decision-making methods that have been developed such as AHP, TOPSIS, DEMATEL, ELECTRE, VIKOR and MOORA. In evacuation process, there are some options that people can use such as an exit point. But in the process, people may not know which exit is the best for them. In this point, there is a multi-criteria decision-making problem, which has some decision points and decision criteria.

Every decision-making method has different formulas. The Analytical Hierarchical Process (AHP) is a decision-making method developed by Thomas L. Saaty. In particular, it is used in the analysis and construction of complex decision-making problems (Hanine et al., 2016). The AHP method is based on human perceptions. The basic inputs in the AHP are variables, decision points and variable importance levels. TOPSIS is one of the multi-criteria decision making methods. Yoon and Wang developed it. TOPSIS determines which distance is the shortest to the positive solution and longest to the negative solution (Singaravel and Selvaraj, 2015). It is based on the ELECTRE method and widely used by the decision makers. TOPSIS is often accepted in the soft system methodology (Mancev, 2016) which deals with soft operational research (OR) methods instead of hard OR. TOPSIS can be used in different areas, different problems. For instance, select the computer integrated manufacturing (Singh and Singla, 2013). VIKOR is used to bring all the decision points to an ideal decision point according to the degree of closeness. The method provides for the creation of a compromise environment by introducing a large number of decision makers into the decision process (Yaralıoğlu, 2010). MOORA, more than two decision points, is carried out simultaneously on the healing process according to certain criteria. It was developed by Brauers and Zavadkas in 2009 (Yaralıoğlu, 2010). Although it is not a very old method, the usage areas are quite common. The ratio method consists of three different approaches, the ratio method, the reference point method and the full multiplicative form. Attri and Grover (2014) indicate that production systems such as a multi-criteria problem, product design, process design, item selection, selection of machine or cutting tool, choice of material transport system, choice of advanced production system can occur at different levels of life cycle (Attri and Grover, 2014). COPRAS sorts and evaluates the decision points step-by-step in terms of importance. It was developed by Zavadskas in 1996 (Yazdani et al., 2017). Apart from many other multi-criteria decision-making techniques, it has many advantages. Some of these are the ability to sort out a large number of alternatives, the ability to achieve results in a shorter time than the AHP and ANP methods, the ability to display simple results and graphically display these results (Parezanovic, Bojkovic, Petrovic & Tarle, 2016). COPRAS can be used according to maximization and minimization criteria (Podvezko, 2011).

Another technology that we use is web-based technologies and programming languages. Web technology is widely used for specific jobs all around the world, especially education, health, business and more. Web is an ideal platform for communication via voice and video (Mehmood and Korika-Pehserl, 2015). Also, web-based distance education system more effective than other methods such as television- or radio-based. Web-based business management systems are web-based software that allows managers to run their jobs and business on the Internet. These conditions bring them effective business process.

Web-based platforms can contain these features

- Content management system
- News, announcement
- Forms
- Photos and videos
- E-commerce
- Workflow
- Statistics (Wikipedia, 2017)

Tab. 1 shows some category and examples about web-based systems.

Tab. 1: Web-Based System Categories and Examples.

Category	Examples
Web pages and blog making	Wordpress, Joomla, Drupal
Information technology and network monitoring	Wireshark, PRTG Network
Education	Sakai, Moodle, Blackboard
Social media	Facebook, Instagram, YouTube
Account systems	Netsis, Xero
Online document tracking system	Bizdoc, ISOFTDOC
Public	GIS, e-Government
e-Commerce	Ali-Express, Amazon

The last technology that we use is computer vision. Computer vision is an inter-disciplinary concept related to artificial intelligence, machine learning, robotics, signal processing and geometry. It provides an inference of the properties of an object's view. Computer vision systems minimize the harmful effects to the environment by reducing the production costs, increasing the product quality, ensuring human safety and renewing the production processes (Klancnik, Ficko, Balic & Pahole, 2015). Computer vision is divided into several subcategories. These categories are listed below.

- Datasets
- Software
- Digital geometry
- Commercial systems
- Feature detection

- Geometry image sensor technology
- Learning
- Morphology
- Motion analysis
- Noise reduction techniques
- Recognition and categorization
- Research infrastructure
- Researches
- Segmentation

There are some computer vision libraries as commercial or open source. Matlab and OpenCV are computer vision libraries. Matlab, with computer vision toolbox, can simulate and design algorithms, functions and video processing systems and computer vision, including feature detection, extraction and matching, object detection and tracking, motion estimation and video processing, 3D computer vision and camera calibration (Mathworks, 2017). The other computer vision library is OpenCV, which was launched in 1998 as a research project at Intel. It may be said that the aim is to provide some tools to help solve computer vision problems (Pulli, Baksheev, Kornyakov & Eruhimov, 2012). OpenCV is free for both commercial and academic use. C, C++, Python and Java can be used in OpenCV. Users prefer Python and Java, especially. In this study, computer vision was used to count people in front of the exit doors.

3. Findings

The application is a web-based platform in which intelligent signages placed in some locations within the building at any time of evacuation of the building are guided by the results of any of the five multi-criteria decision-making techniques to the closest (optimal) exit of people in any location. In the study application, PHP, Javascript, HTML5 and CSS were used as web programming languages. MySQL Database was selected as database. Matlab computer vision library was used as computer vision. And for the mobile application, Android Studio and Java programming language were used. Nine minicomputers, six digital signages and three webcams were used in the study application as hardware. Also, the system runs on a server with 64 core with 2 processors and 256 gigabytes of RAM.

In this study, a building consisting of 2 floors for the application was chosen. Six locations within the building had been identified. Every location was assumed as a decision maker. There were three exit doors in the building. These exit doors were selected as decision points. Distance of the exit doors, width of the exit doors and number of crowd in front of the exit gate were selected as a decision criteria. 0, 0.25, 0.50, 0.75 and 1 were selectable variable importance values. The system is a web-based platform in which there are intelligent digital signages for six locations in the building. Every digital signage has a minicomputer. These computers and digital signages show the best exit route for the user in the building. Since the program is web-based, minicomputers do not need to be very powerful and expensive. Every signage represents its location. According to multi-criteria decision-making techniques such as AHP, TOPSIS, VIKOR, MOORA and CO-PRAS, the system shows best exit route in digital signages to the person who is located in this point. Every digital signage is assumed to be a decision maker and every exit door in the building is considered as a decision point. So the variable values of each digital signage are different (distance from current location to exit doors or crowd values in front of the exit doors, for example). Since each of these screens is placed in certain positions, distance to exits and exit doors width values do not show any change. But crowd values in front of the exit doors can change in some time periods. And it is necessary to continuously detect these values and insert them as a variable in the multi-criteria decision algorithm.

A computer with a webcam was placed at each exit doors to obtain query values via Matlab computer vision library. Every two seconds, data were gathered and these values were sent to database as an input. Management interface imported these variable values from database every two seconds and solved the problem according to using decision method. Finally the algorithm is worked every two seconds and best solution result is shown on the digital signage. The figure below shows this process (Fig. 1).

Fig. 1: System Operation Process.

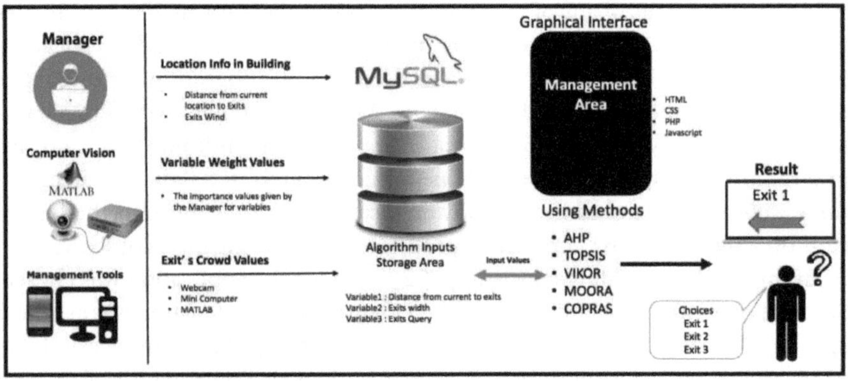

Now, we assume that in location 1 in the building, there are three exit doors and we have to select one of them. There is a digital signage that shows the right way for the users in this point. The system manager attached AHP decision-making method for this location. The following table shows all decision variable values (Tab. 2).

Variable 1 and variable 2 are stable values and it can be calculated before evacuation process. And these values are stored in the database. But variable 3 can take different values continuously. For this reason, the changing values are stored every two seconds. Now, all values of variables are stored in the database and in the final process, the system imports all the values and solves the problem according to the selected method. In this example, AHP is selected. The system has three interfaces. These interfaces are manager, user and mobile interfaces. Through the manager interface, a different multi-criteria decision-making methods and variable weight values can be determined for each location. Specify different variable values through the simulation screen. Results can be seen for each position. The following figure shows that simulation screen of the system (Fig. 2). Also depending on the result, the best exit route can be displayed on the building plan.

Tab. 2: AHP Is Selected Method for Location 1.

Variable 1: Distance to the exit doors		
Exit 1	Exit 2	Exit 3
25m.	8m.	50m.
Variable 2: Width of the exit doors		
Exit 1	Exit 2	Exit 3
2m.	0.80m.	1m.
Variable 3: Number of people in front of the exit doors		
Exit 1	Exit 2	Exit 3
4 people	10 people	3 people
Variable Importance Values		
Variable 1	Variable 2	Variable 3
0.33	0.33	0.33

Fig. 2: Simulation Screen of the System.

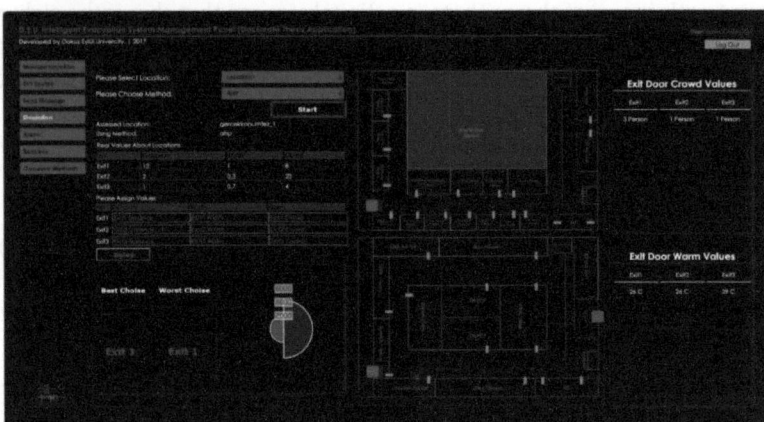

Five decision-making methods results can be compared with the same variable values and importance level values for the same location point. Instant messages, announcements, pictures and videos can be sent to any location. By pressing the emergency button, all the people in the building can be alerted to leave the building immediately. The following figure shows the comparison of the methods, sending messages and multimedia instruments, all the locations and the emergency screen (Fig. 3).

Fig. 3: Comparison of the Methods, Sending Messages and Multimedia Instruments, All Locations and Emergency Screen.

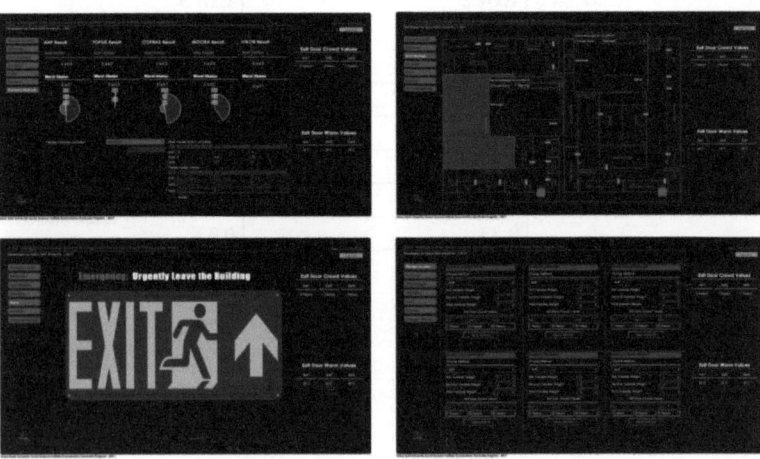

User interface is not open to manage. Users in front of the digital signages watch the screen and adhere to interaction. User interface has some information such as recommended exits, urgent messages, videos, pictures, announcements, distances from current point to all exit doors, reach time to all exits and the best route in building plan. The following figure shows that user and mobile interface (Fig. 4).

Fig. 4: System User and Mobile Interface.

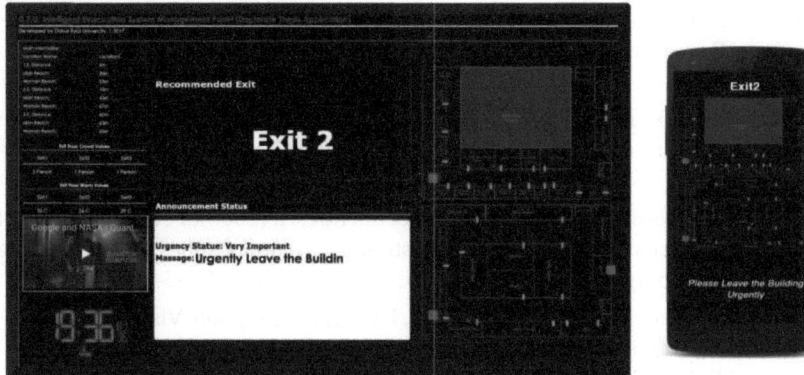

Mobile application is built on android platform. Management tools can be used via mobile interface. Also, user interface can be monitored.

4. Conclusion

Five decision-making methods were used by the system to calculate best exit for users. The building where the system application has been tried is not very complex. The system must be implemented in complex buildings. Electre, Promethee or any multi-criteria decision-making methods may be integrated to the system. In future works, it may be necessary to increase the number of variables. Also, system manager should know multi-criteria decision-making processes or it should make the system ready for use by determining the levels of variable importance. iBeacon technology, one of the new communication technologies, can be integrated into the system. Thus, the system will have a stronger structure.

References

Attri, R., & Grover, S. (2014). Decision Making Over the Production System Life Cycle: MOORA Method. *International Journal of Systems Assurance Engineering*, 5(3), 320–328. DOI 10.1007/S13198-013-0169-2

Hanine, M., Boutkhoum, O., Tikniouine, A., & Agouti, T. (2016). Application of an Integrated Multi-Criteria Decision Making AHP-TOPSIS Methodology for ETL Software Selection. *Springer Open, Springer Plus*, 5, 263.

Helbing, D., Buzna, L., Johansson, A., & Warner, T. (2013). Self-Organized Pedesterian Crowd Dynamics: Experiment, Simulations and Design Solutions ResearchGate. Available: Transportation Science 219:5495–5515. January.

Huovilainen, J., Oy, M., & Finland, V. (2010). Dynamic and Intelligent Evacuation System for Tunnels, *5th International Conference "Tunnel Safety and Ventilation"*, Graz

Kachroo, P., Al-Nasur, S. J., Wadoo, S. A., & Shende, A. (2008), Feedback Control of Crowd Evacuation, Book Title: Pedesterian Dynamics. 2008 Springer-Verlag Berlin Heidelberg. ISBN 978-3-540-75561-6. e-ISBN: 978-3-540-75561-6

Klancnik, S., Ficko, M., Balic, J., & Pahole, I. (2015). Computer Vision-Based Approach to End Mill Tool Monitoring. *International Journal of Simulation Modelling*, 14(4), 571–583.

Krausz, B., & Bauckhage, C. (2011). Automatic Detection of Dangerous Motion Behavior in Human Crowds. *8th IEEE International Conference on Advanced Video and Signal-Based Surveillance,* 30 Aug.-2 Sept. 2011. Klagenfurt, Austria

Lim, G. J., Baharnemati, M. R., & Kim, S. J. (2016). An Optimization Approach for Real Time Evacuation Reroute Planning, Annals of Operations Research. Springer. March 2016, Volume 238, Issue 1–2, pp 375–388.

Mancev, M. D. (2016). The Application of the TOPSIS Method in Selecting the Best Academic Library at the University of Nis. *Canadian Journal of Information and Library Science*, 40(1), 81–96.

Mathworks. (2017). Design and simulate computer vision and video processing systems. https://www.mathworks.com/products/computer-vision.html

Mathworks. (2017). Simulation and Model-Based Design. https://www.mathworks.com/products/simulink.html

Mehmood, R., & Korika-Pahserl, P. (2015). The Web Is Big Business. *Journal of Computing and Information Technology – CIT*, 23(1), 19–27.

Okaya, M., & Takahashi, T. (2015). Evacuation Guidance and Agent Behaviour in Evacuation Simulations. Electronics and Communications in Japan, Vol. 98, No. 10, 2015. Wiley Periodicals, Inc.

Parezanovic, T., Bojkovic, N., Petrovic, M., & Tarle, S. P. (2016). Evaluation of Sustainable Mobility Measures Using Fuzzy COPRAS Method. Management:Journal Of Sustainable Business And Management Solutions In Emerging Economies, 21(78), 53-62. Retrieved from http://management.fon.bg.ac.rs/index.php/mng/article/view/46/38

Podvezko, V. (2011). The Comparative Analysis of MCDA Methods SAW and COPRAS. Inzinerine Ekonomika-Engineering Economics, 22(2), 134–146. ISSN: 1392–2785.

Pulli, K., Baksheev, A., Kornyakov, K., & Eruhimov, V. (2012). Real-Time Computer Vision with OpenCV, Communication of the ACM, 55, 6.

Santos, G., & Aguirre, B. E. (2004). A Critical Review of Emergency Evacuation Simulation Models. In Peacock, R. D., Kuligowski, E. D. (Editors). Proceedings of the Workshop on Building Occupant Movement during Fire Emergencies, June 10–11, 2004. 27–52, Gaithersburg, Maryland: National Institute of Standards and Technology, (forthcoming).

Schultz, M., Rößger, L., Fricke, H., Schlag, B. (2014) Group dynamic behavior and psychometric profiles as substantial driver for pedestrian dynamics. Book: Pedestrian and Evacuation Dynamics 2012, 2014. Springer, Cham

Singaravel, B., & Selvaraj, T. (2015). Optimization of Machining Parameters in Turning Operation Using Combined TOPSIS and AHP Method. Technical Gazette, 22(6), 1475–1480.

Singh, R., & Singla, R. K. (2013). TOPSIS Based Multi-Criteria Decision Making of Feature Selection Techniques for Network Traffic Dataset. International Journal of Engineering and Technology, 5(6), 4598–4604.

Walton, W. D., & Budnick, E. K. (1997). Deterministic Computer Fire Models. NFPA Fire Protection Handbook, 18th ed., Section 11, Chapter 5, National Fire Protection Association (1997).

Xie, C. (2008). Evacuation Network Optimization: Models, Solution Methods and Applications Doctorate, A Dissertation Presented to the Faculty of the Graduate School of Cornell University in Partial Fulfillment of the Requirements for the Degree of Doctor of Philosophy. http://citeseerx.ist.psu.edu/viewdoc/download?doi=10.1.1.917.4120&rep=rep1&type=pdf

Yang, X., & Dong, H. (2015). Effects of Quantity and Position of Guiders on Pedesterian Evacuation 2015 IEEE 18th International Conference on Intelligent Transportation Systems. 15–18 Sept. 2015. Las Palmas, Spain

Yaralıoğlu, K. (2010). *Karar Verme Yöntemleri*, Detay Yayıncılık, Ankara.

Yazdanı, M. (2017). Analysis in Material Selection: Influence of Normalization Tools on Copras-G. *Economic Computation and Economic Cybernatics Studies and Research*, 1, 51.

Zhao, H., & Winter, S. (2015). A Time-Aware Routing Map for Indoor Evacuation, Sensors 2016, 16, 112; doi:10.3390/s16010112. https://www.mdpi.com/journal/sensors

Ahmet Cihat Baktir* and Bilgin Metin

The Technical Challenges of Cloud Computing as a Leading Trend in Business

1. Introduction

Cloud computing is an Information and Communication Technologies (ICT) paradigm which is a leading trend in the business and economic sector, and plays a vital role in the modern business transformation. Actually, this highly capable technology is not a new concept; it has a connection to the grid computing paradigm and other relevant technologies such as utility computing, cluster computing and distributed systems (Foster, Zhao, Raicu & Lu, 2008). Although it has several definitions in the literature, one of the most popular definitions of cloud computing that is made by the NIST explains cloud computing as "a model for enabling ubiquitous, convenient, on-demand network access to a shared pool of configurable computing resources (e.g., networks, servers, storage, applications and services) that can be rapidly provisioned and released with minimal effort or service provider interaction" (Mell & Grance, 2009). This definition includes the most important characteristics and features of the cloud architecture. In addition to these promising infrastructural notions, by using cloud-based system, consumers will be able to access applications and data on demand wherever the consumer is (Buyya, Yeo & Venugopal, 2008).

With the enhancement of the quality of the global Internet connection and trend toward outsourcing the information technology (IT) operations, cloud computing becomes as an essential component of the processes within the organizations. It provides countless advantages through its intrinsic features which result in an increase in the efficiency for the business side. The pay-as-you-go model that is utilized by the cloud technology may be an important factor that helps to decrease the IT infrastructure costs of the companies. The use of this model in addition to all its other technological benefits is generally less noticeable, but in fact it is very important in terms of earnings and new possibilities in business and economics that brings more advantage in the competition. Also, companies that use cloud computing can focus much more on the business side instead of technical issues, and this situation results in an increase in the efficiency of the company.

---

*   Corresponding author: Ahmet Cihat Baktir, cihat.baktir@boun.edu.tr.

The main motivation behind using cloud computing as a "cost decreasing technology" is actually its security and reliability. On the other hand, a significant portion of the companies does not utilize cloud computing because it is still not perceived as a secure technology as much as in-house computing systems. This contrast may be the result of lack of knowledge and unawareness of cloud computing. The concepts of the cloud computing can relatively change as an advantage or disadvantage according to the perception by the organization, and knowledge of users about cloud computing.

The perception of cloud computing is the critical point of deciding on whether utilizing it for the company is advantageous or not. The knowledge and awareness about cloud computing has a direct effect on the perception. However, although it brings essential benefits, it is observed that a number of the companies do not utilize cloud computing due to its challenges. In fact, most of the companies that do not utilize cloud computing due to its challenges do not analyze these points in detail to determine their actual effects. In order to increase the awareness of cloud computing, boost the competition among the companies and enhance the quality of the business output, these challenges should be carefully determined, analyzed and discussed in order to conclude their actual impact. The related studies in the literature discuss a wide range of challenges related to the cloud, which makes it impractical to analyze them all in a single study.

The aim of this study is analyzing the technical challenges of cloud computing and their effects on usage within companies that operate in Turkey. It is believed that the technological side of the cloud challenges has more influence on the usage and this is the main motivation for focusing on the technical challenges. Most of the companies in Turkey are still not eager to adopt cloud computing because of not having an extensive knowledge about the challenges that they may encounter. There are many challenges discussed by the similar studies but none of them provides a complete analysis of depicting the certain effects on the usage within the organizations. In order to analyze the relationship between the technical challenges and the usage of cloud computing, the challenges are gathered through a survey of the literature. These gathered challenges are processed and non-technical ones are eliminated in order to focus on the technical side of the cloud computing technology. To choose the most important ones, the opinions of the cloud experts are received and relatively the less important ones are not taken into consideration for maintaining the quality and homogeneity of the study.

The remainder of this study organizes as follows. The second section briefly discusses the most important technical challenges of cloud computing that are mentioned in the literature, presents the motivation of this study, the definition of the problem and hypotheses and the methodology. The third section initially presents some statistical results about cloud computing usage in Turkey which is then followed by the analysis of the hypotheses provided. The final section concludes the study with the discussion of the results and future work.

2. Methods

Cloud computing is a phenomenon that reshapes the progress of the IT services (Marston, Li, Bandyopadhyay, Zhang & Ghalsasi, 2011). As discussed, it might be perceived as a solution for the ineffective operations and problematic conditions that are associated with the business within a company or an IT department. However, besides its outstanding benefits, utilization of cloud computing may lead to some challenging situations which should be taken into consideration by the companies in order to take a decision of the migration to the cloud technology instead of the legacy in-house systems.

In order to discuss and analyze the effect of the technical challenges of cloud computing on the usage for companies in Turkey, all of the challenges within the literature are gathered together, except the least mentioned ones, and discussed. As a result of this process, there are 9 different technical challenges extracted by clustering 69 different challenges. The most important technical challenges that are used as the main factors of usage in this study are privacy and security, account control, cloud structure, virtualization, monitoring, cloud applications, Internet connection, cloud migration and Service Level Agreement (SLA). These challenges and the related studies that mention them as the main problem for adopting cloud computing as a main business solution are summarized in Tab. 1.

The main objective of this study is extracting and obtaining the effects of these nine technical challenges on the usage of cloud computing. In order to investigate these points, hypotheses should be defined and the research problem should be built upon them. Thus, there are nine hypotheses (H) asserted by this study about the relationships between technical challenges and usage of cloud computing:

- H1: Challenges about privacy and security negatively affect the usage of cloud computing.
- H2: Challenges about account control negatively affect the usage of cloud computing.
- H3: Challenges about cloud structure negatively affect the usage of cloud computing.
- H4: Challenges about virtualization negatively affect the usage of cloud computing.

- H5: Challenges about monitoring negatively affect the usage of cloud computing.
- H6: Challenges about cloud applications negatively affect the usage of cloud computing.
- H7: Challenges about Internet connection negatively affect the usage of cloud computing.
- H8: Challenges about cloud migration negatively affect the usage of cloud computing.
- H9: Challenges about service level agreement (SLA) negatively affect the usage of cloud computing.

Tab. 1: Technical Challenges of Cloud Computing.

Technical Challenges	Related Studies
Privacy and security	(Asaduzzaman, Joseph, Sibaj, & Mohamed, 2012; Garrison, Kim, & Wakefield, 2012; Gonzalez et al., 2012; Greengard, 2010; Jensen, Schwenk, Gruschka, & Iacono, 2009; Marston et al., 2011; Subashini & Kavitha, 2011; Viega, 2009)
Account control	(Kshetri, 2012; Subashini & Kavitha, 2011; Takabi, Joshi, & Ahn, 2010; Viega, 2009; Wang, 2009; Zissis & Lekkas, 2012)
Cloud structure	(Calheiros, Ranjan, & Buyya, 2011; Dillon, Wu, & Chang, 2010; Greengard, 2010; Low, Chen, & Wu, 2011; Masiyev, Qasymov, Bakhishova, & Bahri, 2012; Takabi et al., 2010; Younis & Kifayat, 2013; Zhou, Zhang, Xie, Qian, & Zhou, 2010)
Virtualization	(Gliem & Gliem, 2003; Gonzalez et al., 2012; Popovic & Hocenski, 2010; Subashini & Kavitha, 2011; Younis & Kifayat, 2013; Zhang, Cheng, & Boutaba, 2010)
Monitoring	(Pettey & Tudor, 2010; Rong, Nguyen, & Jaatun, 2013; Zhang et al., 2010)
Cloud applications	(Pettey & Tudor, 2010)
Internet connection	(Baliga, Avre, Hinton, & Tucker, 2011; Greengard, 2010; Grossman, 2009; Henkoğlu & Külcü, 2013; Leavitt, 2009)
Cloud migration	(Masiyev et al., 2012)
Service level agreement (SLA)	(Dillon et al., 2010; Egwutuoha, Schragl, & Calvo, 2013; Moreno-Vozmediano, Montero, & Llorente, 2013; Rong et al., 2013, Takabi et al., 2010; Younis & Kifayat, 2013)

In this study, survey, which is a primary data source, is used as the main instrument and methodology for the data collection. The collected data is used for the verification of hypotheses that are proposed by the designed model. In order to accomplish this objective, an online questionnaire is prepared to conduct it on the IT personnel of the companies and collect the demanded data for the study of the hypotheses. The questionnaire is composed of several sections which include questions about the following points:

- Demographics
- Knowledge about cloud computing
- Usage of cloud computing
- Knowledge about the technical challenges proposed by this study
- Importance given to these technical challenges

All questions except the demographics are prepared with a 5-point Likert Scale. For the questions that aim to measure the level of knowledge IT personnel have about cloud computing questions, 1 represents "far too little" and 5 represents "far too much". Same scaling methodology is used for the questions that are related to the usage of cloud computing. However, for the questions that are asked to reveal whether the cloud technology is given importance as much as necessary, 1 represents "not important" and 5 represents "important".

The questionnaire is conducted online and the target population is IT personnel because it is assumed that IT personnel have the most extensive knowledge about the concept and technical details of cloud computing when they are compared with the rest of the population. After all the methods of distributions, there were a total of 262 responses where 101 of them were partial and eliminated in order to maintain the consistency among the study.

Among the 161 responses, 8 of them were eliminated too because they stated that they do not have any knowledge about the technical challenges of cloud computing. The following section presents and discusses the findings and results of the analysis that are made by using these 153 responses. To use multiple regression analysis for this study, the ratio of observations to the independent variables should not fall below 5 and for more accurate and conservative ratio, according to the study by Miller and Kunce, and Halinski and Feldt, the ratio of observations to the independent variables should be 10 (as cited in Kotrlik & Higgins, 2001). The ratio of observations (153) to the independent variables (9) for this study is seventeen which means that the number of responses is highly accurate for the sample size of this study.

3. Findings

Before the discussion on the effects of technical challenges, an up-to-date status of cloud computing is given as a result of the questionnaire. These statistical results help us to determine the current condition of the organizations in Turkey. Of the participants, 27.5 % work for a company in Bank/Finance sector, which is the first rank among all sectors. Second one in the list is the sector of IT Service Providers with 14.4 %. The sectors that have the least proportion are public sector, health and energy as presented in Tab. 2.

Tab. 2: Sectors of the Companies That Participate in the Study.

Sector	Frequency	Percentage
Bank/finance	42	27.5 %
IT service provider	22	14.4 %
Electronics	17	11.1 %
Other	17	11.1 %
Internet	14	9.2 %
Telecommunication and media	14	9.2 %
Education	9	5.9 %
Industry	6	3.9 %
Automotive	4	2.6 %
Insurance	3	2.0 %
Health	2	1.3 %
Public sector	2	1.3 %
Energy	1	0.7 %

As observed in Tab. 3, more than half of the respondents think that cloud is still not popular enough in Turkey. In fact, this statistic gives us a clue about the utilization of the cloud technology and the general perception. The popularity of cloud computing is still not at the desired level to be commonly used and because of this reason, it should be promoted in order to be acknowledged as an essential instrument in the area of ICT by companies.

Tab. 3: Popularity of Cloud Computing.

Popularity of Cloud Computing	Frequency	Percentage
Far too little	31	20.3 %
Too little	78	51 %
About right	36	23.5 %
Too much	5	3.3 %
Far too much	3	2 %

The respondents' knowledge about cloud computing is another important aspect for this study because of both validating the analysis and extracting the current status in Turkey. There are four different knowledge areas that are asked by the questionnaire:

- General knowledge about cloud computing
- Knowledge about benefits of cloud computing
- Knowledge about principles of cloud computing
- Knowledge about the differences between cloud computing and other systems such as in-house systems.

In Tab. 4, it is seen that IT personnel have an adequate knowledge about the benefits of cloud computing. On the other hand, the least known part is its principles. Also, apart from the table that represents the statistics of knowledge about cloud computing, there are eight respondents who state that they do not know anything about cloud computing which are extracted by the analysis part of the study in advance.

Tab. 4: Knowledge about Cloud Computing.

	General	Benefits	Principles	Other Systems
Far too little	6.5 %	4.6 %	7.2 %	6.5 %
Too little	18.3 %	13.7 %	21.6 %	28.1 %
About right	34 %	32.7 %	40.5 %	27.5 %
Too much	29.4 %	34.6 %	19.6 %	24.8 %
Far too much	11.8 %	14.4 %	11.1 %	13.1 %

After the questions that measure the level of general knowledge about cloud computing and a brief information about the participated organizations, the following discussion is one of the questions related to the analysis of the hypotheses, which is trying to reveal the usage level of cloud computing in companies. According to the results shown by Tab. 5, 24.8 % of the respondents do not use cloud computing and 11.1 % of the respondents use cloud computing far too much. It can be seen that regardless of the usage rate, 75.2 % of the participants use cloud computing for their organizational operations.

Tab. 5: Usage of Cloud Computing.

Usage Amount of Cloud Computing	Frequency	Percentage
Not using	38	24.8 %
Too little	34	22.2 %
About right	34	22.2 %
Too much	30	19.6 %
Far too much	17	11.1 %

It is observed that the popularity of cloud computing in Turkey among the companies is not at the desired level. Most of the IT personnel think that it is currently not a popular alternative for the business operation of the organizations and due to this situation, people do not have extensive knowledge about the technology and its principles. When the statistics are revisited again, it is seen that a quarter of the participants do not use cloud computing in any rate. As stated, the popularity of cloud computing, knowledge about it and the usage rates are all related concepts. Thus, if the awareness of cloud computing and people's knowledge about this technology is increased, it will become a popular solution for the problems within the organizations and usage rates will become higher.

After presenting the current state of cloud computing with respect to its popularity and usage as well as the knowledge about it, the main purpose of this study is investigating the effects of technical challenges on the usage. The importance and effect of the 69 technical challenges is evaluated individually by the participants of the questionnaire and they are clustered under 9 headings as the independent variables of this study.

A correlation analysis is conducted to show the relationship between two variables. According to the bivariate correlation analysis the significantly correlated variables are privacy and security, account control and virtualization. This does not mean if these variables affect or predict the usage of cloud computing. While bivariate correlation analysis shows the strength of a correlation, to consider all technical challenges' effects on usage of cloud computing, multiple regression model analysis is accomplished. The variables' and their correspondence in the formula are shown at the Tab. 6.

Tab. 6: Regression Model Analysis.

Variable Name	Corresponding Variable in Formula
Usage of cloud computing	CU
Privacy and security	PS
Account control	AC
Cloud structure	CS
Virtualization	VT
Monitoring	MT
Cloud applications	CA
Internet connection	IC
Cloud migration	CM
Service level agreement	SLA

With given variables, the regression model of this study is:

$$CU = \alpha + \beta1 \times PS + \beta2 \times AC + \beta3 \times CS + \beta4 \times VT + \beta5 \times MT + \beta6 \times CA + \beta7 \times IC + \beta8 \times CM + \beta9 \times SLA \qquad (1)$$

The result of the regression analysis is presented by Tab. 7 with the coefficient and significance level of each variable. The results show that challenges about cloud structure, monitoring, Internet connection and SLA are not significantly predicting the usage of cloud computing because of their high significance value. Thus, these variables are removed from the regression model no matter what their coefficient values are. With the corresponding coefficients of remaining variables, the latest version of the regression model is:

$$CU = 0.733 + 0.324 \times PS + 0.353 \times AC + 0.459 \times VT - 0.223 \times CA - 0.412 \times CM \qquad (2)$$

Tab. 7: Coefficients and Significance Values of Corresponding Variables.

Variable Name	Coefficient	Significance Value
PS	0.324	0.052
AC	0.353	0.012
CS	0.023	0.780
VT	0.459	0.000
MT	0.037	0.641
CA	-0.223	0.057
IC	0.058	0.479
CM	-0.412	0.088
SLA	0.013	0.871

According to the result of the regression analysis, judgments can be made for all hypotheses. H3, H5, H7 and H9 are rejected at the first step because the corresponding variables' significance values are higher in order to significantly explain the cloud computing usage. For H3, it is found that there is not a significant relationship between the challenges about cloud structure and usage of cloud computing. For H5, it is found that there is not a significant relationship between the challenges about monitoring and usage of cloud computing. For H7, it is found that there is not a significant relationship between the challenges about Internet connection and usage of cloud computing. For H9, it is found that there is not a significant relationship between the challenges about SLA and usage of cloud computing.

H1 asserts that challenges about privacy and security negatively affect the usage of cloud computing. As a result of the regression analysis, it is observed that the importance given to the challenges about privacy and security contribute to the model significantly. However, according to the results, it does not affect the usage in a negative way, thus H1 is rejected. Same situation applies for H2 and H4. Account control and virtualization are significantly predicting the usage of cloud computing but not in a negative way. This means that H1, H2 and H4 are also rejected at the second step due to their positive coefficient.

The remaining hypotheses, which are H6 and H8, are the accepted hypotheses of this study because of their significance and negative effect on the usage of cloud computing. Through Tab. 7 and Equation 2, it is seen that the increase in the importance given to the challenges related to cloud applications decreases cloud computing usage. Also, same thing is observed for the challenges related to cloud migration. These two challenges among 9 challenges in total are the ones that have a negative effect on cloud computing usage. This means, these challenges about cloud migration and cloud applications are seen as a barrier to utilize cloud computing by the companies. On the other hand, it is seen that the challenges about cloud structure, monitoring, Internet connection and SLA do not have any significant effect on cloud computing usage. However, the interesting point of the analysis is about privacy and security, account control and virtualization. These challenges have significant relation with the cloud computing usage but as the importance given to these challenges increases, it is seen that the usage rate also increases or a decrease in importance decreases the usage linearly. It can be stated that a challenge may be perceived as a highly important factor but the benefits of cloud computing can be the dominant parameter for the usage. Through this, even though importance given to the challenges related to privacy and security, account control and virtualization is high, the benefits it provides may be the decision criteria for the companies. If the reverse side is considered, a decrease in the importance given to these challenge decreases the cloud computing usage. For this situation, it can be said that even though these challenges are not important, there may be other criteria that are seen as barriers by the companies. Shortly, even though these challenges are significantly related to the cloud computing usage, there are many other criteria for cloud computing utilization except its challenges.

4. Discussions and Conclusions

H3, H5, H7 and H9 are rejected according to the regression analysis and the prediction of these hypotheses' related variables are not significant. Besides this, H1, H2 and H4 are rejected even though the related variables' prediction are significant. As a result of regression analysis, H1 is rejected because the challenges about privacy and security do not affect the usage negatively, H2 is rejected because the challenges about account control do not affect the usage negatively and H4 is rejected because the challenges about virtualization do not affect the usage negatively.

The accepted hypotheses as regression analysis results in are H6 and H8. H6 states that the challenges about cloud applications negatively affect the usage of cloud computing and H8 states that the challenges about cloud migration negatively affect the usage of cloud computing. These two hypotheses are the accepted ones among the nine hypotheses.

H3, H5, H7 and H9 are rejected due to their insignificant prediction. There may be several reasons for this situation. There are countless benefits of cloud computing systems besides its challenges. The user can decide to use cloud computing mostly because of its benefits. This means that benefits of cloud computing can dominate the challenges of it. While answering the questionnaire, one may consider only the importance given to the challenges while they ignore other factors such as benefits of cloud computing or organization's reasoning but these factors can be the main reason for using or not using the cloud computing.

Privacy and security, account control and virtualization are significantly related to the usage of cloud computing but they do not affect the usage of cloud computing in a negative way. These contrasts can be clarified by several additional explanations. One can give high importance to the challenges about cloud computing but the usage may increase in parallel because the benefits of cloud computing can dominate the challenges about privacy and security, account control and virtualization. Also, company's general approach to cloud computing usage and its challenges can be contrast to the respondents' viewpoint which may result in that the respondents, who are IT personnel, use cloud computing unwillingly within the organization.

The challenges about cloud applications and cloud migration do not have a significant correlation with the usage of cloud computing. On the other hand, regression analysis states that these variables significantly predict the dependent variable with negative coefficient value. None of these two variables significantly correlated with usage of cloud computing individually but when all challenges are composed together, challenges about cloud applications and cloud migration can predict the usage significantly. These situations show that bivariate correlation analysis and regression analysis are almost parallel except significance values because bivariate correlation analysis analyzes each correlation independently.

As further research areas, the challenges about cloud applications and cloud migration need to be studied in depth because there is a perception that they will create huge problems if the business operations are migrated to the cloud data centers. If the challenges under the topic of cloud applications and cloud migration are solved, it may contribute to an increase in the usage of cloud computing in Turkey. The challenges about privacy and security, account control, cloud structure, virtualization, monitoring, Internet connection and SLA do not decrease the usage of cloud computing significantly but they need to be studied together with cloud computing's benefits and other factors in order to find more accurate result.

References

Asaduzzaman, A., Joseph, A. R., Sibai, F. N., & Mohamed, N. (2012, March). Cloud Computing: A Cloudy Future? In *Innovations in Information Technology (IIT)*, 18–20 March *2012 International Conference on* (pp. 78–82). IEEE, Abu Dhabi, United Arab Emirates.

Baliga, J., Ayre, R. W., Hinton, K., & Tucker, R. (2011). Green Cloud Computing: Balancing Energy in Processing, Storage, and Transport. *Proceedings of the IEEE*, 99(1), 149–167.

Buyya, R., Yeo, C. S., & Venugopal, S. (2008, September). Market-Oriented Cloud Computing: Vision, Hype, and Reality for Delivering It Services as Computing Utilities. In *High Performance Computing and Communications*, 25–27 September *2008. HPCC'08. 10th IEEE International Conference on* (pp. 5–13). IEEE, Dalian, China.

Calheiros, R. N., Ranjan, R., & Buyya, R. (2011, September). Virtual Machine Provisioning Based on Analytical Performance and QoS in Cloud Computing Environments. In *Parallel Processing (ICPP), 13–16 September 2011 International Conference on* (pp. 295–304). Taipei, Taiwan, IEEE.

Dillon, T., Wu, C., & Chang, E. (2010, April). Cloud Computing: Issues and Challenges. In *Advanced Information Networking and Applications (AINA), 20–23 April 2010 24th IEEE International Conference on* (pp. 27–33). Perth, WA, Australia, IEEE.

Egwutuoha, I. P., Schragl, D., & Calvo, R. (2013). A Brief Review of Cloud Computing, Challenges and Potential Solutions. *Parallel & Cloud Computing, 2*(1), 7.

Foster, I., Zhao, Y., Raicu, I., & Lu, S. (2008, November). Cloud Computing and Grid Computing 360-Degree Compared. In *Grid Computing Environments Workshop, 12–16 November 2008. GCE'08* (pp. 1–10). Austin, TX, USA, IEEE. *Central & Easter Europe and Middle East & Africa Quarterly Server Virtualization Tracker.* (2009). IDC CEMA

Garrison, G., Kim, S., & Wakefield, R. L. (2012). Success Factors for Deploying Cloud Computing. *Communications of the ACM, 55*(9), 62–68.

Gliem, J. A., & Gliem, R. R. (2003). *Calculating, Interpreting, and Reporting Cronbach's Alpha Reliability Coefficient for Likert-Type Scales.* Midwest Research-to-Practice Conference in Adult, Continuing, and Community Education. 8–10 October 2013, Columbus, OH, USA.

Gonzalez, N., Miers, C., Redígolo, F., Simplício, M., Carvalho, T., Näslund, M., & Pourzandi, M. (2012). A Quantitative Analysis of Current Security Concerns and Solutions for Cloud Computing. *Journal of Cloud Computing, 1*(1), 1–18.

Greengard, S. (2010). Cloud Computing and Developing Nations. *Communications of the ACM, 53*(5), 18–20.

Grossman, R. L. (2009). The Case for Cloud Computing. *IT Professional, 11*(2), 23–27.

Henkoğlu, T., & Külcü, Ö. (2013). Evaluation of Conditions Regarding Cloud Computing Applications in Turkey, EU and the USA. *In International Symposium on Information Management in a Changing World (pp. 36–42). Springer, Berlin, Heidelberg.*

Jensen, M., Schwenk, J., Gruschka, N., & Iacono, L. L. (2009, September). On Technical Security Issues in Cloud Computing. In *Cloud Computing, 21–25 September 2009. CLOUD'09. IEEE International Conference on* (pp. 109–116). IEEE. Bangalore, India.

Kotrlik, J. W. K. J. W., & Higgins, C. C. H. C. C. (2001). Organizational Research: Determining Appropriate Sample Size in Survey Research Appropriate Sample

Size in Survey Research. *Information Technology, Learning, and Performance Journal*, 19(1), 43.

Kshetri, N. (2012). Cloud Computing in Developing Economies. *Kshetri, Nir (2010)"Cloud Computing in Developing Economies", IEEE Computer*, 43(10), 47–55.

Leavitt, N. (2009). Is Cloud Computing Really Ready for Prime Time. *Growth*, 27(5), 15–20.

Low, C., Chen, Y., & Wu, M. (2011). Understanding the Determinants of Cloud Computing Adoption. *Industrial Management & Data Systems*, 111(7), 1006–1023.

Marston, S., Li, Z., Bandyopadhyay, S., Zhang, J., & Ghalsasi, A. (2011). Cloud Computing—The Business Perspective. *Decision Support Systems*, 51(1), 176–189.

Masiyev, K. H., Qasymov, I., Bakhishova, V., & Bahri, M. (2012, October). Cloud Computing for Business. In *Application of Information and Communication Technologies (AICT)*, 17–19 October *2012 6th International Conference on* (pp. 1–4). Tbilisi, Georgia, IEEE.

Mell, P., & Grance, T. (2009). The NIST Definition of Cloud Computing. *National Institute of Standards and Technology*, 53(6), 50.

Moreno-Vozmediano, R., Montero, R. S., & Llorente, I. M. (2013). Key Challenges in Cloud Computing: Enabling the Future Internet of Services. *Internet Computing, IEEE*, 17(4), 18–25.

Pettey, C., & Tudor, B. (2010). *Gartner Says Worldwide Cloud Services Market to Surpass $68 Billion in 2010*. Gartner Inc., Stamford, Press Release.

Popovic, K., & Hocenski, Z. (2010, May). Cloud Computing Security Issues and Challenges. In *MIPRO*, 24–28 May *2010 Proceedings of the 33rd International Convention* (pp. 344–349). Opatija, Croatia, IEEE.

Rong, C., Nguyen, S. T., & Jaatun, M. G. (2013). Beyond Lightning: A Survey on Security Challenges in Cloud Computing. *Computers & Electrical Engineering*, 39(1), 47–54.

Subashini, S., & Kavitha, V. (2011). A Survey on Security Issues in Service Delivery Models of Cloud Computing. *Journal of Network and Computer Applications*, 34(1), 1–11.

Takabi, H., Joshi, J. B., & Ahn, G. J. (2010). Security and Privacy Challenges in Cloud Computing Environments. *IEEE Security & Privacy*, 8(6), 24–31.

Viega, J. (2009). Cloud Computing and the Common Man. *Computer*, 42(8), 106–108.

Wang, C. (2009). Forrester: A Close Look at Cloud Computing Security Issues. *CSO Security and Risk*.

Younis, M. Y. A., & Kifayat, K. (2013). Secure Cloud Computing for Critical Infrastructure: A Survey. *Liverpool John Moores University*, United Kingdom, Tech. Rep.

Zhang, Q., Cheng, L., & Boutaba, R. (2010). Cloud Computing: State-of-the-Art and Research Challenges. *Journal of Internet Services and Applications*, 1(1), 7–18.

Zhou, M., Zhang, R., Xie, W., Qian, W., & Zhou, A. (2010, November). Security and Privacy in Cloud Computing: A Survey. In *Semantics Knowledge and Grid (SKG)*, 1–3 November *2010 Sixth International Conference on* (pp. 105–112). Beijing, China, IEEE.

Zissis, D., & Lekkas, D. (2012). Addressing Cloud Computing Security Issues. *Future Generation Computer Systems*, 28(3), 583–592.

Abide Coşkun Setirek* and Aysun Bozanta

Industry 4.0 and Key Technologies: A Review

1. Introduction

The latest advancement in information and communication technologies (ICT) enables the increase in productivity and efficiency in industrial manufacturing. Countries with developed economies have been seeking to monitor and even create those new technologies to improve their wealth of nations.

"Industry 4.0," which is a new industrial revolution, is the recent hotspot for most global industries and the information industry. There are three main aspects of the Industry 4.0: the Smart Product, the Smart Machine and the Augmented Operator (Weyer, Schmitt, Ohmer & Gorecky, 2015). The smart product is capable of holding the information about operational data and requirements as an individual building plan. Therefore, it can determine required resources and manage the production process (Stock & Seliger, 2016; Weyer, Schmitt, Ohmer & Gorecky, 2015). The smart machine implies the cyber-physical production systems (CPPS) which are able to communicate and collaborate with the surrounding physical devices, production modules and products through open networks and semantic descriptions (Monostori, 2014; Weyer, Schmitt, Ohmer & Gorecky, 2015). The third aspect of Industry 4.0 is the augmented operator who is able to adapt to challenging work environment of highly modular production systems (Weyer, Schmitt, Ohmer & Gorecky, 2015). The worker who will face with a large variety of jobs ranging from specification and monitoring to verification of production strategies should be the most flexible part of the CPPS (Gorecky, Schmitt, Loskyll & Zühlke, 2014).

There are several technologies involved in this global trend such as the Internet of Things (IoT), cyber-physical systems (CPSs), cloud manufacturing, wireless technologies, RFID technologies, big data analytics and artificial intelligence (AI), robotics, 3D printing (3DP), augmented reality (AR) and simulation (Posada et al., 2015). These technologies are essential for successful Industry 4.0 application. Especially, the IoT and CPSs, which can be defined as transformative technologies for managing interconnected systems between its physical assets and computational capabilities (Lee, Bagheri & Kao, 2015), are central to the Industry 4.0 vision (Posada et al., 2015).

---

*   Corresponding author: Abide Coşkun Setirek, abide.coskun@boun.edu.tr, Boğaziçi University Hisar Campus Bebek, +90 212 359 6933.

This study aims to thoroughly review the research performed on Industry 4.0 and its key technologies. Based on the research objective defined, it was aimed to produce an updated snapshot of the existing knowledge relevant to Industry 4.0 and to propose influential suggestions for the future research. Although there are an increasing number of studies about Industry 4.0, a comprehensive review study is rare in the literature.

2. Literature Review

In literature, Industry 4.0 concept which refers to the fourth industrial revolution and be called by smart industry, smart factory, smart manufacturing and smart production have been studied also as advance manufacturing, smart manufacturing, digital manufacturing, computer-integrated manufacturing, agile manufacturing, cellular manufacturing, additive manufacturing and discrete manufacturing. According to Rüßmann et al. (2015), horizontal and vertical system integration, the IoT, the cloud, big data analytics, simulation, additive manufacturing (3DP), AR and robot are the technological areas that underpin Industry 4.0. Industrial automation systems are becoming smarter with these technologies. In literature, there are many research focused on developing and using these technologies for industrial automation systems. All these technologies were studied from Industry 4.0 perspective.

Industry 4.0 concept is based on integration of both IoT and CPSs (Pisching, Junqueira, Santos Filho & Miyagi, 2015). Wollschlaeger and his colleagues (2017) reviewed the impact of IoT and CPSs on industrial automation from an Industry 4.0 perspective. Ravi and Wu (2016) studied the implication of IoT and services for the chemical manufacturing industry. Molano et al. (2017) analyzed the use of IoT in manufacturing industry. Wan and his colleagues (2016) proposed a software-defined industrial Internet of Things (IIoT) architecture in the context of Industry 4.0. On the other hand, Thramboulidis (2015) presents a CPSs-based approach for industrial automation systems, and a CPSs architecture for Industry 4.0-based manufacturing systems was proposed by Lee, Bagheri and Kao (2015). PLC-based control of industrial automation systems (Mok, 2015), and SOA-based industrial automation systems (Jammes et al., 2014) were also studied in the context of Industry 4.0. In addition, Leitão, Colombo & Karnouskos (2016) implemented an industrial automation based on CPSs technologies and handled the challenges of it.

IoT and CPSs are converging to the Internet of Services which uses the cloud-based manufacturing (Pisching, Junqueira, Santos Filho & Miyagi, 2015).

Givehchi, Trsek and Jasperneite (2013) provide an overview of the latest concepts of cloud computing technology for industrial automation in their study. In literature, there are also studies which implement cloud-based industrial automation systems (Hegazy & Hefeeda, 2015; Peake et al., 2015; Wenger et al., 2015). Yen et al. (2014) presented an advanced manufacturing solution to Industry 4.0 trend through cloud computing technologies. Manufacturers need to use emerging technologies, such as advanced analytics and CPSs-based approaches which improve efficiency and productivity in order to become more competitive (Lee, Lapira, Bagheri & Kao, 2013). According to them, data has become more accessible and ubiquitous with the IoT contributing to the big data environment. In literature, researchers studied recent advances and trends (Lee, Lapira, Bagheri & Kao, 2013; Lee, Kao & Yang, 2014; Yin & Kaynak, 2015; Zhong, Xu, Chen & Huang, 2017); analytics (Lee, Kao & Yang, 2014; Zhang, Ren, Liu & Si, 2017); frameworks (Zhang et al., 2015); and challenges (Yin & Kaynak, 2015) in big data for Industry 4.0.

AI is another key technology for smart manufacturing in Industry 4.0 and authors handled this technology from Industry 4.0 perspective. Martinez et al. (2017) developed an open platform for the Intelligent Industrial Internet (I3) targeting industrial connectivity and sensing deployments. Ramezan and Jassbi (2017) implement a hybrid expert decision support system based on artificial neural networks in process control from an Industry 4.0 perspective. Saldivar et al. (2016) used a clustering genetic algorithm in order to identify smart design attributes for Industry 4.0 customization.

RFID is an important part of CPS, IoT or Industry 4.0 (Yu, Sriram, Alfnes & Strandhagen, 2016). Some research used RFID for an intelligent application or system. For example, Mansour and Jelassi (2014) have used RFID technology in order to gather data from shop floor for an intelligent manufacturing system. Some of the researchers studied the efficiency of RFID technologies. Balog, Szilágyi, Dupláková and Minďaš (2016) studied on effect verification of external factor to readability of RFID transponder.

With the advances in AR and virtual reality (VR), these approaches have been started to integrate into the industry environment (Malý et al., 2016). Malý et al. studied on AR experiments with industrial robot in Industry 4.0 environment. Ugarte et al. (2016) focused on bringing the Markerless AR technology to assist in manufacturing processes. Paelke (2014) presents an AR system that supports human workers in a rapidly changing production environment. Caricato et al. (2014) provided reliable decision support system for analysing the application of AR technologies in manufacturing.

Additive manufacturing (AM) processes utilize from a computer-aided design file for taking information and converting it to a stereolithography file (Wong & Hernandez, 2012). With intensive research, significant progress has been made in the development of AM processes and practical applications (Guo & Leu, 2013). Rapid prototyping and 3DP technologies have been studied mainly. Several applications (Krupke et al., 2015; Välimäki, 2017), processes, process designs and models (Chen, Zhou & Lao, 2011; Ghazanfari, Li, Leu & Hilmas, 2016; Ponche et al., 2014) for AM were proposed in literature.

Flexible and reconfigurable systems through robotics are studied by Miller (1985) and Coppola (2014). Bahrin, Othman, Nor and Azli (2016) reviewed the advances of robotic and automation technology in achieving Industry 4.0. Park (2014) tried to identify the core technologies for robots to proliferate into new manufacturing settings. Gustavsson (2016) tried to enable human–robot collaboration in assembly manufacturing using speech recognition, haptic control and AR.

Simulation optimization-based tools are also important in Industry 4.0. They are used for complex industrial systems and automation technologies and Industry 4.0 literature includes some simulation studies (Fleischer et al., 2006; Melouk, Freeman, Miller & Dunning, 2013; Dunke & Nickel, 2015).

According to Cammin et al. (2016), wireless networks (WNs) have many advantages and play a key role in the Industry 4.0 as they will facilitate the deployment of novel industrial applications for smart factories and intelligent manufacturing systems. For this reason, WNs have been attracting more attention from academic communities (Cammin et al., 2016; Sepulcre et al., 2016). | According to Sepulcre et al. (2016), significant advances are still necessary for reliable industrial wireless networks (IWNs) and they studied on the reliability of IWNs. Some researchers have discussed IWN features and related techniques, provide a new architecture and propose some applications for IWNs and IWN standards (Li et al., 2015).

3. Methodology

3.1 Database Selection

This study aims to review the research performed on Industry 4.0 and to propose influential suggestions for future research. Based on the research objective defined, it was aimed to produce an updated snapshot of the existing knowledge relevant to Industry 4.0. There are many different databases available for searching scholarly articles, and it is important that the appropriate databases are included. For this purpose, six databases including journals whose subject of interests are engineering, economics and business namely Scopus, ProQuest, EbscoHost, Web of Knowledge, IEEE, Science Direct were searched with the keyword "Industry 4.0" without any limitation. This search resulted in the following numbers of papers: Scopus: 967, ProQuest: 915, EbscoHost: 700, Web of Knowledge: 550, IEEE: 219, Science Direct: 146. Therefore, Scopus, including 967 publications about Industry 4.0, which is more than all other databases, was chosen as a target database for this study.

3.2 Keyword Selection

After first search, key technologies were specified among the most used keywords of resulting publications. Keywords which have similar meanings or same meaning with a different writings and key technologies that were created from those keywords are presented in Tab. 1.

Tab. 1: Key Technologies Related with Industry 4.0.

Key Technologies	Related Keywords and Their Frequencies
Cyber- Physical Systems (CPSs)	Cyber Physical Systems (CPSs) (112), Cyber-physical Systems (43), Cyber-physical Systems (CPS) (42), Cyber Physical System (37), Cyber Physicals (15), Cyber Physical Systems (14), CPS (13), Cyber-Physical Systems (13), Cyber-physical System (9)
Internet of Things (IoT)	Internet Of Things (101), Internet Of Things (IoT) (28), IoT (19), Internet Of Things (IoT) (9), Internet Of Things (IoT) (10)
Big Data	Big Data (77), Data Analytics (8)
AI	Artificial Intelligence (32), Intelligent Agents (12), Intelligent Systems (9)
Robotics	Robots (24), Robotics (21), Industrial Robots (10)
VR-AR	Virtual Reality (26), Augmented Reality (19)
Wireless	Wireless Sensor Networks (17), Wireless Telecommunication Systems (15)
Cloud	Cloud Computing (22), Cloud Manufacturing (13)
3D Printers	3D Printers (11)
RFID	RFID (9), Radio Frequency Identification (RFID) (9)
Simulation	Simulation (7)

3.3 Sample Selection

The final search, which yielded 310 publications, was conducted according to following limitations: Document type (article and conference paper), source type (journals and conference proceedings), language (English), keywords (in Tab. 1 and "Industry 4.0"). In Fig. 1, the distribution of publications according to the key technologies is presented. There can be overlap between key technologies which means a specific paper can be related to more than one technology.

Fig. 1: The Distribution of the Number of Publications According to the Key Technologies.

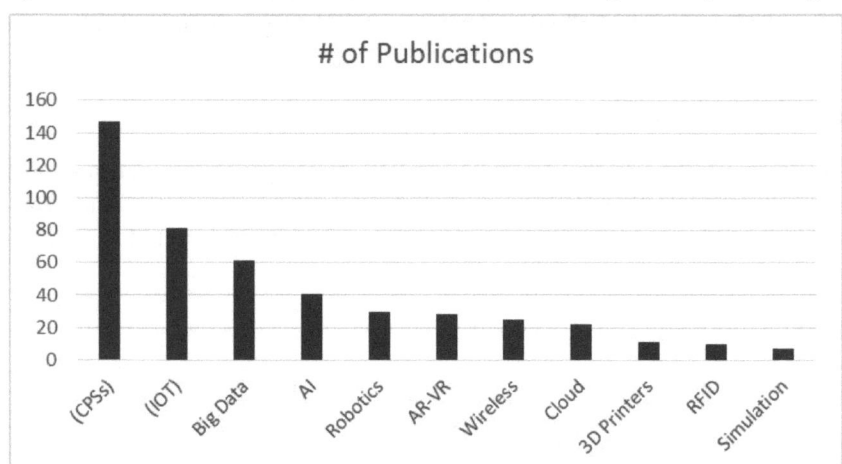

After the key technology classification, the publications were examined according to their publications years, classified in accordance with the research type classification as empirical or conceptual (reference) and the research design classification as quantitative, qualitative or mixed (reference). In addition, publication countries are also specified for each article.

4. Results

In this phase, the publication year of the "Industry 4.0" articles was examined. The study interested in how the research about the issue "Industry 4.0" had evolved with key technologies until today. Research type, research design and publication country were also handled. The following topics were explored: the number of "Industry 4.0" publications published per year, the percentage of the publications by research type as empirical or conceptual, the percentage of the publications by research design as qualitative, quantitative or mixed research design, a list of technologies and frequencies of them by publication year, the frequencies of the themes by publication year and authors' countries.

4.1 Industry 4.0 Research Years

When the distribution of articles was examined by publication year, it was found that 310 key technologies–related Industry 4.0 publications have been published until mid-2017. While there were only 7 papers published in 2013, 21 and 47 papers were published, respectively, in 2014 and 2015 (Fig. 2). However, there is a great increase in the number of publications that have been published until mid-2017. In the mid of 2017, the number of publications is 68.

Fig. 2: Number of Key Technologies–Related Industry 4.0 Publications by Years.

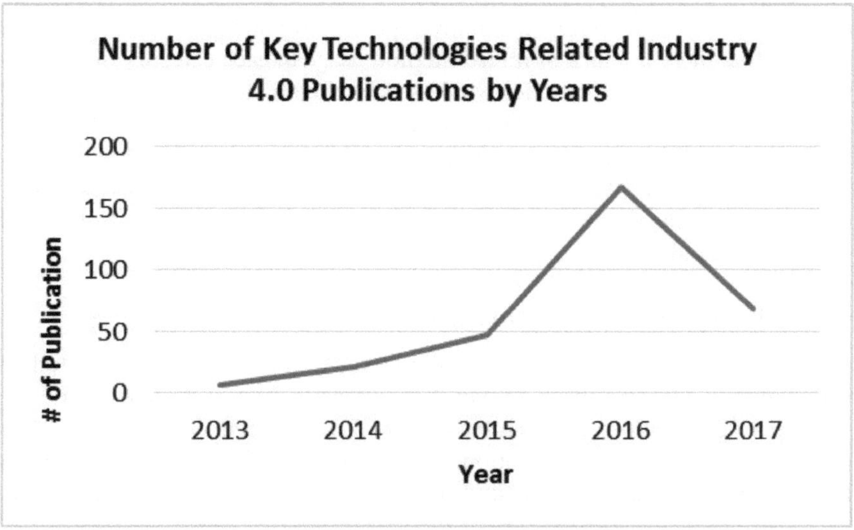

4.2 Industry 4.0 Key Technology Research by Research Type and Research Design

In Fig. 3, the number of conceptual and empirical publications and the number of qualitative, quantitative and mixed methods used in publications are presented according to each key technology. It can be observed from Fig. 2 that the number of empirical research is more than the number of conceptual research. Especially, studies about CPS, big data, AI, robotics, VR and wireless technologies are mostly empirical. The lack of conceptual studies on those fields may be the future research directions for the researchers.

Fig. 3: Research Types and Designs of Publications According to Key Technologies.

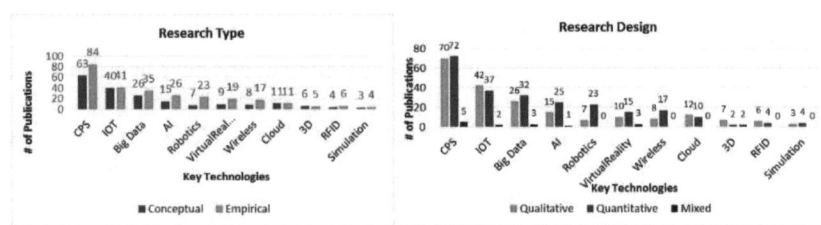

Publications in each key technology category were examined in accordance with their research design namely qualitative, quantitative and mixed methods. It is obvious that there is a lack of mixed method usage in each category of publications. In the categories of robotics, wireless technologies, cloud technologies, RFID and simulation, there is no publication which was conducted by using the mixed method. In addition, qualitative design was slightly used in CPS, big data, AI, robotics, VR, wireless technologies and simulation-related publications. Quantitative design was slightly applied in IoT, cloud technologies, 3D printers and RFID-related publications.

4.3 Industry 4.0 Key Technology Research by Years

As a result of the literature review, totally 11 key technologies which are studied with Industry 4.0 were specified. Fig. 4 summarizes these technologies and the number of publications for each technology. It shows that the highest number of publication, with 147 publications, belongs to "CPSs". In addition, 81 papers focuses on "IoT", while "big data and analytics"–related papers follow it with 61.

Fig. 4: The Number of Publications by Key Technologies.

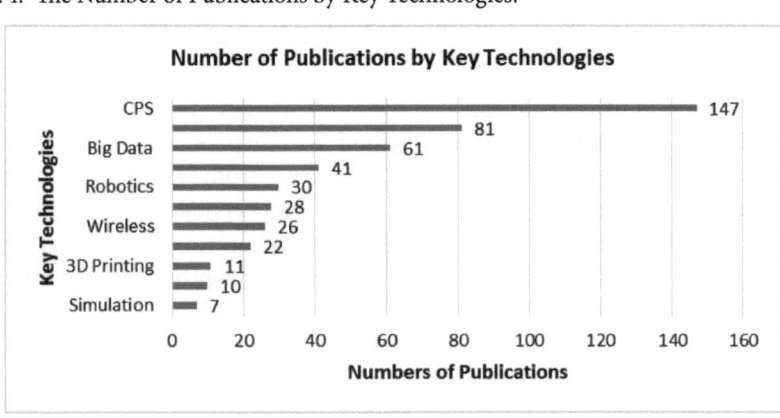

Fig. 5 shows the number of technology-related Industry 4.0 papers according to years. The number of papers published between the years 2013 and 2017 has steadily increased and most of the articles for each key technology were published in 2016. According to Fig. 5, almost all categories have exponential growth in the number of papers being published during the period; however, between 2014 and 2015, the growth in the category "AR-VR" is logarithmic. On the other hand, the increase in the category "CPSs" and "IoT" are considerably higher than the increase in other categories for all periods. In addition, "Big data and analytics" seems to be the most attention-getting area during the period.

Fig. 5: Changes in the Number of Publications by Years.

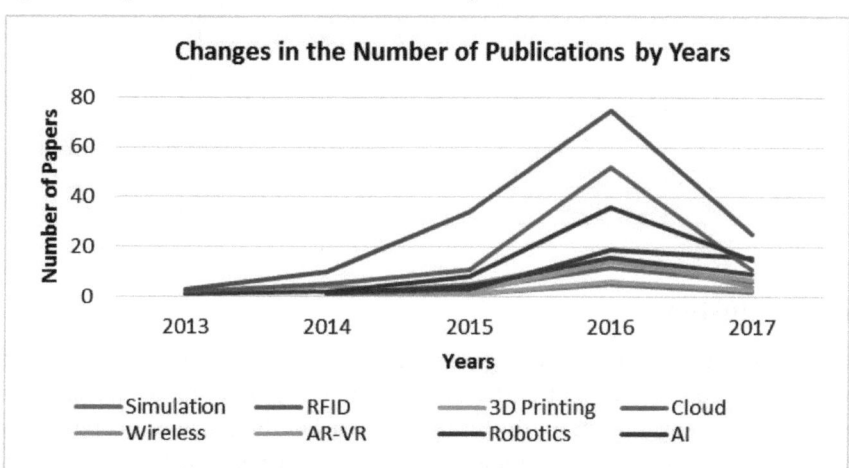

4.4 Industry 4.0 Key Technology Research by Countries

When Industry 4.0 key technology research is examined according to authors' country, it is obvious that the highest number of research belongs to Germany for almost every technology except big data and analytics, cloud computing and RFID technologies. In these areas, China has the highest number of research (Tab. 2).

Tab. 2: Industry 4.0 Key Technology Research by Countries.

Country	CPS	Big Data	IoT	AI	Wireless	Robotics	AR-VA	Cloud	3DP	RFID	Simulation
Germany	60	13	20	10	7	13	16	2	4		1
China	16	14	8	2	5	4	2	11	2	3	
US	10	6	3	2	3	1	1	1			
Spain	9	4	4	1	1	1	2			1	1
UK	7	7	4	1	1		1		1	1	
Italy	7	3	3	1	1		1			1	1
Taiwan	5	3	2	1	2	1		2	1		
Austria	8	1	1	1		3	1	1			
Hungary	5	5	2	2							1
Czech	5	2	2	1		2	1				1
Others	58	22	28	19	14	5	3	3	3	4	2

5. Conclusion

The chapter contributes to the ongoing discussion centering on Industry 4.0 by presenting updated snapshot of the existing knowledge relevant to Industry 4.0 and to propose suggestions for the future research. For this purpose, Scopus database was chosen as a target database for this study. Keywords which have similar or same meaning with a different writings of key technologies were determined. The publications in Scopus database was examined by using keywords which represent the key technologies of Industry 4.0 with resource type (conference proceedings and journals), document type (conference paper and article) and language (English) limitations. The publications were examined according to their publications years, classified in accordance with the research type classification as empirical or conceptual and the research design classification as quantitative, qualitative or mixed.

As a result, 310 key technologies–related Industry 4.0 publications were analyzed. According to findings, a great increase was seen in the number of papers in 2016 with 167 publications. In addition, more empirical research were studied rather than conceptual research, especially in studies about CPS, big data, AI, robotics, VR and wireless technologies. Moreover, findings showed that there is a lack of mixed method usage in each category of publications. More qualitative design was used in CPS, big data, AI, robotics, VR, wireless technologies and simulation-related publications while quantitative design was dominated in IoT, cloud technologies, 3D printers and RFID related publications. Another finding is that most of the research includes the "CPSs", "IoT" and "big data and analytics" concepts. Each key technology was published mostly in 2016 and almost all categories have exponential growth in the number of papers during the period. Only between 2014 and 2015, the growth in the category "AR-VR" is logarithmic. The increase in the category "CPSs" and "IoT" are considerably higher than the increase in other categories for all periods. Moreover, "big data and analytics" seems to be the most attention getting area during the given period. According to country classification, Germany has the highest number of research in almost each technology category except big data and analytics, cloud computing and RFID technologies. China has the highest number of research in these categories.

5.1 Future Research Directions

- According to research type: CPS, big data, AI, robotics, VR and wireless technologies studies are mostly empirical. The lack of conceptual studies on those fields may be the future research directions for the researchers.
- According to research design: there is a lack of mixed method usage in each key technology category of publications. In the categories of robotics, wireless technologies, cloud technologies, RFID and simulation, there is no publication conducted by using mixed method.
- Qualitative design was slightly used in CPS, big data, AI, robotics, VR, wireless technologies and simulation-related publications.
- Quantitative design was slightly applied in IoT, cloud technologies, 3D printers and RFID-related publications.
- The mostly studied key technologies are IoT, CPS and big data analytics as expected. However, there are still research opportunities for cloud manufacturing, wireless technologies, RFID technologies, AI, robotics, 3DP, AR and simulation.

5.2 Limitations

The main limitations of this study can be the usage of one data source – Scopus. Although the Scopus compounds many publications in the field, other databases can be added to the study to make it more comprehensive. On the other hand, amore structured approach can be used to analyze publications, for instance bibliometric analysis.

References

Bahrin, M. A. K., Othman, M. F., Nor, N. H., & Azli, M. F. T. (2016). Industry 4.0: A Review on Industrial Automation and Robotic. *Journal Teknologi (Sciences & Engineering)*, 78(6–13), 137–143.

Balog, M., Szilágyi, E., Dupláková, D., & Minďaš, M. (2016). Effect Verification of External Factor to Readability of RFID Transponder Using Least Square Method. *Measurement*, 94, 233–238.

Cammin, C., Krush, D., Heynicke, R., Scholl, G., Schulze, C., Thiede, S., & Herrmann, C. (2016, September 6–9). Coexisting Wireless Sensor Networks in Cyber-Physical Production Systems. In *Emerging Technologies and Factory Automation (ETFA), 2016 IEEE 21st International Conference on* (pp. 1–4)), Berlin, Germany, New York: IEEE.

Caricato, P., Colizzi, L., Gnoni, M. G., Grieco, A., Guerrieri, A., & Lanzilotto, A. (2014). Augmented Reality Applications in Manufacturing: A Multi-Criteria Decision Model for Performance Analysis. *IFAC Proceedings Volumes*, 47(3), 754–759.

Chen, Y., Zhou, C., & Lao, J. (2011). A Layerless Additive Manufacturing Process Based on CNC Accumulation. *Rapid Prototyping Journal*, 17(3), 218–227.

Coppola, G. (2014). *Advances in Parallel Robotics for Flexible and Reconfigurable Manufacturing* (Doctoral dissertation). University of Ontario Institute of Technology, Oshawa, Canada.

Dunke, F., & Nickel, S. (2015). Simulation-Based Optimisation in Industry 4.0. In M. Rabe & U. Clausen (Eds.), Simulation in Production and Logistics 2015 (pp. 69–78). Stuttgart, Germany: Fraunhofer IRB Verlag.

Fleischer, J., Ender, T., & Wienholdt, H. (2006). A Simulation Based Optimization Concept for Production Systems. ZWF Zeitschrift für wirtschaftlichen Fabrikbetrieb, 101(9), 480–485.

Ghazanfari, A., Li, W., Leu, M. C., & Hilmas, G. E. (2016). A Novel Extrusion-Based Additive Manufacturing Process for Ceramic Parts. In *Solid Free* (pp. 1509–1529). Fabr. Symp, Austin, TX.

Givehchi, O., Trsek, H., & Jasperneite, J. (2013, September 10–13). Cloud Computing for Industrial Automation Systems—A Comprehensive Overview. In *Emerging Technologies & Factory Automation (ETFA), 2013 IEEE 18th Conference on* (pp. 1–4) Cagliari, Italy, New York : IEEE.

Gorecky, D., Schmitt, M., Loskyll, M., & Zühlke, D. (2014, July 27–30). Human-Machine-Interaction in the Industry 4.0 Era. In *Industrial Informatics (INDIN), 2014 12th IEEE International Conference on* (pp. 289–294) Porto Alegre, Brazil, New York: IEEE.

Guo, N., & Leu, M. C. (2013). Additive Manufacturing: Technology, Applications and Research Needs. *Frontiers of Mechanical Engineering*, 8(3), 215–243.

Gustavsson, P. (2016). Using Speech Recognition, Haptic Control and Augmented Reality to Enable Human-Robot Collaboration in Assembly Manufacturing: (Unpublished PhD thesis research proposal). University of Skövde, School of Engineering ScienceThe Virtual Systems Research Centre, Skövde, Sweden. Retrieved October 11, 2018, from http://www.diva-portal.org/smash/record.jsf?pid=diva2%3A956517&dswid=-7816.

Hegazy, T., & Hefeeda, M. (2015). Industrial Automation as a Cloud Service. *IEEE Transactions on Parallel and Distributed Systems*, 26(10), 2750–2763.

Jammes, F., Karnouskos, S., Bony, B., Nappey, P., Colombo, A. W., Delsing, J., ... & Bangemann, T. (2014). Promising Technologies for SOA-Based Industrial Automation Systems. In A. Colombo et al. (Eds.), Industrial Cloud-Based Cyber-Physical Systems (pp. 89–109). Cham, Switzerland: Springer International Publishing.

Krupke, D., Wasserfall, F., Hendrich, N., & Zhang, J. (2015). Printable Modular Robot: An Application of Rapid Prototyping for Flexible Robot Design. Industrial Robot: An International Journal, 42(2), 149–155.

Lee, J., Bagheri, B., & Kao, H. A. (2014, July 27–30). Recent Advances and Trends of Cyber-Physical Systems and Big Data Analytics in Industrial Informatics. In *International Conference on Industrial Informatics (INDIN)* (pp. 1–6) Porto Alegre, Brazil, New York: IEEE.

Lee, J., Bagheri, B., & Kao, H. A. (2015). A Cyber-Physical Systems Architecture for Industry 4.0-Based Manufacturing Systems. *Manufacturing Letters*, 3, 18–23.

Lee, J., Kao, H. A., & Yang, S. (2014). Service Innovation and Smart Analytics for Industry 4.0 and Big Data Environment. *Procedia Cirp*, 16, 3–8.

Lee, J., Lapira, E., Bagheri, B., & Kao, H. A. (2013). Recent Advances and Trends in Predictive Manufacturing Systems in Big Data Environment. *Manufacturing Letters*, 1(1), 38–41.

Leitão, P., Colombo, A. W., & Karnouskos, S. (2016). Industrial Automation Based on Cyber-Physical Systems Technologies: Prototype Implementations and Challenges. *Computers in Industry*, 81, 11–25.

Li, X., Li, D., Wan, J., Vasilakos, A. V., Lai, C. F., & Wang, S. (2015). A Review of Industrial Wireless Networks in the Context of Industry 4.0. *Wireless Networks,*, 1(23) 23–41.

Maly, I., Sedlácek, D., & Leitão, P. (2017 July 19–21). Augmented Reality Experiments with Industrial Robot in Industry 4.0 Environment. In Industrial Informatics (INDIN), 2016 IEEE 14th International Conference on (pp. 176–181), Poitiers, France. New York: IEEE.

Mansour, W., & Jelassi, K. (2014, November 3–6). RFID Technology to Control Manufacturing Systems Using OPC Server. In *Electrical Sciences and Technologies in Maghreb (CISTEM), 2014 International Conference on* (pp. 1–4) Tunis, Tunisia, New York: IEEE.

Martinez, B., Vilajosana, X., Kim, I. H., Zhou, J., Tuset-Peiró, P., Xhafa, A., … & Lu, X. (2017). I3Mote: An Open Development Platform for the Intelligent Industrial Internet. *Sensors*, 17(5), 986.

Melouk, S. H., Freeman, N. K., Miller, D., & Dunning, M. (2013). Simulation Optimization-Based Decision Support Tool for Steel Manufacturing. *International Journal of Production Economics*, 141(1), 269–276.

Miller, S. M. (1985). Industrial Robotics and Flexible Manufacturing Systems. In *P.R. Kleindorfer (Ed), The Management of Productivity and Technology in Manufacturing (pp. 9–55). New York: Plenum Press.*

Mok, J. H. (2015). Image Processing and PLC Based Conveyor Belt System for Industrial Automation and Control. (Project Report). UTeM, Melaka, Malaysia. Retrieved October 11, 2018, from http://eprints.utem.edu.my/17859/.

Molano, J. I. R., Lovelle, J. M. C., Montenegro, C. E., Granados, J. J. R., & Crespo, R. G. (2017). Metamodel for Integration of Internet of Things, Social Networks, the Cloud and Industry 4.0. *Journal of Ambient Intelligence and Humanized Computing*, 1–15.

Monostori, L. (2014). Cyber-Physical Production Systems: Roots, Expectations and R&D Challenges. *Procedia CIRP*, 17, 9–13.

Paelke, V. (2014, September). Augmented Reality in the Smart Factory: Supporting Workers in an Industry 4.0 Environment. In A. Grau, H. Martínez (Eds), Emerging Technology and Factory Automation (ETFA), 2014 IEEE (pp. 1–4). New York: IEEE.

Park, F. (2014, December 10–12). Robotics and Manufacturing. In *Control Automation Robotics & Vision (ICARCV), 2014 13th International Conference on* (pp. 1–1) Singapore, Singapore, New York: IEEE.

Peake, I. D., Vuyyuru, A., Blech, J. O., Vergnaud, N., & Fernando, L. (2015, December 14–17). Cloud-Based Analysis and Control for Robots in Industrial Automation. In *Parallel and Distributed Systems (ICPADS), 2015 IEEE 21st*

International Conference on (pp. 837–840) Melbourne, VIC, Australia, New York: IEEE.

Pisching, M. A., Junqueira, F., Santos Filho, D. J., & Miyagi, P. E. (2015, April). Service Composition in the Cloud-Based Manufacturing Focused on the Industry 4.0. I In L. Camarinha-Matos et al. (Eds), Technological Innovation for Cloud-Based Engineering Systems. DoCEIS 2015. IFIP Advances in Information and Communication Technology 450 (pp. 65–72). Cham, Switzerland: Springer International Publishing.

Ponche, R., Kerbrat, O., Mognol, P., & Hascoet, J. Y. (2014). A Novel Methodology of Design for Additive Manufacturing applied to Additive Laser Manufacturing Process. Robotics and Computer-Integrated Manufacturing, 30(4), 389–398.

Posada, J., Toro, C., Barandiaran, I., Oyarzun, D., Stricker, D., deAmicis, R., Pinto, E., Eisert, P., Döllner, J., & Vallarino, I. (2015). Visual Computing as a Key Enabling Technology for Industrie 4.0 and Industrial Internet. *IEEE Computer Graphics and Applications*, 35(2), 26–40.

Ramezani, J., & Jassbi, J. (2017, May 3–5). A Hybrid Expert Decision Support System Based on Artificial Neural Networks in Process Control of Plaster Production–An Industry 4.0 Perspective. In L. Camarinha-Matos, M. Parreira-Rocha, J. Ramezani (Eds), Technological Innovation for Smart Systems, , 8th IFIP WG 5.5/SOCOLNET Advanced Doctoral Conference on Computing, Electrical and Industrial Systems, DoCEIS 2017 (pp. 55–71), Costa de Caparica, Portugal. Springer, Cham.

Ravi, R., & Wu, L. C. M. (2016). *Demystifying Industry 4.0: Implications of Internet of Things and Services for the Chemical Industry* (Doctoral dissertation, PhD thesis). Malaysia Institute for Supply Chain Innovation, Selangor, Malaysia.

Rüßmann, M., Lorenz, M., Gerbert, P., Waldner, M., Justus, J., Engel, P., & Harnisch, M. (2015). *Industry 4.0: The Future of Productivity and Growth in Manufacturing Industries*. Boston USA: Boston Consulting Group, 14.

Saldivar, A. A. F., Goh, C., Li, Y., Chen, Y., & Yu, H. (2016, September 7–8). Identifying Smart Design Attributes for Industry 4.0 Customization Using a Clustering Genetic Algorithm. In *Automation and Computing (ICAC), 2016 22nd International Conference on* (pp. 408–414) Colchester, UK, New York: IEEE.

Sepulcre, M., Gozalvez, J., & Coll-Perales, B. (2016). Multipath Qos-Driven Routing Protocol for Industrial Wireless Networks. Journal of Network and Computer Applications, 74, 121–132.

Stock, T., & Seliger, G. (2016). Opportunities of Sustainable Manufacturing in Industry 4.0. *Procedia CIRP*, 40, 536–541.

Thramboulidis, K. (2015). A Cyber-Physical System-Based Approach for Industrial Automation Systems. *Computers in Industry*, 72, 92–102.

Ugarte, R. J., Barrena, N., Diez, H. V., Alvarez, H., & Oyarzun, D. (2016, September, 13–16). Augmented Reality System to Assist in Manufacturing Processes. In *Proceedings of the XXVI Spanish Computer Graphics Conference* (pp. 75–82) Salamanca, Spain, Delft, Netherland: Eurographics Association.

Välimäki, E. (2017). *Modelling, Simulation and Validation of CMT Process: An Application for Additive Manufacturing.* Master's Thesis. Tampere University of Technology, Tampere, Finland.

Wan, J., Tang, S., Shu, Z., Li, D., Wang, S., Imran, M., & Vasilakos, A. V. (2016). Software-Defined Industrial Internet of Things in the Context of Industry 4.0. *IEEE Sensors Journal*, 16(20), 7373–7380.

Wenger, M., Zoitl, A., Blech, J. O., Peake, I., & Fernando, L. (2015, December 14–17). Cloud Based Monitoring of Timed Events for Industrial Automation. In *Parallel and Distributed Systems (ICPADS), 2015 IEEE 21st International Conference on* (pp. 827–830), Melbourne, VIC, Australia, New York: IEEE.

Weyer, S., Schmitt, M., Ohmer, M., & Gorecky, D. (2015). Towards Industry 4.0-Standardization as the Crucial Challenge for Highly Modular, Multi-Vendor Production Systems. *IFAC-PapersOnLine*, 48(3), 579–584.

Wollschlaeger, M., Sauter, T., & Jasperneite, J. (2017). The Future of Industrial Communication: Automation Networks in the Era of the Internet of Things and Industry 4.0. *IEEE Industrial Electronics Magazine*, 11(1), 17–27.

Wong, K. V., & Hernandez, A. (2012). A Review of Additive Manufacturing. *ISRN Mechanical Engineering*, 2012.

Yen, C. T., Liu, Y. C., Lin, C. C., Kao, C. C., Wang, W. B., & Hsu, Y. R. (2014, August 18–22). Advanced Manufacturing Solution to Industry 4.0 Trend through Sensing Network and Cloud Computing Technologies. In *Automation Science and Engineering (CASE), 2014 IEEE International Conference on* (pp. 1150–1152) Taipei, Taiwan, New York: IEEE.

Yin, S., & Kaynak, O. (2015). Big Data for Modern Industry: Challenges and Trends [Point of View]. *Proceedings of the IEEE*, 103(2), 143–146.

Yu, Q., Sriram, P. K., Alfnes, E., & Strandhagen, J. O. (2016). RFID Integration for Material Management Considering Engineering Changes in ETO Industry. In *In Nääs I. et al. (Eds), Advances in Production Management Systems. Initiatives for a Sustainable World. APMS 2016. IFIP Advances in Information and Communication 488 (pp. 501–508). Cham, Switzerland: Springer.*

Zhang, Y., Ren, S., Liu, Y., & Si, S. (2017). A Big Data Analytics Architecture for Cleaner Manufacturing and Maintenance Processes of Complex Products. *Journal of Cleaner Production*, 142, 626–641.

Zhang, Y., Zhang, G., Wang, J., Sun, S., Si, S., & Yang, T. (2015). Real-Time Information Capturing and Integration Framework of the Internet of Manufacturing Things. *International Journal of Computer Integrated Manufacturing*, 28(8), 811–822.

Zhong, R. Y., Xu, C., Chen, C., & Huang, G. Q. (2017). Big Data Analytics for Physical Internet-Based Intelligent Manufacturing Shop Floors. *International Journal of Production Research*, 55(9), 2610–2621.

Kerem Kayabay*, Mehmet Ali Akyol, Mert Onuralp
Gökalp, Altan Koçyiğit and P. Erhan Eren

Current Research Topics in Industry 4.0 and an Analysis of Prominent Frameworks

1. Introduction

The fourth industrial revolution is expected to change production in the near future. This concept promises comprehensive transformations for manufacturing plants in which ordinary machines interact with each other to create context-aware, self-sufficient and even conscious production systems. Enabled by advancements in a series of technology domains, the next industrial revolution builds upon the third industrial revolution that was driven by advancements in electronics and information technology (IT) that allowed automation of complex tasks in mass production systems. Since we already utilize IT in manufacturing today, we need a clearer understanding of Industry 4.0 in order to distinguish between today's ordinary factory and the next-generation smart factory.

Manufacturing organizations have certain motivations to invest in and make the transition into the next industrial era. For example, they want much more **flexibility** in their production line. In the next-generation factories, throughout the same production line, reconfigurable robots allow variations in products and even single unique items can be produced according to individual customer needs. As we move on to the next-generation of manufacturing, mass production now becomes **mass customization** (European Parliamentary Research Service, 2015).

Resilient production is an important motivation and it is yet to be realized. When machines are self-aware and clustered in machine groups, they can now track equipment wear and tear, and compare outcomes with same similar type of machines. In other words, they can do health assessments and predictions on when they are going to require maintenance (Lee, Kao & Yang, 2014). Taking this one step further, machines can automatically organize the production chain to minimize maintenance operations hence decreasing disruptions in production line. **Productivity** is improved when machine failures are omitted as much as

---

*   Corresponding author: Kerem Kayabay, kayabay@metu.edu.tr, METU Informatics Institute, Universiteler Mahallesi, Dumlupınar Bulvarı, No: 1, 06800, Ankara, Turkey, +90 312 210 7880.

possible and disruptions in production are planned in advance (European Parliamentary Research Service, 2015).

Production optimization is taken to a new level when machines can flexibly adapt the production chain according to changing requirements in real time. Sensors allow inventory monitoring and process automation, enabling data driven supply-chain management (Industrial Internet Consortium, 2016). Configuration of machinery is now managed by intelligent assistance systems improving human-machine cooperation (Posada et al., 2015). Simulation allows operators to test machine settings in a virtual world before the physical changeover (Rüßmann et al., 2015). Data-driven supply-chain scheduling, improved human-machine interaction and virtual modeling of manufacturing processes in the production floor decrease product time-to-market (European Parliamentary Research Service, 2015).

Quality of products is not just monitored throughout the manufacturing process, but real-time data is gathered from field once products are deployed by customers (Bosch, 2014). Insight into shop floor visibility enhances **quality control** (Industrial Internet Consortium, 2016). Instead of monitoring samples from small batches, data from every piece can now be collected and analyzed. The result of the analyses can be used to detect problems in production line so that adjustments to machine configurations can be made, and all of these can be done in real-time (European Parliamentary Research Service, 2015).

Moving toward Industry 4.0, there are a number of business opportunities to be discussed. Flexibility in production encourages innovation as prototyping is made easier since setup times decrease in production lines. Innovation and the ability to produce customer-driven products provoke **new products, services and business models** (European Parliamentary Research Service, 2015). As routine works are abstracted away in the production floor, labor is used much more efficiently. This does not necessarily mean job losses, but it means job optimizations, in other words **streamlined human resources**. Continuous risk assessments and process coordination take place, so there are fewer service calls, unplanned maintenance and remote monitoring and intervention (Industrial Internet Consortium, 2016). Industry 4.0 is expected to change the interaction of businesses with customers. A product becomes an agent providing multiple services over a lifetime and transmits data back to the factory. Together with the flexibility in production, Industry 4.0 brings new opportunities to **customer relationship management** and **marketing** techniques increasing customer satisfaction (Bosch, 2014). Optimizations in labor, supply-chain and production line, enhanced quality control and increased customer involvement in the production floor leads to **cost savings** and **revenue increase** at the same time, giving the business an advantage in the competitive marketplace (Industrial Internet Consortium, 2016).

Industry 4.0 is a comprehensive transformation in production (European Parliamentary Research Service, 2015), and it promises to change processes, operations and outcomes in a manufacturing plant. To guide this transformation and define new standards for the new era, engineering organizations, researchers in academia, leading technology innovators and government organizations work together toward architectural frameworks. These frameworks provide higher level abstractions and integrations for isolated research dimensions to make the transition faster and smoother. This chapter is intended to be a beacon to explain the transition into the next-generation of automation in production. After discussing the enabler technology stack that make Industry 4.0 possible, the most important research dimensions toward the transformation effort are classified and explained. After that, prominent architectural frameworks are introduced and whether these frameworks can satisfy each research dimension is discussed.

2. Method

Industry 4.0 vision is enabled by a set of advancements in multiple areas of IT domain. Advancements in the enabler technology stack drive the academia and the industry to put up new research efforts toward realization of Industry 4.0 goals. In this section, after presenting the enabler technologies, research dimensions in this domain are comprehensively classified and explained.

In order to classify the ongoing research efforts toward Industry 4.0, we have conducted a literature review. Our review process started with examining white and technical papers published by Industrial Internet Consortium, European Parliamentary Research Service and the Boston Consulting Group. White papers we found in Industrial Internet Consortium lead us to other complimentary resources published by Bosch, Pega Manufacturing and World Economic Forum. Investigating these papers, we obtained the goals of Industry 4.0, and the business motivations of investing into next-generation manufacturing systems. We also gathered keywords to search for like flexibility in production, predictive maintenance, collaboration in manufacturing, software defined manufacturing and so on. Searching for these keywords, we identified academic papers. In these chapters, we separated enabler technologies from research activities. To give an example, Machine Learning is an enabler technology of predictive maintenance. Then we found overlapping areas in these research activities, which we present in Research Dimensions. Even though classified in different dimensions, introduced research dimensions have cross cutting enabler technologies and advancements in a research area can significantly accelerate the advancement in another research area.

2.1 Enabler Technologies

The Industry 4.0 concept promises high potential impact on the industry at a relatively low cost with the latest automation and digital technologies. In order for Industry 4.0 concept to come true with its full potential, meshing the physical world of machines with the digital world of data and analytics is required. For this purpose, a technology stack that includes critical enabler of the next Industrial Revolution to be utilized in a variety of contexts is needed. Although, each of these enabling technologies in the stack has its own disruptive impact on the industry, when these technologies are integrated together, they promise a significant capability that has never been possible by now. Moreover, with the integration of enabling technologies, the Industry 4.0 concept will become more simple, convenient and accurate to production. In order to understand Industry 4.0 concept better, the following technologies which are also depicted in Fig. 1 and their contributions to the next industrial revolution are investigated: cloud computing, Internet of Things (IoT), distributed systems, big data, machine learning, virtual reality, augmented reality, business process management/process mining, networking and cybersecurity and semantic technologies.

Fig. 1: Industry 4.0 Technology Stack.

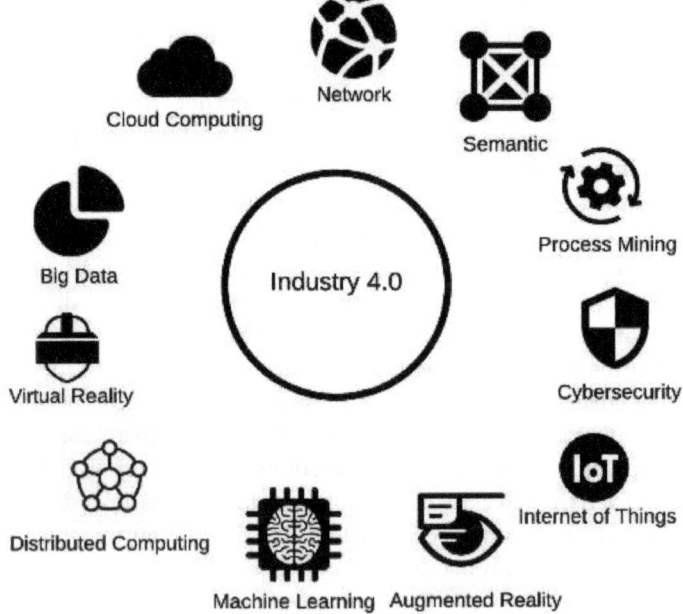

The smart factories are based on **cyber-physical systems (CPS)**, which are connected with **Internet of Things (IoT)** devices. The technological advancements in IoT domain enable us to shift toward to Industry 4.0 by monitoring, controlling and optimizing the production process from procurement of raw material to distribution of finished products with the intelligent connectivity of IoT devices. The increasing digitalization manufacturing processes with a wide range of IoT devices results in exponential growth in machine-to-machine (M2M) data communication over a non-reliable network. Developing a reliable data communication strategy is essential for secure and resilient network infrastructure to protect crucial industrial devices and manufacturing production lines from **cybersecurity** threats. Therefore, industrial network requires flexible infrastructures for effective data exchange and enhanced reliability and scalability. **Software defined networking (SDN)** technology can provide a viable solution, which exploits abstraction and inexpensive hardware advancements in an effort to build more flexible and reliable network systems.

The heterogeneous IoT devices are generating data at a high rate, which needs to be processed in a timely manner. Businesses need to process these data timely to convert valuable information for their decision making and process-optimization activities. The main challenge in the industry is to develop new models and distributed tools to handle these **big data.** Moreover, utilizing advanced **machine learning** algorithms on the collected big data from smart production systems can provide insights to enable preventative maintenance. Early warnings for anomalies and system failures eliminate unnecessary inventory, increase the product quality and optimize the production process. The collected big data also constitutes a valuable asset to apply **business process management** and **process mining** techniques that can be used for observing and diagnosing inefficiencies in business processes. The results of process mining help organizations to improve overall performance and efficiency of their businesses processes. However, **distributed computing** is required to process these collected big data and scale up the system when the overall workload increases. At this point, **cloud computing** is a significant enabler for the Industry 4.0 by providing the tight integration of computing resources and application development platforms with low costs. The cloud computing provides a scalable infrastructure which is capable of dealing with the significant workloads across the Industry 4.0 devices created. Moreover, it also helps us to build a development platform to implement applications for analyzing data collected from industrial devices to gain insight for automating and optimizing production processes. The Predix (General Electric, 2017) provided by General Electric is one of the important platform as a service models to enable industrial-scale analytics for management of asset performance and optimization of operations.

Virtual reality (VR) and **augmented reality (AR)** are other key enabler technologies for human-machine-interaction of Industry 4.0. They enhance customer experience by visualization of customized products and improve workforce training by allowing workers to experience the work environment and to get familiar with special equipment in a simulated environment.

Semantics technologies play a crucial role regarding the management of things, devices and services to provide flexible, adaptable and interoperable service creation in manufacturing process (Thoma, Braun, Magerkurth & Antonescu, 2014; Wahlster, 2014). Machine-understandable content enables sharing information between humans and manufacturing software agents when automating processes to refine production lines.

2.2 Research Dimensions

The proliferation of Industry 4.0 requires new research fields to be investigated in more detail. We have categorized Industry 4.0 research dimensions classifying the ongoing studies in academia. Even though the contribution level of each research dimension might differ, each research dimension has crucial effects on goals of Industry 4.0. According to our taxonomy, the research fields that need to be investigated are individualized production, collaborative manufacturing, end-to-end digital integration, reconfigurable manufacturing systems and automation in manufacturing.

One of the key issues in customer satisfaction and rapid response to changing market conditions is to provide diversity of personalized products serving varying needs and offering them at affordable prices. There is a paradigm shift from producing standardized products for homogeneous markets to creating variety of products and customization. **Mass customization** (MC) harnesses the mass production and customization principles into mass production of individually customized products and services (Davis, 1989; Pine, 1993). Low-cost production of high quantities and the high-cost but low volume production of individualized products is one of the dilemmas faced by the industry today. Mass production focuses on producing greater output and increasing throughput to lower the prices so that everyone can afford them. On the other hand, in mass customization, the objective is to produce a greater variety of products or services at costs and in time comparable to mass production so that everyone can have what they want. This leads to individualized production, which requires increased flexibility, agility and re-configurability in production through mass customization. Hence, the digital transformation plays a very crucial role in **individualized production** and we depend on new technology to have greater variety and shorter development cycles in a more economical way. One of the potential technologies that can be used for mass customization is rapid manufacturing which is derived from rapid prototyping that aims quick creation of prototypes by addition or subtraction of material. Rapid manufacturing can be defined as "the use of a computer aided design (CAD)-based automated additive manufacturing process to construct parts that are used directly as finished products or components" (Hopkinson, Hague & Dickens, 2006).

Collaborative manufacturing requires the collaboration of stakeholders which include organizations, suppliers, customers and business partners to work together (Yan, Ye, Wang & Hua, 2010). With Industry 4.0, **factory-to-factory interoperability**, **supply networks** and **supply chain scheduling** assist to reduce the risks and combine the available resources to expand into different markets without further manufacturing investments (Chien & Kuo, 2013). Companies need to evolve their manufacturing processes to work collaboratively by making their production lines interoperable with other factories, creating a network of adaptive supply chains. Thus, manufacturing can be more efficient and costs can be reduced over time. The important enabler technology for collaborative manufacturing to prosper is IoT. As the IoT technologies become widespread in the industrial environments, manufacturing functionalities can be served as a service, which enables different manufacturing facilities to be utilized collaboratively for different manufacturing purposes.

End-to-end digital integration is another area that requires attention in terms of Industry 4.0 development. With the idea of smart factories in Industry 4.0, integration between the factories constitutes a crucial requirement in order to create an efficient set of production lines. In order to achieve this, end-to-end digital integration should be established between the factories. There are different types of integrations that need to be realized. These digital integration types are **connections between real and virtual worlds, horizontal and vertical integrations, data integrations** and **process virtualization integrations**. The goal of end-to-end digital integration is to close the gap between manufacturing, product design and the customer. With horizontal and vertical integrations, the information can be managed and accessed by different agents in the manufacturing process. Technologies such as virtual reality help to connect real and virtual worlds (Wang, Törngren & Onori, 2015). Moreover, visual computing enables creating more integrated and complete manufacturing processes with technologies including user interfaces, computer graphics, image processing, 3D and visualization.

With the Industry 4.0, same manufacturing facilities can be used for different manufacturing purposes. In order to have a factory that can be used to manufacture different types of products, the manufacturing systems in the factories should be easily reconfigurable to produce different products. This leads to concept of **reconfigurable manufacturing systems**, which allows the utilization of manufacturing systems for different purposes. In order to realize the reconfigurable manufacturing systems, the following technologies have a crucial impact: **software-defined systems**, **plug & work architectures**, **automated assembly** and **cloud-based design and manufacturing**. It has been demonstrated (Wan et al., 2016) that software-defined Industrial Internet of Things (IIoT) for Industry 4.0 will accelerate the implementation of Industry 4.0 through configurable information exchange mechanisms with customizable networking protocols. Plug & work approach is also one of the most crucial requirements for industrial automation (Jung, Song, Watson & Usländer, 2017). In this approach, components in manufacturing processes can be easily integrated without manual changes within the remaining manufacturing system (Schleipen et al., 2015). Plug & work capabilities are also required to build an automated assembly line where assembly line can be reconfigured for different production purposes. Cloud based design and manufacturing, virtual industry clusters, hybrid cloud manufacturing and clusters of 3D printing are needed in order to create a reconfigurable manufacturing system in the Industry 4.0 era (Zhang et al., 2014).

Even though **Automation in Manufacturing** has started with the third industrial revolution, it is still a fundamental research dimension for the Industry 4.0 revolution (Stock & Seliger, 2016). Without the full Automation in Manufacturing, transition into Industry 4.0 cannot be realized. In order to improve the automation in manufacturing in the Industry 4.0, manufacturing processes need to be more intelligent and adaptive. Eventually, Industry 4.0 should allow companies to make production with little to no human intervention. In order to achieve the full automation, some of the previously mentioned research areas such as reconfigurable manufacturing and plug & work systems need to be realized. Additionally, automation in manufacturing reduces the errors resulted from human mistakes, reduces the **waste in manufacturing**, decreases the cost of maintenance with **intelligent monitoring & diagnostics** and enables the machines to be able to self-optimize the manufacturing process, thus increasing the efficiency in the manufacturing. Machine learning is one of the important enabler technologies to study on. Automation requires adaptive learning. With machine learning algorithms, manufacturing processes can be optimized to be more efficient. The goal is to create fully autonomous production without human intervention.

3. Findings

So far we introduced the technology stack which enables the next industrial revolution, and classified the ongoing research activities that address the goals of Industry 4.0. These research activities typically target single and isolated aspects of production, therefore they do not take modularity and exchangeability into account (Kolberg & Zühlke, 2015). In order for the industry to adopt these solutions and apply them to production line, these solutions should be integrated together. Moreover, the introduced enabler technologies are recently hot topics in industry and academia. For example, in big data domain, there is already an abundance of tools and services for processing big data. Nevertheless, new tools continue to emerge especially for processing stream data. For the industry to adopt these tools into business processes, ongoing research now focuses on higher level programming models which can abstract users away from platform specific languages, APIs, protocols and resource management algorithms (Gokalp et al., 2016).

The complete transformation of an ordinary factory into a smart factory requires integration of the introduced research areas providing higher level abstractions. To this end, engineering organizations try to establish standards and some guidelines to facilitate the transition of ordinary factories (Infosys, 2016). The Industrial Internet Consortium (IIC) brings together academicians, leading technology innovators and government organizations to implement global development standards necessary to transform industrial systems (Industrial Internet Consortium, 2017a). First published in June 2015, the Industrial Internet Reference Architecture (IIRA) is a standards-based open architecture and provides a common framework at a high level of abstraction in order to identify most important issues and patterns in many use cases (Industrial Internet Consortium, 2017b). IIRA provides multiple viewpoints for an Industrial Internet Architecture. We focus on functional and implementation viewpoints for the purpose of this discussion. Functional viewpoint decomposes a typical IIoT system into five functional domains which are control, operations, information, application and business. In the implementation viewpoint, these functional viewpoints are mapped to a three-tier-architecture where the tiers are edge, platform and enterprise. Control domain stands at the edge tier, and **end-to-end digital integration** starts here as connectivity and entity abstractions are found in this domain. Connectivity abstraction encapsulates the underlying communication technologies while entity abstraction provides virtual representations for sensors and actuators. Elements of functional domain are mapped to corresponding tiers throughout the organization in the implementation viewpoint. IIRA addresses **automation**

in production in the operations domain where provisioning and deployment, monitoring and diagnostics, optimizations and prognostics take place. Currently, IIRA currently does not comprehensively address issues related with **collaborative manufacturing, individualized production** and **reconfigurable manufacturing systems**.

The Reference Architectural Model Industrie 4.0 (RAMI 4.0) is a result of joint collaboration between several institutions. RAMI 4.0 is a three dimensional structure to cover all crucial aspects of Industry 4.0 (VDI, 2015). Its first horizontal dimension serves to define hierarchy levels within an organization where each hierarchy level defines different functionalities performed by the business. The second horizontal dimension defines product and facility life cycle distinguishing between prototyping and manufacturing phases. The vertical dimension is a layered axis that can be considered as a decomposition of a complex system split up into manageable parts from an IT perspective. Climbing up the layers, we move from real world to the digital world and from physical things up to business processes. Layers in the vertical axis altogether fulfill the requirements of **end-to-end digital integration**. Communication layer standardizes communication of physical machines found in the asset layer. In the integration layer, technical processes can be manipulated by computers, human machine interaction takes place and events are produced by the assets. In the information layer, data integrity is assured before preprocessing and consistent data integration takes place. Vertical integration is satisfied since events generated in the integration layer can be used to advance business processes in the business layer. Some aspects **of individualized production** are covered in the product and life cycle dimension since there is a distinction between prototypes and products. RAMI 4.0 architecture may be inadequate to satisfy the requirements of **automation in production** fully. Data preprocessing and execution of rules are handled only in information layer, which is the third layer from the physical level. Quick changes in configuration of individual machines may not be possible when data is preprocessed so far away from the edge devices. In the product and facility life cycle axis, a product from sales phase can report back improvements, however, automation to minor adjustments in production are not included in the architecture. RAMI 4.0 does not include other research dimensions, which are **collaborative manufacturing** and **reconfigurable manufacturing systems**.

IBM has also published a three layered distributed architecture for Industry 4.0. Its three layers are called edge, plant and enterprise. In this architecture, information emerges from the edge and as it flows through each layer, data analytics are performed (IBM, 2017). IBM's architecture addresses **end-to-end digital**

integration by communicating production events from edge to plant and enterprise using tools, services, standard protocols and representations such as OPC, MQTT, TCP/IP, JSON and XML. Furthermore, at the edge, the cyber-physical systems feedback loop can do edge analytics to perform quick adjustments in production. Some results are also forwarded to plant and enterprise if necessary. **Automation in production** is covered since autonomous adjustments in production can be done according to data analytics especially at the edge layer. The manufacturing processes can also be dynamically configured as results of insights gained from plant and enterprise analytics are passed back to the edge layer. **Collaborative manufacturing** is briefly covered as enterprise level analytics may lead to changes in cross-plant schedules. However, IBM's Industry 4.0 architecture currently does not address issues related with **individualized production** and **reconfigurable manufacturing systems**.

Apart from introduced architectures, leading technology companies has also published reference architectures especially for IoT domain. These architectures include GE Predix Architecture (General Electric, 2015), Intel IoT Architecture (Intel, 2015), Bosch IoT Suite (Bosch, 2016) and KoçSistem Platform 360 (KoçSistem, 2017). However, these architectures do not set guidelines for the next industrial transformation. Therefore, we do not analyze them in terms of how they satisfy the research dimensions of Industry 4.0.

4. Discussion and Conclusions

In the next-generation manufacturing plant, there are real-time interactions between products, machines, systems and services. These real-time interactions are already materializing today as the advancing enabler technologies are increasingly adopted by the industry. Yet, in order for Industry 4.0 to be distinguishable and its goals to become true, a more comprehensive transformation needs to happen. The definition of Industry 4.0 is not still clear and exact; however, introduced research fields are crucial in terms of realization of Industry 4.0 goals.

Advancements in introduced research areas are supported by the developments in reference architectural frameworks. In this article, after introducing enabler technologies and research dimensions of Industry 4.0, we analyze three prominent architectural frameworks for whether they address all research efforts in this domain. Among the classified research dimensions, all three frameworks try to address end-to-end digital integration and automation in production. That being said, further studies are needed for these frameworks to extensively cover other research areas which are classified under individualized production, collaborative manufacturing and reconfigurable manufacturing systems. These architectural frameworks are in active development, and we believe as these frameworks are updated they will act as guides for industrial organizations in the next industrial revolution.

References

Bosch. (2014). Creating Connected Manufacturing Operations in the Internet of Things. Retrieved from Accessed on 31.06.2017http://www.mcrockcapital.com/uploads/1/0/9/6/10961847/20140901_bosch_software_innovations_connectedmanufacturing_white_paper_final.pdf.

Bosch. (2016). The Bosch IoT Suite. Retrieved from Accessed on 31.06.2017 https://www.bosch-si.com/media/en/bosch_si/iot_platform/bosch-iot-suite_product-brochure.pdf.

Chien, C.-F., & Kuo, R.-T. (2013). Beyond Make-or-Buy: Cross-Company Short-Term Capacity Backup in Semiconductor Industry Ecosystem. *Flexible Services and Manufacturing Journal*, 25(3), 310–342.

Davis, S. M. (1989). From "Future Perfect": Mass Customizing. *Planning Review*, 17(2), 16–21. https://doi.org/10.1108/eb054249.

European Parliamentary Research Service. (2015). Briefing Industry 4.0 Digitalisation for productivity and growth. Retrieved from http://www.europarl.europa.eu/RegData/etudes/BRIE/2015/568337/EPRS_BRI(2015)568337_EN.pdf Accessed on 31.06.2017.

General Electric. (2015). Predix Architecture and Services. Retrieved from https://www.predix.com/sites/default/files/ge-predix-architecture-r092615.pdf Accessed on 31.06.2017.

General Electric. (2017). Predix. Retrieved from https://www.predix.io/ Accessed on 31.06.2017.

Gokalp, M. O., Kayabay, K., Akyol, M. A., Eren, P. E., & Kocyigit, A. (2016). Big Data for Industry 4.0: A Conceptual Framework. In 15–17 Dec *2016 International Conference on Computational Science and Computational Intelligence (CSCI)* (pp. 431–434). IEEE. https://doi.org/10.1109/CSCI.2016.0088 Las Vegas, NV, USA, Publisher location: US.

Hopkinson, N., Hague, R., & Dickens, P. (2006). *Rapid Manufacturing: An Industrial Revolution for the Digital Age*. John Wiley & Sons, US.

IBM. (2017). IBM Cloud Garage Method: Internet of Things—IoT Architecture: Industrie 4.0. Retrieved May 29, 2017, from https://www.ibm.com/devops/method/content/architecture/iotArchitecture#industrie_40?cm_sp=Blog-_-blogcta-_-ArchCenter.

Industrial Internet Consortium. (2016). Smart Factories: A Symphony of the Industrial Internet in Action. Retrieved 31.06.2017 from http://www.iiconsortium.org/IIC-Smart-Factory-1-pager.pdf.

Industrial Internet Consortium. (2017a). Industrial Internet Consortium. Retrieved June 4, 2017, from http://www.iiconsortium.org/.

Industrial Internet Consortium. (2017b). The Industrial Internet of Things Volume G1: Reference Architecture. Retrieved 31.06.2017 from http://www.iiconsortium.org/IIC_PUB_G1_V1.80_2017-01-31.pdf.

Infosys. (2016). Interoperability between IIC Architecture & Industry 4.0 Reference Architecture for Industrial Assets. Retrieved 31.06.2017 from https://www.infosys.com/engineering-services/white-papers/Documents/industrial-internet-consortium-architecture.pdf.

Intel. (2015). The Intel® IoT Platform. *Architecture Specification White Paper: Internet of Things (IoT)*, 1–11.

Jung, J., Song, B., Watson, K., & Usländer, T. (2017). Design of Smart Factory Web Services Based on the Industrial Internet of Things. In Proceedings of the 50th Hawaii International Conference on System Sciences.

KoçSistem. (2017). KoçSistem IoT. Retrieved June 4, 2017, from https://iot.kocsistem.com.tr/frame-comm-init.

Kolberg, D., & Zühlke, D. (2015). Lean Automation Enabled by Industry 4.0 Technologies. *IFAC-PapersOnLine*, 48(3), 1870–1875. https://doi.org/10.1016/j.ifacol.2015.06.359.

Lee, J., Kao, H. A., & Yang, S. (2014). Service Innovation and Smart Analytics for Industry 4.0 and Big Data Environment. *Procedia CIRP*, 16, 3–8. https://doi.org/10.1016/j.procir.2014.02.001.

Pine, B. J. (1993). Mass *Customization: The New Frontier in Business Competition*. Harvard Business Press, US.

Posada, J., Toro, C., Barandiaran, I., Oyarzun, D., Stricker, D., De Amicis, R., … & Vallarino, I. (2015). Visual Computing as a Key Enabling Technology for Industrie 4.0 and Industrial Internet. *IEEE Computer Graphics and Applications*, 35(2), 26–40. https://doi.org/10.1109/MCG.2015.45.

Rüßmann, M., Lorenz, M., Gerbert, P., Waldner, M., Justus, J., Engel, P., & Harnisch, M. (2015). *Industry 4.0: The Future of Productivity and Growth in Manufacturing*. Boston Consulting Group, US.

Schleipen, M., Lüder, A., Sauer, O., Flatt, H., & Jasperneite, J. (2015). Requirements and Concept for Plug-and-Work. *At-Automatisierungstechnik*, 63(10), 801–820.

Stock, T., & Seliger, G. (2016). Opportunities of Sustainable Manufacturing in Industry 4.0. *Procedia CIRP*, 40, 536–541. https://doi.org/10.1016/j.procir.2016.01.129.

Thoma, M., Braun, T., Magerkurth, C., & Antonescu, A.-F. (2014). Managing Things and Services with Semantics: A Survey. In 5–9 May *2014 IEEE Network Operations and Management Symposium (NOMS)* (pp. 1–5). IEEE. Krakow, Poland, Publisher location: US https://doi.org/10.1109/NOMS.2014.6838366

VDI. (2015). Status Report: Reference Architecture Model Industrie 4.0. Retrieved 31.06.2017 from https://www.zvei.org/fileadmin/user_upload/Presse_und_Medien/Publikationen/2016/januar/GMA_Status_Report__Reference_Archtitecture_Model_Industrie_4.0__RAMI_4.0_/GMA-Status-Report-RAMI-40-July-2015.pdf.

Wahlster, W. (2014). *Semantic Technologies for Mass Customization* (pp. 3–13). Springer International Publishing. https://doi.org/10.1007/978-3-319-06755-1_1.

Wan, J., Tang, S., Shu, Z., Li, D., Wang, S., Imran, M., & Vasilakos, A. (2016). Software-Defined Industrial Internet of Things in the Context of Industry 4.0. *IEEE Sensors Journal*, 1–1. https://doi.org/10.1109/JSEN.2016.2565621.

Wang, L., Törngren, M., & Onori, M. (2015). Current Status and Advancement of Cyber-Physical Systems in Manufacturing. *Journal of Manufacturing Systems*, 37(Part 2), 517–527.

Yan, J., Ye, K., Wang, H., & Hua, Z. (2010). Ontology of Collaborative Manufacturing: Alignment of Service-Oriented Framework with Service-Dominant Logic. *Expert Systems with Applications*, 37(3), 2222–2231.

Zhang, L., Mai, J., Li, B. H., Tao, F., Zhao, C., Ren, L., & Huntsinger, R. C. (2014). Future Manufacturing Industry with Cloud Manufacturing. In *Cloud-Based Design and Manufacturing (CBDM)* (pp. 127–152). Springer, Germany.

Ceyda Ünal*,Cihan Çılgın and Vahap Tecim

Designing of Manufacturing Process with Mobile-Based Smart Systems

1. Introduction

Before looking at Industry 4.0, it is necessary to remember the previous industrial revolutions: In 1698, the first commercial steam machine developed by the British engineer Thomas Savery, and then steam technology developed by James Watt in 1763, became the starting point of the first industrial revolution. With the first industrial revolution, the first transition was made from manpower production to machine production (mass production).

Fig. 1: Evolution of Industry 4.0. (Source:Authors)

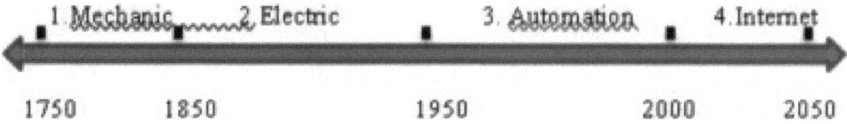

1. Mechanic	2. Electric	3. Automation	4. Internet
1750	1850	1950	2000	2050

The industrial revolution means shifting from a production style based on human and animal power to a production style dominated by machine power. This type of production emerged in England in the 18th century especially in the weaving sector, then spread to other areas. With the transition of machine-based production has increased in shape and quantity (Yediyıldız, 1994).

As seen in Fig. 1, the beginning of the use of electrical energy in the 18th century and the introduction of steel into the production process led to the development of transportation, logistics, the mass production bands and thus the second industrial revolution. The growth of the automotive sector with the development of petrol-based internal combustion engines has been the decisive factor in the industrial revolution.

The emergence of digitally based microchips and computers and the use of these technologies in production enterprises to increase production has launched the third industrial revolution.

---

*	Corresponding author: Ceyda Unal, ceyda.unal@deu.edu.tr, Dokuz Eylul University Department of Management Information Systems.

The rapid growth of the technologies that have played a role in the development of the third industrial revolution and with becoming the Internet widespread supporting the developments in information and communication technologies; new technologies and new concepts have emerged with the result of increasing bandwidth and data transfer speed on the Internet network. Concepts such as big data, cloud computing, Internet of Things that emerged in the light of these developments constituted the infrastructure of the fourth industrial revolution. The technological devices and devices used by the industry are becoming increasingly easier to access the wired or mobile Internet networks, and they can process, analyze and deduce these data on these servers by communicating with each other by providing data flow between them, storing them on the remote servers via sensors or various platforms brought out the fourth industrial revolution. The Industry 4.0 concept was first used in the Hannover Fair in Germany in 2011 (Ferber, 2012). According to Alçın (2016), Industry 4.0 now shows the emergence of a new paradigm in production.

The aim of Industry 4.0 is to connect and integrate traditional industries, especially production areas, to realize flexibility, adaptability and productivity and to enhance effective communication between producers and consumers and their internal dynamics (Gorecky, Schmitt, Loskyll & Zuhlke, 2014). Industry 4.0 is able to customize production to meet consumer needs better by making it possible to perform dynamic simultaneous production planning instead of traditional prediction-based production systems, as well as strengthening collaboration between producers and suppliers. The methods and techniques used in the production process are based on the existing structure, such as quality, efficiency and productivity, depending on the human factor. This makes it difficult to plan an objective development process.

The execution, monitoring or supervision of the business processes in production enterprises in a dynamic, flexible and efficient structure, in particular, increases the quality, productivity and cooperation considerably. Ideas, concepts and technologies emerging along with the development of the Industry 4.0 concept ensure that the industry is more productive when managing business processes.

Wahlster (2012) propounds that the availability of cyber systems as Internet-accessible solutions has resulted in a more cost-effective and efficient operation. At the same time, they have helped to realize new business models and intelligent control systems. In this context, the management and supervision of the business processes and elements with an integrated system, or the communication of the production elements to each other, indirectly improves productivity, quality and efficiency by enabling the enterprises to use human resource and all other production elements more effectively and efficiently.

In the final report of Acatech's Industry 4.0 Forum (2013), the following are some of the distinguishing innovations that this new era brings: Global interaction of machines with storage systems and resources, the development of unique intelligent products with location data, realization of smart factories, adapting to product specifications, providing resource optimization, realization of new business models, new social infrastructure in the workplace for employees and business structure sensitive to individual differences, better work/life balance, response to individual consumer requests and intelligent software developed for immediate engineering and instantaneous response to problems.

The aim of this research is to increase the cooperation of production machinery, production personnel and management level under the heading of Industry 4.0, especially in manufacturing enterprises and to activate management of production process with intelligent devices connected to the Internet. In this context, not only the products produced but also the devices that produce these products must be included in a smart system. In this way, the industry should use the Internet of mobile-based smart devices, cloud systems and the Internet of Things as a collaboration tool in order to increase the concepts such as flexibility, speed and productivity which are the reasons for the emergence of Industry 4.0.

In the scope of the study, thanks to the developed mobile and web-based information system, the abovementioned mobile-based intelligent devices, cloud systems and the Internet of Things are used as a collaboration tool. With the developed information system, personnel and machine inspection are made more effective at the same time and machine efficiency is presented as an output by monitoring machine usage instantly.

2. Method

The main element of this chapter is to provide instant communication between production machines and other production elements such as human resources to enable efficient and accurate production. The mobile-based application designed was developed by using hybrid and native mobile application development methodologies together. Thus, better performance is aimed with the hybrid mobile application development methodology in terms of providing a more user-friendly interface and instant data updates; and besides that with the native mobile application development methodology, fast and secure access is implemented.

Platforms and software used for the purpose of content preparation in the research are HTML5, Android Studio, PHP, JavaScript and Google Cloud Services. User-friendly interfaces designed with HTML5 have been used in both the mobile application and the admin panel. Using Android Studio, the necessary events, procedures, classes and designs were implemented with the Java software language. In addition, while developing web-based information system as admin panel, PHP software language for database connections, Bootstrap CSS framework and Javascript for design are used as Google Cloud Service, Google Cloud Messaging are used (GCM) to send notifications. With GCM, information and notifications can be sent to the mobile device from the server without the mobile devices being open or closed. GCM, which is used to send notifications in this research, is also frequently used in daily life by messaging services, banks, online services and many other applications.

In addition, MySQL and SQLite databases are used for database operations in mobile application, preventing data loss due to instant Internet interruptions and MySQL is used as database for web-based information system. Using the MySQL database on a web server, the data can be updated and displayed instantly by the users.

Fig. 2: Methodology of the Application. (Source:Authors)

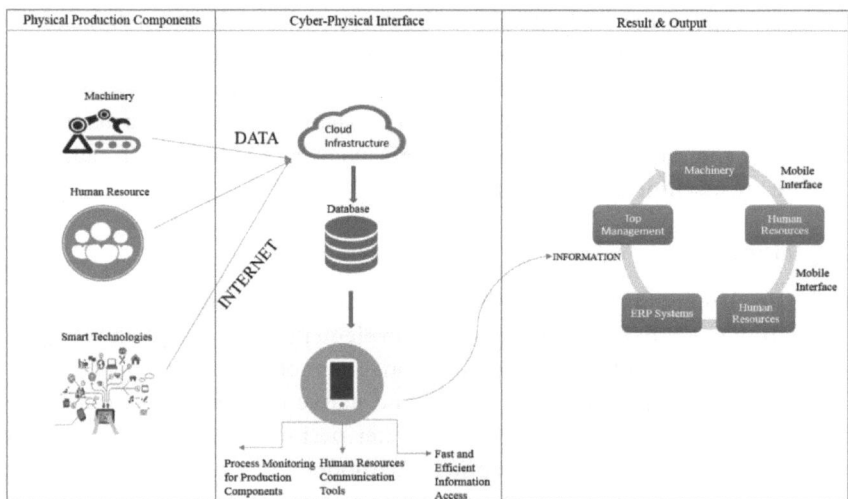

As seen in Fig. 2, applications developed, data collected from machine, human resources and intelligent technologies increase the communication on the mobile devices by storing it on the cloud infrastructure via the Internet, thus enhancing efficiency, effectiveness, speed and flexibility. In this way, human–human communication and human–machine communication are provided more effectively. In addition, production personnel are instantly informed by making processes instant.

Fig. 3: Application Form. (Source:Authors)

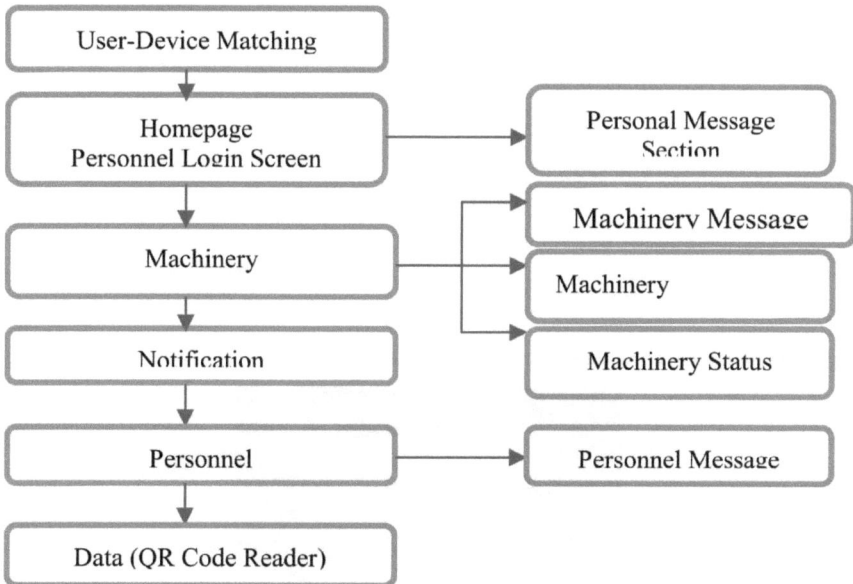

As shown in Fig. 3, the flow of application as a form is explained in detail. To explain the operation of the application in this context, as shown in Fig. 4 and Fig. 5, a personnel ID is required to match the device ID with the user ID at application startup. In this way, necessary relationships are established between the used smart device and the personnel using it. In the homepage section opened after the login screen, user login is requested. In order to be able to enter the related fields, the personnel must be logged in on this login screen.

Fig. 4: Login Screen. (Source:Authors)

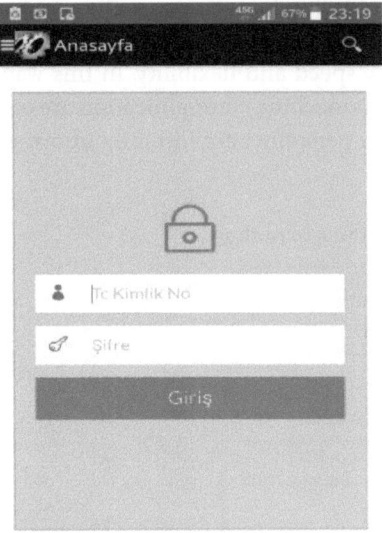

Fig. 5: Personnel Secure Login Screen. (Source:Authors)

Fig. 6: Menu. (Source:Authors)

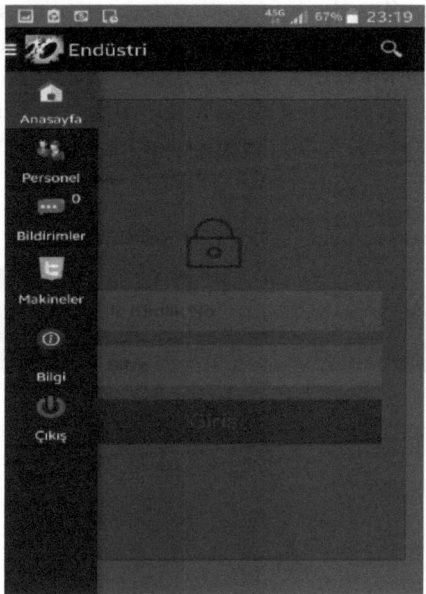

Fig. 7: Personnel List. (Source:Authors)

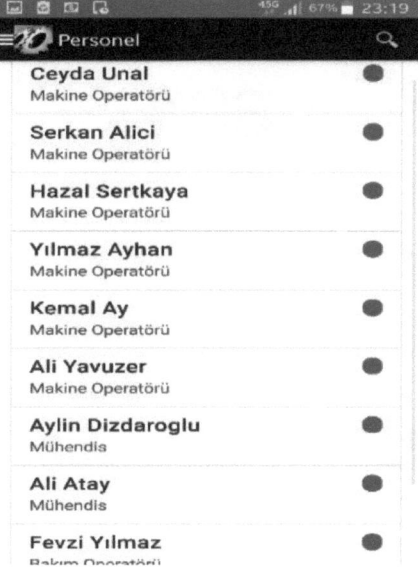

In Fig. 6, the related modules included in the application are designed as menus. The personnel section lists the information and active/passive status of all the personnel except the user who entered the user. Users who have logged in to the application are shown with a green icon in this section. Detailed information about any personnel on the staff list is shown in Fig. 7. In addition, any staff member can write a message from the "Message" section in this section when he/she wants to write a message to the person concerned.

Fig. 8: Detailed Information about Personnel. (Source:Authors)

Fig. 9:   Machine Status. (Source:Authors)

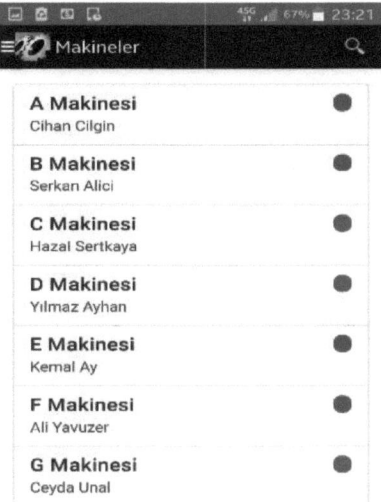

Fig. 10:   Detailed Information about Machine. (Source:Authors)

Fig. 11: Machine Message Screen. (Source:Authors)

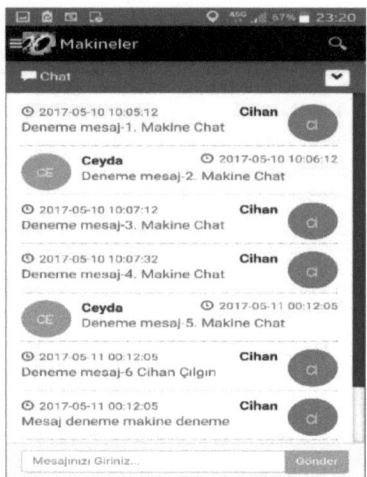

In Figs 8 and 9, detailed information about personnel chosen and machine status as active/passive meaning is displayed. As shown in Fig. 10, the details of the selected machine, message and status change fields are displayed. Messages left by authorized users in the chat area logged in from this area are shown in Fig. 11. Fig. 12 shows the details of the notifications sent from the admin panel to the user.

Fig. 12: Notifications Sent to Users. (Source:Authors)

In addition to the mobile-based application, the associated web-based information system has been developed to ensure that the design is user-friendly and up-to-date. Besides, information sharing has been made more effective by using various visual tools in the reporting of data.

Fig. 13: Admin Panel. (Source:Authors)

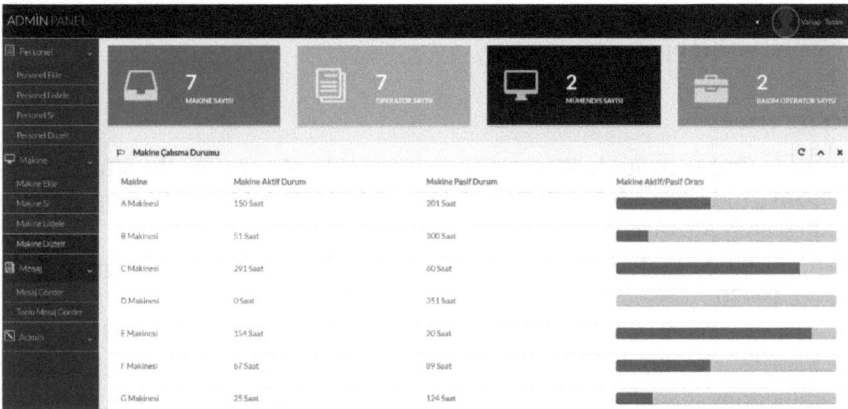

As shown in Fig. 13, the home page of the application's web-based management panel shows the instantaneous data of user machines, operators and other production personnel.

In addition, as shown in the machine operation status table, the active and passive operation states of the machines are shown as a graphical representation of the performance of the machines related to the dynamically received data. In addition, in the drop-down menu on the left in Fig. 13, new records and updates about personnel and machinery can be made and detailed information can be accessed via this menu. The notification messages to be sent to the personnel can be sent to the person individually or collectively if requested from the "Messages" section in the menu.

3. Discussion and Conclusion

Looking at the pre- and post-use situations of the mobile and web-based application designed within the scope of this study, it is observed that the performance of communication between the human resource, machine and management stage in the production process increases and collaboration becomes effective. As a consequence, improvements in business processes provide rigorous process innovation by improving the way things are done, accelerated and made more productive.

Thanks to the designed web-based management panel, this work can be easily integrated into existing Enterprise Resource Planning systems and can provide a more dynamic and flexible structure. The application developed is currently a prototype and in testing phase. In further studies, it will be tested on different manufacturing operations.

References

Acatech. (2013). Acatech: Recommendations for Implementing the Strategic Initiative Industrie 4.0, Final Report of the Industry 4.0 Working Group, Acatech Research Alliance.

Alçın, S. (2016). Endüstri 4.0 ve İnsan Kaynakları. *Popüler Yönetim Dergisi*, 63(47), pp. 19–30.

Ferber, S. (2012). *Industry 4.0—Germany Takes First Steps toward the Next Industrial Revolution. Bosch ConnectedWorld Blog.* http://blog.bosch-si.com/categories/manufacturing/2012/10/industry-4-0-germany-takes-first-steps-toward-the-next-industrial-revolution/. Accessed on May 25, 2017.

Gorecky, D., Schmitt M., Loskyll M., & Zuhlke, D. (2014). Human-Machine-Interaction in the Industry 4.0 Era. In *12th IEEE International Conference on Industrial Informatics (INDIN)*, pp. 289–294 Jan 1, 2014, ,Porto Alegre, Brazil. IEEE.

Wahlster, W. (2012). *Industry 4.0: Towards the 4th Industrial Revolution*, Luxembourg–Kirchberg: Forum Business Meets Research.

Yediyıldız, B. (1994). *Tarih* (2 volumes), İstanbul: MEB Press.

M. Hanefi Calp*, Ahmet Doğan, Aslıhan Tüfekci and
Türksel Kaya Bensghir

Readiness of MIS Undergraduate Programs in Turkish Universities to Industry 4.0

1. Introduction

Rapid developments in science and technology support the field of production considerably in today's world. Germany, which has revolutionized the production area, has played a leading role in the field of industrial information technology (IT). Many industrialized countries have begun to design own industrial infrastructure to meet the requirements of Industry 4.0 (Coşkun, Gençay & Kayıkcı, 2016).

The content of "Industry 4.0" first appeared in an article which was published by the German government in November 2011 as a high-technology strategy for 2020. The fourth stage of the industrialization is called as "Industry 4.0" thereafter mechanization, electrification and information. The term of "Industry 4.0" reappeared at an industrial exposition in Hannover Germany and quickly became a German national strategy in April 2013. "Industry 4.0" has been commonly discussed and has become an important matter for most global industry and information industry in recent years (Zhou, Liu & Zhou, 2015).

The aims of Industry 4.0 are to provide a higher level of operational efficiency and a higher level of automation. Five important features of Industry 4.0 are digitalization, optimization and privatization of production, automation and adaptation, the interaction of human and computer, value-added services and businesses, automatic data exchange and communication. These features are not only related to Internet technologies and advanced algorithms, but also refers that Industry 4.0 is a value-added accumulation of knowledge and process. Besides the dynamic nature of Industry 4.0, there is no systematic and extensive examination of Industry 4.0 yet (Posada et al., 2015; Roblek, Meško & Krapež, 2016; Thames & Schaefer, 2016; Yang, 2017).

In this study, the examination of MIS departments of undergraduate curriculum with Industry 4.0 and its sub-departments that are located in Turkey's state

---

* Corresponding author: Dr. M. Hanefi Calp, mhcalp@ktu.edu.tr, Karadeniz Technical University, Faculty of Economics & Administrative Sciences, Management Information Systems.

universities and the status of readiness is aimed. First, the information about the term Industry 4.0 is given briefly. Later, the methods are given in the second part and results of the study are included in the third part, in this context. Finally, the conclusions and recommendations obtained from the study are presented in the fourth part.

1.1 Industry 4.0

The first industrial revolution began with the introduction of mechanical manufacturing equipment and then it continued with a second industrial revolution that includes the mass production of goods. The third revolution (digital revolution) is using the electronic and computer technology from the early 1970s to the present in the automation and control of manufacturing processes by increasing. The inclusion of IoT (Internet of Things) technology into the production environment has taken the industrialization to the fourth stage (Kagermann, Wahlster & Helbig, 2013; Shrouf, Ordieres & Miragliotta, 2014; The Economist Intelligence Unit, 2013). According to the results of a survey made by the American Society for Quality (ASQ) in 2014, the productivity of 82 % of the organizations declared that they have implemented smart manufacturing applications, 49 % had fewer product flaws and 45 % had increased customer satisfaction (PRWeb, 2017). Besides, Economist Intelligence Unit predicted the future use of IoT by organizing a survey at the global trade community and 38 % of the participants believe that most of the IoT will have a great influence on the market and the industry. Three years after the survey was implemented, 96 % of the participants expect businesses to use IoT, 63 % believe that the companies in late integration will fall behind and 45 % believe that IoT will make companies eco-friendlier (The Economist Intelligence Unit, 2013).

Pan and the others (2015) think that the activities in the Eco Industrial Park in Singapore should be re-updated with Industry 4.0. Kolberg & Zülhke (2015) mentioned the importance of implementing automation with Industry 4.0 first. Then, they indicated the advantages and disadvantages of production and discussed the implementation of the production with Industry 4.0. In this context, they indicated that Industrial 4.0, which is thought to have a complicated structure, will have a less risky, standard, transparent and efficient structure by implementing with lean systems.

Paelke (2014) has developed an augmented reality system that supports employees in a rapidly changing manufacturing environment. The system directs the user to unknown tasks (e.g., collection of new products) and visualizes the information directly in the relevant spatial context by providing the spatially saved information directly to the user's view. It is shown that the results are very positive by having many real tests.

Gorecky and the others (2014) presented solutions for technological help to employees that provides the cyber-physical world and its interactions to be implemented in the form of smart user interfaces. They draw attention to the necessity of adequate strategies to create the necessary interdisciplinary understanding for Industry 4.0 in their study, in addition to their technological means.

When studies related to Industry 4.0 are examined, it arises that the purpose of Industry 4.0 is to provide more efficient and effective design, production and marketing environment by integrating industry with more information technology rather than destroying human focused workforce in factories.

2. Method

In the study, the relationship of MIS undergraduate curriculum with Industry 4.0 in Turkey's state universities and the status of readiness to Industry 4.0 was examined. First of all, the list of state universities with MIS undergraduate departments was determined and the curriculum data of all the MIS departments of these universities were collected in this process. Then, a literature review on Industry 4.0 was made, the studies in the literature were analyzed, and the sub-departments of Industry 4.0 that form the basis of the study were identified. The mentioned ones are classified under nine groups as IoT, smart factories, cyber-physical systems, big data, autonomous robots, simulation, system integration, cloud computing and augmented reality system. The 3D printer, which is one of the new technologies, is predicted that it will be a decisive factor in all of these areas. Finally, all lectures and their contents were examined by the research group and revealed the relationship with the sub-departments identified in Industry 4.0. The universe and sampling of the study, data collection, analysis and interpretation of data and findings headings are included in this part.

2.1 The Universe and Sample

MIS undergraduate departments in the state universities in Turkey that have educational activities create the universe of the research and have reached all of the departments.

2.2 Data Collection Process

The state universities which have MIS undergraduate departments in Turkey and the curriculums of the courses in this section are used in the analysis process within this study. First of all, for this purpose, the list of the universities which have MIS undergraduate program is obtained from YÖK Atlas (https://yokatlas.

yok.gov.tr/). Course curriculums of the relevant departments, any course content or ECTS information, are taken from the web addresses of the universities as of April 2017. | In addition, the number of faculty members of the departments in question and the titles of whether students graduated or not, were determined and all collected data were recorded within the scope of the research.

2.3 Analysis of Data

Each course given in the MIS departments of the universities was examined in detail with the research group, and the level of relations with the scope of Industry 4.0 was determined in this process. The curriculums of the courses and the contents of nine sub-heading previously identified under Industry 4.0 were taken into account when determining the relationships. Information systems courses (information society and e-state, information system strategies, management information systems, e-commerce etc.) and computer science courses (object oriented programming, web programming, internet programming, artificial intelligence, computer networks, computer hardware and system software etc.) have been taken into consideration and the courses in which management and information subjects are integrated in general (technology management, information management, database management systems, production and operations management, etc.) are included. Math, History, English, etc., have not been included in the analysis as well as general cultural courses. Then, the courses, which are included into the analysis, can be related with the nine Industry 4.0 sub-departments identified previously or to which extent they cover the specified departments of these courses have been attempted to determine. A course can take place in one or more of the departments under the nine designated heading at this point. For example, management information systems, technology management, innovation management, etc., have been included in all of these departments within the scope of the analysis, as the courses are also concerned with nine main technology areas that form the components of Industry 4.0.

3. Results

The results obtained from the analysis of collected data and the comments made on these results are included in this part. First of all, it has been determined that there are a total of 19 state universities which have MIS undergraduate department in Turkey; this situation is shown with red color in Fig. 1 in some provinces. Fig. 1 was prepared by the research group using the collected data as of April 2017.

It is seen that MIS departments are located in the western provinces where the industry is generally more intense in the country and only in Sivas and Erzurum in the east when the Fig. 1 is examined. When the value of MIS department of the industry, the economy and therefore the country is considered, it is thought that the number of departments needed to be enhanced both quantitatively and qualitatively in general in this context.

Fig. 1: The Provinces Where the State Universities Have MIS Departments in Turkey (Since April 2017).

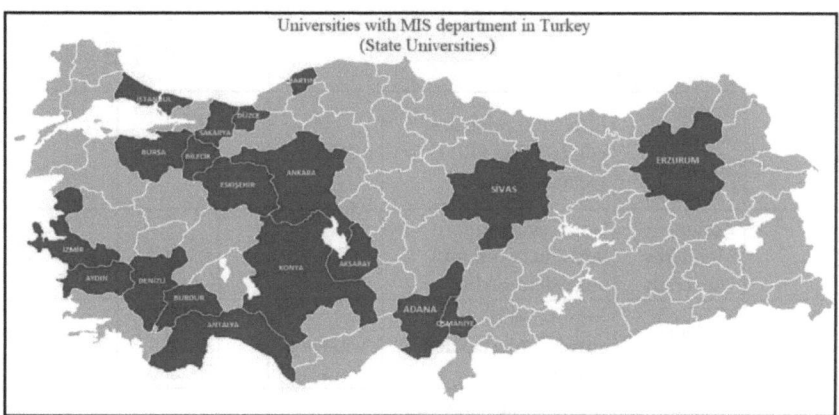

In Tab. 1, the column headings state the names of the universities, the access links of MIS departments, the provinces they are located in, the numbers of faculty members and whether they graduated or not. First of all, it can be said that MIS undergraduate department is still in the development process in Turkey when the number of faculty members of the universities is observed. Besides, the universities with the lowest faculty members are Adnan Menderes University, Akdeniz University, Bartın University, Uludag University, Mehmet Akif Ersoy University and Necmettin Erbakan University; the university with the most faculty members is Bogazici University and it has 71 faculty members in total Turkey wide. However, the number of assistant professors (Asst. Prof.) (45) is more than associate professor (Assoc. Prof.) (16) and professor (Prof.) (10) and it can be interpreted that academic staff that grows under this staff can reach a level that can meet the needs of the department in the mentioned universities in the near future.

Besides, the workload is too much for academic staff who carry out educational activities in Turkey's state universities and the support such as lecturers from other departments can be interpreted when the number of faculty members given in

Tab. 1 is low and the number of courses is also considered high. Finally, the number of graduate students despite the number of low faculty members in Bartın University and Mehmet Akif Ersoy University draws attention.

Tab. 1: MIS Departments, Number of Faculty Members and Graduation Situations According to Universities (As of April, 2017)

No	University's Name	Access Link	City	Instructor Number				Graduation Situation
				Prof.	Assoc. Prof.	Asst. Prof.	TOTAL	
1	Adana Science and Technology University	http://mis.adanabtu.edu.tr/PagesTR/DefaultTR.aspx	Adana	1	1	3	5	-
2	Adnan Menderes University	http://akademik.adu.edu.tr/bolum/sokeisletme/ybs/	Aydın	1	0	2	3	-
3	Akdeniz University	http://yonetimbilisimsistemleri.akdeniz.edu.tr/	Antalya	0	0	3	3	-
4	Aksaray University	http://ybs.aksaray.edu.tr/tr	Aksaray	0	2	3	5	-
5	Atatürk University	https://atauni.edu.tr/yonetim-bilisim-sistemleri-bolumu-1	Erzurum	1	1	5	7	+
6	Bartın University	http://iibf.bartin.edu.tr/bolumAnasayfa.aspx?bolumId=7	Bartın	0	2	1	3	+
7	Bilecik Şeyh Edebali University	http://w3.bilecik.edu.tr/iibf/bolumler/ybs/	Bilecik	0	1	4	5	-
8	Cumhuriyet University	http://iibf.cumhuriyet.edu.tr/index.php?f=5	Sivas	0	2	2	4	+
9	9 Eylül University	http://iibf.deu.edu.tr/akademik-birimler/yonetim-bilisim-sistemleri/	İzmir	2	1	2	5	-
10	Düzce University	http://www.duzce.edu.tr/yonetim-bilisim-sistemleri/	Düzce	1	1	4	6	-

No	University's Name	Access Link	City	Instructor Number				Graduation Situation
				Prof.	Assoc. Prof.	Asst. Prof.	TOTAL	
11	Osmaniye Korkut Ata University	http://ybs.osmaniye.edu.tr/	Osmaniye	1	1	3	5	+
12	Pamukkale University	http://www.pau.edu.tr/ybs	Denizli	0	1	3	4	-
13	Sakarya University	http://www.ybs.sakarya.edu.tr/tr	Sakarya	2	1	4	7	-
14	Uludag University	http://inif.uludag.edu.tr/TR/Default.aspx?kat=sayfa&id=27	Bursa	0	0	3	3	-
15	Yıldırım Beyazıt University	http://mis.academy/index.php?title=Main_Page	Ankara	1	2	3	6	-
16	Anadolu University	https://www.anadolu.edu.tr/acikogretim/turkiye-programlari/acikogretim-siste-mindeki-programlar/yonetim-bilisim-sistemleri	Eskişehir	0	0	0	0	-
17	Bogazici University	http://www.mis.boun.edu.tr/	İstanbul	5	6	2	13	+
18	Mehmet Akif Ersoy University	http://ztyo.mehmetakif.edu.tr/?page=bolumlerYbs	Burdur	0	1	2	3	+
19	Necmettin Erbakan University	https://konya.edu.tr/yonetimbilisimsistemleri	Konya	0	0	3	3	-
TOTAL				10	16	45	71	

Tab. 2 lists the number of courses given in MIS undergraduate departments at state universities. Looking at the number of courses offered, it is seen that Mehmet Akif Ersoy University has the lowest number (15 courses) and Adnan Menderes University and Aksaray University have the highest number (54 courses).

Tab. 2: The Number of Courses Given in MIS Undergraduate Departments at State Universities.

University's Name	City	Number of Courses
Adana Science and Technology University	Adana	42
Adnan Menderes University	Aydın	54
Akdeniz University	Antalya	33
Aksaray University	Aksaray	54
Atatürk University	Erzurum	22
Bartın University	Bartın	45
Bilecik Şeyh Edebali University	Bilecik	40
Cumhuriyet University	Sivas	29
9 Eylül University	Izmir	39
Düzce University	Düzce	32
Osmaniye Korkut Ata University	Osmaniye	30
Pamukkale University	Denizli	41
Sakarya University	Sakarya	31
Uludag University	Bursa	37
Yıldırım Beyazıt University	Ankara	30
Anadolu University	Eskisehir	19
Boğazici University	Istanbul	21
Mehmet Akif Ersoy University	Burdur	15
Necmettin Erbakan University	Konya	24

In Fig. 2, the relation of Industry 4.0 with curricula of the courses opened in MIS undergraduate departments of the state universities are given on university basis. According to Graphic 1, the general relationship levels of Düzce University, Osmaniye Korkut Ata University, Atatürk University, Adnan Menderes University, Bilecik Sheikh Edebali University, Yıldırım Beyazıt University, 9 Eylül University and Bogazici University are higher and Mehmet Akif Ersoy University is lower; but it is noteworthy that this university has a high rate of 42.9 %, especially in the field of "big data". Furthermore, when each of the figures is examined separately on a university basis, it appears that the levels of the relationships of the sub-areas of Industry 4.0 with the curriculum are not balanced. This can be said to be due to the fact that the qualifications of the academic staff in the relevant department are taken into account when the courses opened in the departments are determined.

Graphic 1: Relationship Level of MIS Undergraduate Curricula in Universities with Industry 4.0.

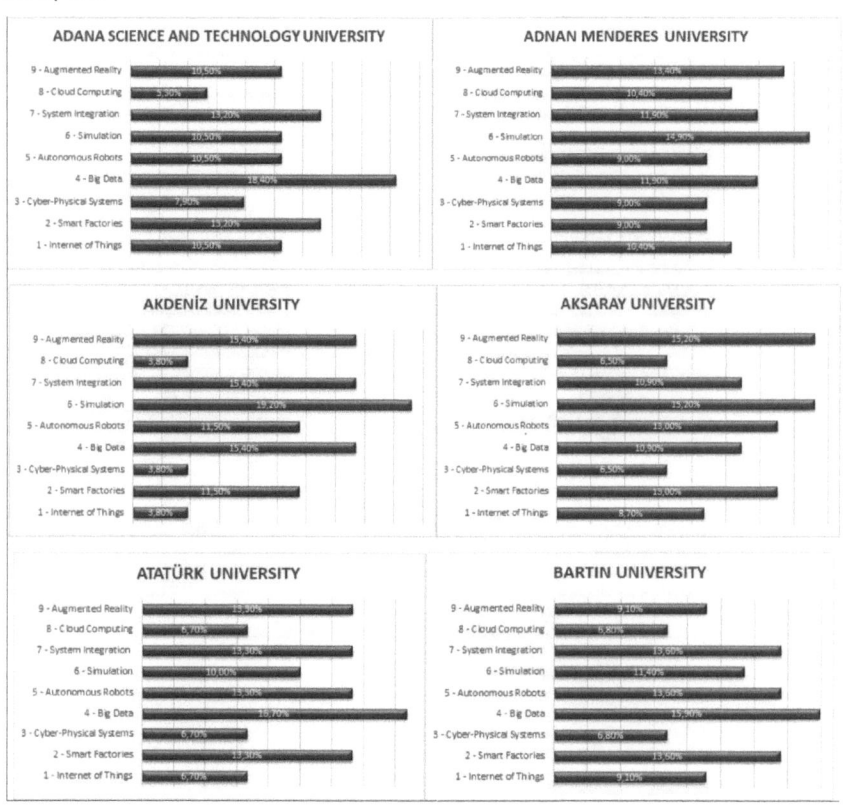

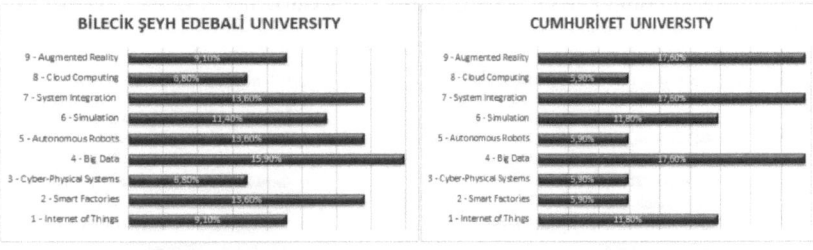

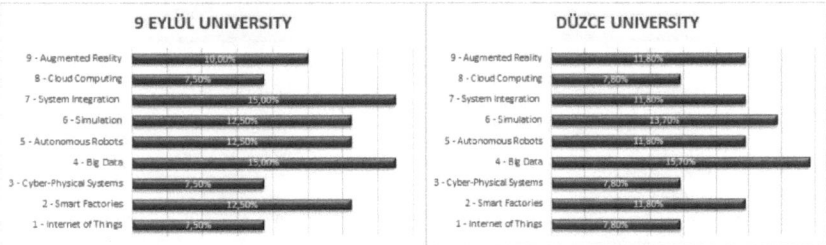

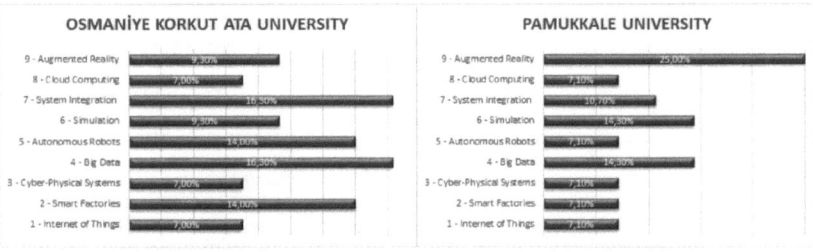

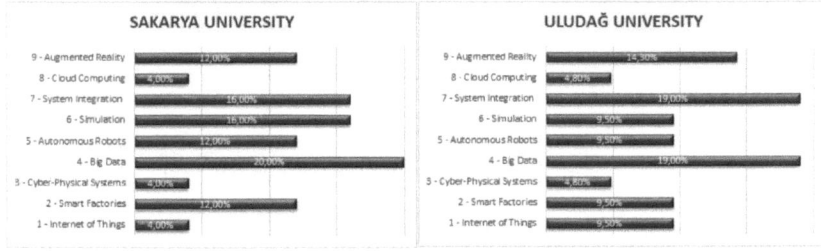

In Tab. 3, according to universities, there are levels of relationship (lowest and highest) between the courses and Industry 4.0 sub-fields. When Tab. 3 is examined, it is observed that the level of readiness of the universities especially in the fields of "IoT, cyber-physical systems and cloud computing" is low and "system integration and big data" is high. In addition, another remarkable point in Tab. 3 is that Mehmet Akif Ersoy University does not have any courses related to the subfields of "IoT, cyber-physical systems, simulation, cloud computing and augmented reality"; but the relationship between the "big data" field and the course curriculum is high at 42.90 %.

Tab. 3: The Lowest/Highest Related Sub-Areas Between Course Curricula and Industry 4.0 According to Universities

University's Name	Industry 4.0			
	Lowest		Highest	
	Field Name	(%)	Field Name	(%)
Adana Science and Technology University	Cloud Computing	5,30	Big Data	18,40
Adnan Menderes University	Smart Factories, Cyber-Physical Systems Autonomous Robots	9,00	Simulation	14,90
Akdeniz University	Internet of Things, Cyber-Physical Systems, Cloud Computing	3,80	Simulation	19,20
Aksaray University	Cyber-Physical Systems, Cloud Computing	6,50	Simulation, Augmented Reality	15,20
Atatürk University	Internet of Things, Cyber-Physical Systems, Cloud Computing	6,70	Big Data	16,70
Bartın University	Cyber-Physical Systems, Cloud Computing	6,80	Big Data	15,90
Bilecik University	Cyber-Physical Systems	6,10	Big Data	16,30
Cumhuriyet University	Smart Factories, Cyber-Physical Systems, Autonomous Robots, Cloud Computing	5,90	Augmented Reality, System Integration, Big Data	17,60
Dokuz Eylül University	Internet of Things, Cyber-Physical Systems, Cloud Computing	7,50	System Integration, Big Data	15,00
Düzce University	Internet of Things, Cyber-Physical Systems, Cloud Computing	7,80	Big Data	15,70
Osmaniye Korkut Ata University	Internet of Things, Cyber-Physical Systems, Cloud Computing	7,00	System Integration, Big Data	16,30
Pamukkale University	Internet of Things, Smart Factories, Cyber-Physical Systems, Autonomous Robots, Cloud Computing	7,10	Augmented Reality	25,00

| University's Name | Industry 4.0 | | | |
| | Lowest | | Highest | |
	Field Name	(%)	Field Name	(%)
Sakarya University	Internet of Things, Cyber-Physical Systems, Cloud Computing	4,00	Big Data	20,00
Uludag University	Cyber-Physical Systems, Cloud Computing	4,80	System Integration, Big Data	19,00
Yıldırım Beyazıt University	Internet of Things, Cyber-Physical Systems, Cloud Computing	5,90	Smart Factories, Big Data, Autonomous Robots, Simulation	14,70
Anadolu University	Internet of Things, Cyber-Physical Systems, Autonomous Robots, Simulation, Cloud Computing	7,10	System Integration, Big Data	21,40
Bogazici University	Cyber-Physical Systems, Cloud Computing	4,80	System Integration, Big Data	19,00
Mehmet Akif Ersoy University	Internet of Things, Cyber-Physical Systems, Simulation, Cloud Computing, Augmented Reality	0,00	Big Data	42,90
Necmettin Erbakan University	Internet of Things, Cyber-Physical Systems, Simulation, Cloud Computing	5,90	Smart Factories, Big Data, Autonomous Robots	17,60

4. Conclusions and Recommendations

In this study, the relationship of Industry 4.0 with the MIS undergraduate curriculum in Turkey was examined. In the research process, first of all, when we look at the number of lecturer of the MIS department in Turkey, these sections are still in development, and there is a need for teaching staff; but recently it has been concluded that the staff will be strengthened with the participation of newly educated lecturers.

It is seen that Düzce University is the highest level of readiness and Mehmet Akif Ersoy University is the lowest level when we look at the relations of universities with Industry 4.0 in general. However, Mehmet Akif Ersoy University attracts attention with a high rate of 42.90 % in the field of big data.

In addition, it can be said that, in the direction of the results obtained, and considering that Industry 4.0 is a new concept, the awareness of Industry 4.0 in universities is still occurring. Panel, symposium, congress etc. must be realized to bring more agenda and to raise awareness about the concept of Industry 4.0 in universities.

It is also important that the Industry 4.0 subfields are integrated into the curriculum and that these fields are transferred practically to undergraduate students because human power needed by the digital economy can be trained with the required qualifications and skills.

As a result, it is clear that investments in the scope of Industry 4.0 in universities will contribute to the economy of the country on a global scale. Nationally and internationally, in cooperation with relevant actors and with the support of all sections of the university, necessary steps must be taken to ensure that industry 4.0 revolution takes place more in Turkey industrialization strategies.

References

Coşkun, S., Kayıkcı, Y. & Gençay, E., (2017). Adapting Engineering Education to Industriie 4.0 Vision. *CoRR,* abs/1710.08806, url: https://arxiv.org/abs/1710.08806.

Gorecky, D., Schmitt, M., Loskyll, M., & Zühlke, D. (2014, July). Human-Machine-Interaction in the Industry 4.0 Era. In *Industrial Informatics (INDIN), 27–30 July 2014, 2014 12th IEEE International Conference on* (pp. 289–294). IEEE Porto Alegre, Brazil.

Kagermann, H., Wahlster, W., & Helbig, J. (2013, April), "Recommendations for Implementing the Strategic Initiative Industrie 4.0," *Final report of the Industrie 4.0 Working Group, acatech – National Academy of Science and Engineering.*

Kolberg, D., & Zühlke, D. (2015). Lean Automation Enabled by Industry 4.0 Technologies. *IFAC-PapersOnline,* (pp. 1870–1875).

Paelke, V. (2014, September). Augmented Reality in the Smart Factory: Supporting Workers in an Industry 4.0. Environment. In *Emerging Technology and Factory Automation (ETFA), 16–19 September 2014, IEEE, (pp.1–4), Barcelona, Spain.*

Pan, M., Sikorski, J., Kastner, C., Akroyd, J., Mosbach, S., Lau, R., & Kraft, M. (2015). Applying Industry 4.0 to Jurong Island Eco Industrial Park. *Energy Procedia. 75 (2015), 1536 – 1541. doi: 10.1016/j.egypro.2015.07.313.*

Posada, J., Toro, C., Barandiaran, I., Oyarzun, D., Stricker, D., de Amicis, R., ... & Vallarino, I. (2015). Visual Computing as a Key Enabling Technology for Industrie 4.0 and Industrial Internet. *IEEE Computer Graphics and Applications*, 35(2), 26–40.

Process engineering. "Brave new world: Industry 4.0 technology", Retrieved from http://processengineering.co.uk/article/2017121/brave-new-world-indu , Access Date: 18.11.2017.

PRWeb. Milwaukee, WI (PRWEB) December 19, 2013, Retrieved from www.prweb. com/releases/2013/12/prweb11430148.htm. Access Date: March 10, 2017.

Roblek, V., Meško, M., & Krapež, A. (2016). A Complex View of Industry 4.0. *SAGE Open*, 6(2), 2158244016653987.

Shrouf, F., Ordieres, J., & Miragliotta, G. (2014, December). Smart Factories in Industry 4.0: A Review of the Concept and of Energy Management Approached in Production Based on the Internet of Things Paradigm. In *Industrial Engineering and Engineering Management (IEEM), 2014 IEEE International Conference on* (pp. 697–701). 9–12 Dec. 2014. Bandar Sunway, Malaysia. IEEE.

Thames, L., & Schaefer, D. (2016). Software-Defined Cloud Manufacturing for Industry 4.0. *Procedia CIRP*, 52, 12–17.

Witchalls, C., & Chambers, J. (2013). The internet of things business index: A quiet revolution gathers pace. The Economist Intelligence Unit. Retrieved from http://www. economistinsights. com/analysis/internet-things-business-index. Access Date: 20 December 2017.

YangLu, (2017), Industry 4.0: A Survey on Technologies, Applications and Open Research Issues, *Journal of Industrial Information Integration,* doi: 10.1016/j. jii.2017.04.005

Zhou, K., Liu, T., & Zhou, L. (2015, August). Industry 4.0: Towards Future Industrial Opportunities and Challenges. In *Fuzzy Systems and Knowledge Discovery (FSKD), 2015 12th International Conference on* (pp. 2147–2152). IEEE. 15–17 Aug. 2015. Zhangjiajie, China.

Ebru Gökalp*, Mert Onuralp Gökalp and P. Erhan Eren

Industry 4.0 Revolution in Clothing and Apparel Factories: Apparel 4.0

1. Introduction

The clothing and apparel industry, which is one of the main sectors of the industrialization movement for developing countries, involves the use of fabric and materials to produce men's, women's and children's clothing, knitted underwear and outerwear. This sector has contributed significantly to Turkey's economy in recent years. It has 16 % of Turkey's total exports in 2016, which is the second largest share after the automotive industry (TEA, 2017). Turkey ranks fifth in the world, based on apparel exports. Since the 1950s, clothing and apparel industry is Turkey's leading sector in terms of employment. The number of apparel companies in Turkey is around 35,000, and about 500,000 people are employed at these companies.

With the removal of exportation quotas applied to China in 2005, the clothing and apparel sector has started having difficulties with competing in global markets, especially against Chinese firms. In order to survive in this tough global competitive market, apparel manufacturers need to transform their manufacturing processes toward having a more flexible production system to meet the rapid changes in the global market, delivering orders to customers as early as possible to meet increased customer expectations, using the human workforce more efficiently to achieve high productivity levels and utilizing all resources effectively. In today's rapidly globalizing business world, the information technology (IT) industry is continuously coming up with new technologies. Thus, it is important to invest in these emerging technologies for businesses to attain competitive advantage and improve their operational efficiency by generating valuable insights to enhance the decision-making process, develop new business models and drive new revenue streams.

After three important industrial revolutions, as mechanical, electrical and digital revolution, the world is now witnessing the fourth industrial revolution that integrates emerging IT concepts, including cyber-physical systems (CPS), Internet of

---

* Corresponding author: Ebru Gökalp, egokalp@metu.edu.tr, METU Informatics Institute, Ankara/Turkey, 0312 210 7881.

Things (IoT) and big data (Srivastava, 2016). The latest industrial revolution is called "Industrie 4.0" in Germany, and "Industrial Internet of Things" (IIoT) in the United States. It is defined by Acatech as "the technical integration of CPS into manufacturing and logistics and the use of the Internet of Things and Services in industrial processes. This will have implications for value creation, business models, downstream services and work organization" (Kagermann, Helbig, Hellinger & Wahlster, 2013). It is stated that there is an expected increase of 23 % (78.77 billion Euros) in Germany's Gross Domestic Product (GDP) from 2013 to 2025 based on the realization of the Industry 4.0 concept (Bauer & Horváth, 2015). The Joint Apparel Union Forum has stated that these developments provide a solution for the future problem of workers' inadequacy and for minimizing the human impact at all stages of production to increase productivity. Therefore, there is a need for clothing and apparel factories to align their corporate strategies with the fourth industrial revolution.

Within the scope of this study, a conceptual smart apparel factory, called "Apparel 4.0", is proposed, and the benefits and challenges of "Apparel 4.0" are analyzed. "Apparel 4.0" includes innovative technologies provided by the fourth industrial revolution. The impacts of these emerging technologies on the production system and managerial activities in a clothing and apparel company are investigated as well.

The rest of the chapter is organized as follows. Section 2 provides background information about industrial revolutions as well as clothing and apparel production systems in the scope of this study. The proposed conceptual smart factory Apparel 4.0 is explained in Section 3. The main benefits and challenges of Apparel 4.0 are investigated in Section 4. Finally, we conclude the chapter and state directions for future research in Section 5.

2. Background

2.1 Industrial Revolutions in the Clothing and Apparel Sector

The first industrial revolution began with the discovery of the steam engine in England in 1712 and later spread to Europe and America. The first mechanical weaving loom was developed in 1785 by British inventor Edmund Cartwright. Progress in the textile sector during this revolution underlies the adoption of textile consumption as a basic need, as Maslow would later explain.

The second industrial revolution started in 1870 when electricity began to be used in the industrial field. The serial production was first realized in 1910 by Henry Ford as part of the second industrial revolution. The impact of this revolution in the clothing and apparel sector is that the sewing machine began to be produced in a serial manner. Despite being discovered in the past, Isaac Singer patented the first sewing machine in 1851, and with this development, clothing production and consumption gained momentum. Later, sewing machines began to be used in other production areas such as shoes (McNeil, 2002).

The third industrial revolution, also called the Digital Revolution, began with the use of the first programmable management system in 1969. With this revolution, information and communication technology (ICT) has started to be used in the industry and the transition from analogue to digital technology has been achieved, thanks to integrated systems obtained in the light of developments in microprocessors, software, fiber optic cables and telecommunication domains.

The fourth industrial revolution was first announced by the German Federal Government at the Hannover Fair in 2011. Within the context of this revolution, which is also called Industry 4.0, CPS and IoT can communicate with each other and people in real time (Kagermann, Helbig, Hellinger & Wahlster, 2013; Monostori, 2015). A copy of the physical world is created in the virtual environment and it is provided that decentralized decisions are made by machines. Thus, it is aimed to develop new service and business models providing efficiency, productivity, transparency, monitoring of systems, ease of fault detection, reduced costs and increased flexibility (Porter & Heppelmann, 2014, 2015).

Fig. 1: Industrial Revolutions.

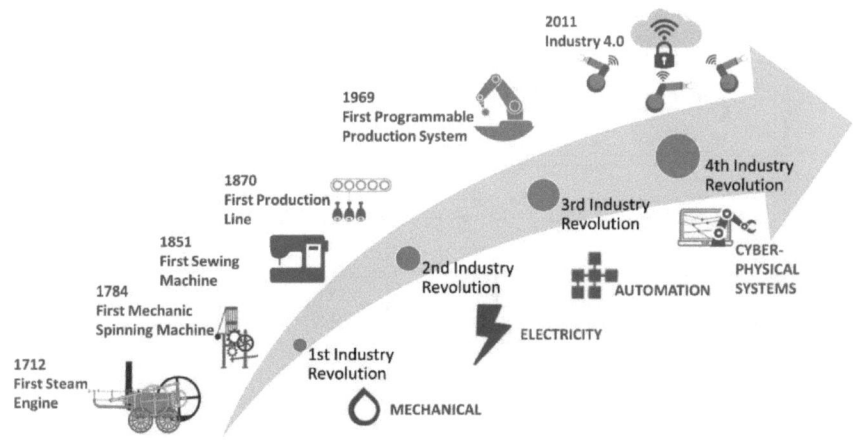

The key features of Industry 4.0 are categorized as follows:

- **Interoperability:** Cyber-physical systems are able to communicate with each other and people in real time through the use of IoT and Internet of Services.

- **Virtualization**: Sensor data is connected through simulation models in a virtual environment.
- **Autonomous Management**: Cyber-physical systems make their own decisions by determining optimum decisions based on the collected and processed big data.
- **Real-Time Management**: Collecting and processing data in a real-time fashion with distributed processing and big data approaches.
- **Internet of Services**: Cyber-physical systems deliver their services to people and other systems via Internet of Services.
- **Modular structure**: Providing a modular structure to adapt to rapidly changing requirements.

2.2 Clothing and Apparel Production System

The basic flow of the production system processes in a clothing and apparel factory is shown in Fig. 2. The apparel factories' production is on the make-to-order basis instead of make-to-stock, and there is no option of an inventory of goods that can be sold at another time. Therefore, the initial step of the production process is the receipt of a new order from the customer in an apparel factory.

Fig. 2: Clothing and Apparel Production System.

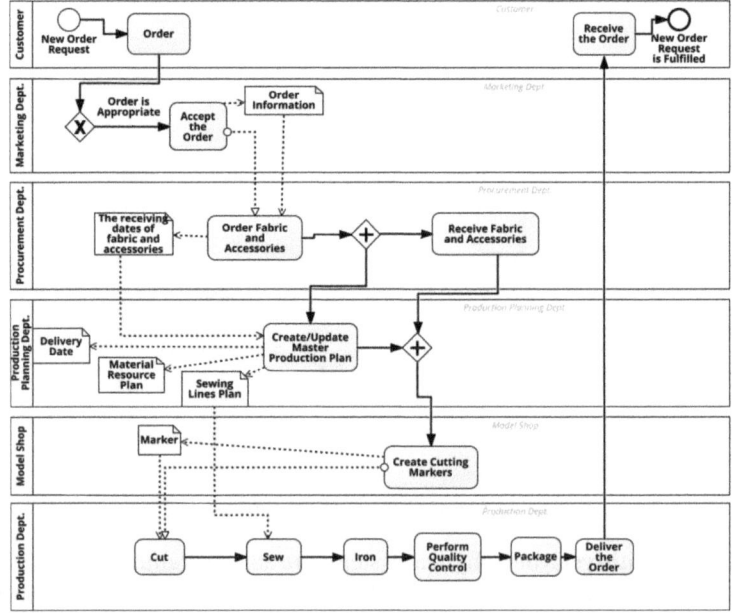

The customer notifies the marketing department about the order information such as price, order quantity, body size, material samples and desired delivery date. If the order is appropriate, the marketing unit accepts the order and delivers the order information to the relevant units. After determining the unit cloth consumption and unit material requirement, the procurement unit identifies the most suitable fabric and material supplier and initiates the procurement process. The production planning department develops the master production plan and determines the delivery date of the order based on the information of the operation analysis information, factory capacities as well as fabric and material delivery dates. After receiving the fabrics and materials from the warehouse connected to the purchasing unit, the cutting markers prepared by the model house department are transferred to the cutting department and the first step of the production process is started, then sewing, ironing, quality control and packaging are done. During the sewing process, the sewing workers are placed next to each other, and a sewing line is created so that each worker can make one or more sewing steps of the product. If the product is required to be washed, such as jeans, the washing process is performed, followed by ironing. During the quality control process after ironing, fabric and production defects are checked. Lastly, in the packaging process, related labels are added to the product and then packaged. As a final step, shipment of the order to the customer is done.

3. The Conceptual Smart Apparel Factory Proposal: Apparel 4.0

Emerging information technology services enable the development of cheaper and more efficient production processes through efficient integration of smart device hardware with cyber-physical systems. In this study, a holistic view of the innovative approaches that can be formed by the fourth industrial revolution in the clothing and apparel factories is proposed as a conceptual smart apparel factory system, called Apparel 4.0, as shown in Fig. 3.

Apparel 4.0 is examined next, under two main headings: Production and Management.

3.1 Apparel 4.0 for Production

In apparel factories, production processes include cutting, sewing, buttoning, ironing, quality control, packaging and shipment. The proposed innovative approaches for production are as follows:

Fig. 3: Apparel 4.0.

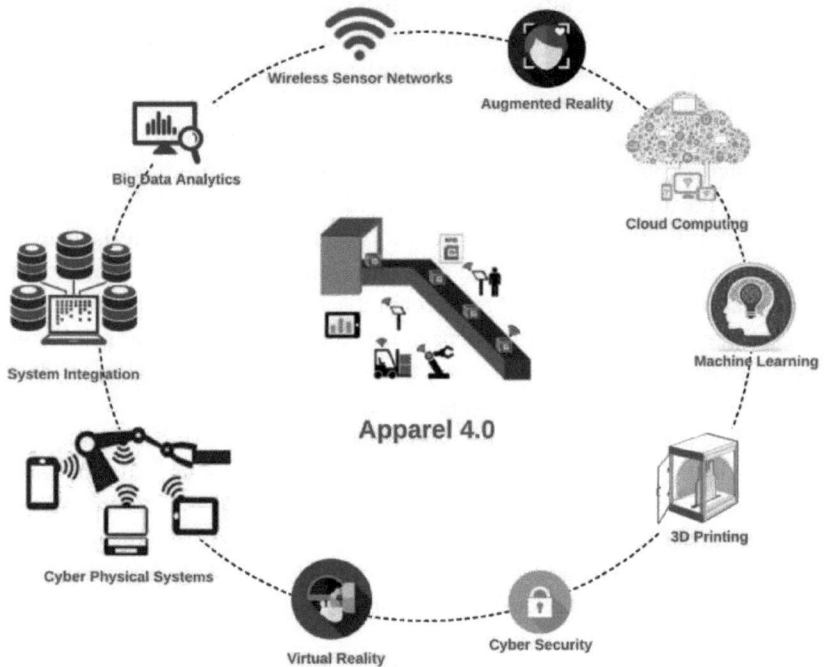

- **Digital Information Transfer**
  - o Modeling/drawing the garment sketch in 3D format.
  - o Examining the model in a digital environment supported by virtual reality technologies.
  - o Sending the drawings/markers of the product to the cutting system via a wireless network in a digital environment using the cloud technology infrastructure.
  - o Protecting industrial devices against cybersecurity threats, while implementing digital information transfer by establishing a reliable data communication on a secure and flexible network infrastructure.
- **Predictive Maintenance**
  - o *Predictive maintenance* comprises a variety of data analytics and statistical techniques to uncover hidden patterns and capture relationships among devices. It mainly aims to predict possible device or equipment failures and to define a maintenance strategy accordingly, in order to decrease failure rate and increase device utilization (Lee, Kao & Yang, 2014).

o *Cyber-physical systems* equipped with sensors, actuators and processors are intelligent electronic systems with Internet connectivity. They comprise extracting optimized decisions to preserve the capability and the functionality of the system by controlling problems of devices from large data-streams in real time. They can make self-optimizing decisions by anticipating errors and quality problems that can occur.

- **Human-Robot Technology Collaboration in Cutting Department**
  o Carrying fabrics from the warehouse to the cutting room by cyber-physical systems.
  o Spreading the fabric on the cutting table by cyber-physical systems and robots.
  o Completing the cutting operation through laser systems with a minimum-level of human interaction.
- **Intelligent Manufacturing**
  o RFID tags placed on the garments contain information on how to make, iron, button, wash and pack. It is a method for automatic recognition of individual objects using radio frequency.
  o RFID readers placed in each production station are integrated with the system. By reading the RFID tag of a product, the information on where the buttons are to be placed or the type of the button is received and the button operation is performed via the cyber-physical systems based on this information.
  o By reading the RFID tag of a garment, the information about the water temperature for washing operation and the optimum temperature for ironing operation is obtained, and the machines set the degrees automatically for washing and ironing operations without any human-machine interaction.
- **Robotic Quality Control**
  o To satisfy customer requirements, the final product must meet certain quality criteria that are predetermined before delivery to the customer. These quality standards include criteria such as the accuracy of the product's body measurements, the quality of the fabric and manufacturing operations. Today, quality control is done by humans, but this sometimes causes incorrect results. In the context of Apparel 4.0, *computer-aided quality control systems* have been established to speed up the quality control process, increase its success and collect production-related data regularly.
  o *Advanced image processing methods* and *machine learning approaches* provide the capability for easily reporting quality problems of the final product.

- **Packaging with Cyber-Physical Systems**
  - o The packaging process is done according to the information contained in the RFID label, so that there is no human interaction with the *cyber-physical systems*.
  - o The *RFID labels* are put on the products to make the production information accessible by the customer, hence increasing the transparency.

3.2 Apparel 4.0 for Management

The managerial activities include receiving orders from customers, prototype modeling, customer approval, procurement of required fabrics and materials, production planning, enterprise resource management, employee performance management and production management activities. Innovative approaches within the scope of Apparel 4.0 are summarized as follows:

- **End-to-End Digital Integration**
  - o The *IoT* that provides the communication network in which physical things are interconnected with each other or with larger systems.
  - o *Big data applications* which provide storage, retrieval, cleaning, visualization, analysis and interpretation of the large volume, the abundant variety and the fast incoming data obtained from IoT technologies.
  - o *Cloud systems* provide storage and accessibility of these data.
  - o The *integration of cloud infrastructures* involves both vertical and horizontal integration of the physical environment by creating a copy of the physical environment in the virtual environment.
  - o Thus, it is ensured that the data and information are accessible and manageable by administrative systems such as enterprise resource planning and production systems.
- **Wireless Sensor Networks**
  - o Wireless sensor networks, consisting *of IoT Technologies as RFID tags, RFID readers, AutoID sensors, Wi-Fi, GSM or low-power near-field communication technologies* that provide wireless communication as well as automatic data collection, synchronization and analysis at every stage, from storage to the end user (Keung Kwok & Wu, 2009; Stoppa & Chiolerio, 2014; Zhong, Dai, Qu, Hu & Huang, 2013)
- **3D Product Design**
  - o With the ability to examine garments in 3D with *virtual reality technologies*, the preliminary prototype product can be obtained in the virtual environment, and the customer can receive and print the model out of a 3D

printer. Accordingly, the rate of human errors is reduced in the production preparation phase and the customer approval period is shortened.

- **Customer's Real Time Order Tracking**
  - o Tracking the location and status of orders with *RFID tags* and allowing the customer to monitor the product status instantly. Accordingly, transparency and customer satisfaction are increased.
- **Real Time Production Planning**
  - o There are *sensors* in all automation devices operating in the manufacturing system.
  - o The data collected from the products' RFID tags in production processes are processed with real time *big data processing approaches*.
  - o The factors affecting the production speed are monitored.
  - o *Batch processing approaches* are used to process collected data in order to improve product quality and customer satisfaction by developing *business intelligence* solutions.
- **Real-Time Product Tracking**
  - o With the *RFID tags* placed on each product, the information about the stage of the product can be collected in real time.
  - o With the *digitalized production line*, the information about each production step can be collected in real-time.
  - o Thus, the production planning department can be informed in a timely manner when necessary, in order to optimize the main production plan. It has been determined that the integration of *RFID technology* and *big-data* in a production monitoring and decision-making system is able to increase the overall flexibility, capability and agility of the production systems (Cooray & Rupasinghe, 2015).
- **Real Time Employee Performance Management**
  - o *Real-time data* is used to balance the production line, which is formed by placing the predecessor and successor workers in the sewing line, to perform each sewing operation. The RFID tag on the part that will form the product is used to collect information about how long the activities are carried out by each worker, in real time.
  - o It is aimed to increase workers' productivity by generating employee performance reports covering the entire working time obtained through the utilization of *big data processing approaches*.
- **Real Time Supplier Performance Management**
  - o In the clothing and apparel sector, it is a very important issue to get the proper manufacturing materials in optimal quantities and at the proper

time in order to increase the capability of responding to customer needs as quickly as possible. The manufacturing of a single garment involves many materials such as fabric, threads, buttons and zippers. To this end, increasing decision-making capability of clothing and apparel factories for materials and suppliers management is a very crucial process. For this purpose, the *integration of IoT and big data technologies* can be utilized to analyze data about supplier requests, delivery dates and quality control results.

o The collected data from IoT devices can be processed in a real time fashion with big data processing approaches to manage the rapidly changing raw material demand by monitoring the customer demands and raw materials in the warehouse in a timely manner.

- **Production Line Balancing**
  o In clothing and apparel factories, there is a need to balance the production line in order to increase the overall productivity. Therefore, the overall production flow can be monitored to detect the bottlenecks in sewing line and develop an optimization solution with the digitized data to increase the agility of the factory.

- **Human–Robot Technology Collaboration in Warehouse Management**
  o The *RFID tags* can be placed on raw materials such as fabrics and accessories to store information about the order, quantity and location of raw materials in the warehouse. Thus, the fabric and materials are automatically fetched when they are required, via robots and/or cyber-physical systems without any human intervention. Thus, in Apparel 4.0, it is possible to reduce the production and inventory costs as well as human errors to a minimum level, while increasing the production speed.

- **Real Time Warehouse Management**
  o The utilization of wireless sensor networks in warehouses is crucial to collect real-time data about production materials (Prasanna & Hemalatha, 2012; Zhang, Huang, Sun & Yang, 2014). With the emerging real-time processing approaches, it has become possible to observe and manage materials at every stage of the production line. Therefore, problems associated with materials management can be avoided at an early stage.

- **Sewing Training with Augmented Reality**
  o With the emerging *augmented reality* technology, the factory workers are able to gain experience on the sewing machines by training in the digital environment. Thus, the productivity and efficiency can be increased and the manufacturing defects which are caused by human faults can be prevented in the production operations where human workforce is needed.

According to the studies in the literature (Sääski et al., 2008), the productivity of workers can be increased by up to 50% and the rate of human errors can be decreased with the use of augmented reality.

- **Training Robots with Kinect Technology**
  - o In clothing and apparel factories, the sewing is a very complicated and difficult process to remove human-machine interaction completely. Therefore, the *Kinect technology* can be utilized to teach cyber-physical systems how human workers perform sewing processes for each type of product. The Kinect technology has been developed with the aim of tracking the movements of users. It includes red-green-blue-depth camera sensor to perform background subtraction, blob detection and pose estimation in real-time (Pham, 2009). In the literature, there are also some studies which utilize Kinect to train cyber-physical systems (Chen, Wang & Lin, 2014; Gil, Mateo & Torres, 2014; Lisboa et al., 2013).

4. Discussion

The main benefits of Apparel 4.0 are as follows:

- *Agility:* The collected data from the factory can be processed to monitor and analyze the ongoing production process in a real-time fashion which allows us to detect and solve problems on the production line in the shortest time period.
- *Transparency:* With the customer's access to the integrated production system, the real-time monitoring of suppliers, and the real-time report generation about production processes, the transparency of the apparel factories can be increased easily.
- *Increased quality:* We are able to increase product quality through reduced human errors by the virtue of automated production process with the cyber-physical systems and robotics technologies. The integration of image processing and machine learning approaches also allows us to increase quality control precision, with the automated quality control process. Moreover, the sewing training with augmented reality also increases the overall product quality by reducing human errors.
- *Increased productivity:* The integration of cyber-physical systems and human-robot technologies, real-time production line management, real-time performance management and predictive maintenance enable us to improve overall productivity and agility.

- *Reduced operational costs:* Automation of processes has paved the way for minimizing human intervention on production as well as reducing overall labor costs. Moreover, the end-to-end digital integration and automated worker performance reporting also help us to reduce the managerial activities and operational costs.
- *Reduced order delivery time:* The utilization of 3D product design, real-time supplier management, real-time production management, automated processes, cyber-physical systems and human-robot technologies allow us to reduce order delivery time.
- *Increased customer satisfaction:* With the increased product quality, transparency and decreased order delivery time, it is possible to increase overall customer satisfaction.

The main challenges of the proposed smart factory of Apparel 4.0 are as follows:

- *Initial investment cost:* Cyber physical systems, robotics, virtual reality, 3D product design, wireless sensor networks, big data infrastructures and their integrations are very costly. Moreover, most apparel companies are small-and-medium sized enterprises. To this end, the companies need to inspect the cost and benefits of the innovative approaches that are proposed in the context of Apparel 4.0, in order to determine a detailed roadmap to initiate their smart factory transition processes.
- *Privacy and security:* Even though there are many studies in the literature that investigate the privacy and security of digital data, it is still an important issue for many organizations. The collected data from smart factory environment is very critical for confidentiality and security of the factories. To this end, the communication among smart devices should be provided with secure machine-to-machine protocols to protect data.
- *Technical challenges:* The proposed technologies in the scope of Apparel 4.0 are relatively new technologies, therefore the shortage of experienced workers in these technologies is a serious constraint. For this reason, national policies should be developed and supported to increase the quantity of experienced workers.
- *Lack of a global standard:* The lack of a global standard developed for Industry 4.0 causes an important difficulty. Therefore, a standardized Industry 4.0 maturity model needs to be developed with the aim of establishing a standard roadmap that will guide the firms and practitioners (Gökalp, Şener & Eren, 2017).

- *Social difficulties:* With Industry 4.0, the number of required low-skilled labor force will shift toward more high-skilled complex jobs which require a more intense focus on emerging technologies. This brings an important social problem, unemployment. It is predicted that the Industry 4.0 revolution will cause 5 million people to quit their jobs in 15 developed economies over the next 5 years (World Economic Forum, 2016). Therefore, there is a need to develop a roadmap to train the labor force on emerging technologies.

5. Conclusion

Today, we are on the verge of a new industrial revolution named as the fourth industrial revolution, and enterprises that successfully adopt this new revolution can survive in the competitive global market. Within the scope of this study, Apparel 4.0, which consists of the innovative approaches resulting from the fourth industrial revolution in the clothing and apparel sector, is proposed. In addition, the benefits and challenges of Apparel 4.0 have been analyzed.

As a future work, performing cost-benefit analysis of Apparel 4.0 and conducting pilot studies to determine necessary updates in Apparel 4.0 are planned. Thus, Apparel 4.0 is intended to be a guide for clothing and apparel producers and practitioners in terms of the applicability of proposed innovative approaches.

References

Bauer, W., & Horváth, P. (2015). Industrie 4.0 – Volkswirtschaftliches Potenzial für Deutschland. *Controlling*, 27(8–9), 515–517. https://doi.org/10.15358/0935-0381-2015-8-9-515

Chen, T., Wang, Y.-C., & Lin, Z. (2014). Predictive Distant Operation and Virtual Control of Computer Numerical Control Machines. *Journal of Intelligent Manufacturing*, Volume 28, Issue 5, pp 1061–1077.

Cooray, P., & Rupasinghe, T. (2015). A Real Time Production Tracking and a Decision Support System (PTDSS): A Case Study from an Apparel Company. (December 22, 2015) 12th International Conference on Business Management (ICBM) 2015, Faculty of Management Studies and Commerce, University of Sri Jayewardenepura, Sri Lanka.

Gil, P., Mateo, C., & Torres, F. (2014). 3D Visual Sensing of the Human Hand for the Remote Operation of a Robotic Hand. *International Journal of Advanced Robotic Systems*, 11(2), 26.

Gökalp, E., Şener, U., & Eren, P. E. (2017). Development of an Assessment Model for Industry 4.0: Industry 4.0-MM. *Communications in Computer and Information Science* (770). https://doi.org/10.1007/978-3-319-67383-7_10

Kagermann, H., Helbig, J., Hellinger, A., & Wahlster, W. (2013). Recommendations for Implementing the Strategic Initiative INDUSTRIE 4.0: Securing the Future of German Manufacturing Industry; Final Report of the Industrie 4.0 Working Group. Forschungsunion. Retrieved from: https://www.din.de/blob/76902/e8cac883f42bf28536e7e8165993f1fd/recommendations-for-implementing-industry-4-0-data.pdf (Accessed on: 22.02.2018)

KeungKwok, S., & Wu, K. K. W. (2009). RFID-Based Intra-Supply Chain in Textile Industry. *Industrial Management & Data Systems*, 109(9), 1166–1178.

Lee, J., Kao, H.-A., & Yang, S. (2014). Service Innovation and Smart Analytics for Industry 4.0 and Big Data Environment. *Procedia Cirp*, 16, 3–8.

Lisboa, H. B., de Oliveira Santos, L. A. R., Miyashiro, E. R., Sugawara, K. J., Miyagi, P. E., & Junqueira, F. (2013). 3D Virtual Environments for Manufacturing Automation. In *22nd International Congress of Mechanical Engineering (COBEM 2013)*, University of São Paulo, Brazil, 3–7 November 2013 (pp. 3–7).

McNeil, I. (2002). *An Encyclopedia of the History of Technology*. Routledge New York.

Monostori, L. (2015). Cyber-Physical Production Systems: Roots from Manufacturing Science and Technology. *At-Automatisierungstechnik*, 63(10), 766–776.

Pham, A. (2009). E3: Microsoft Shows Off Gesture Control Technology for Xbox 360. *Los Angeles Times*, 1. Retrieved from: http://latimesblogs.latimes.com/technology/2009/06/microsofte3.html. (Accessed on: 22.02.2018)

Porter, M. E., & Heppelmann, J. E. (2014). How Smart, Connected Products Are Transforming Competition. *Harvard Business Review*, 92(11), 64–88.

Porter, M. E., & Heppelmann, J. E. (2015). How Smart, Connected Products Are Transforming Companies. *Harvard Business Review*, 93(10), 96–114.

Prasanna, K. R., & Hemalatha, M. (2012). RFID GPS and GSM Based Logistics Vehicle Load Balancing and Tracking Mechanism. *Procedia Engineering*, 30, 726–729.

Sääski, J., Salonen, T., Liinasuo, M., Pakkanen, J., Vanhatalo, M., & Riitahuhta, A. (2008). Augmented Reality Efficiency in Manufacturing Industry: A Case Study. In *DS 50: Proceedings of NordDesign 2008 Conference*, Tallinn, Estonia, August 21–23, 2008.

Srivastava, S. K. (2016). Industry 4.0. *Lucknow: BHU Engineer's Alumni*. pp. 23–24.

Stoppa, M., & Chiolerio, A. (2014). Wearable Electronics and Smart Textiles: A Critical Review. *Sensors*, 14(7), 11957–11992.

TEA. (2017). Economy and Foreign Trade Report. Retrieved 22.02.2018 from: http://www.tim.org.tr/.

World Economic Forum. (2016). The Future of Jobs: Employment, Skills and Workforce Strategy for the Fourth Industrial Revolution. World Economic Forum, Geneva, Switzerland. January 2016.

Zhang, Y., Huang, G. Q., Sun, S., & Yang, T. (2014). Multi-Agent Based Real-Time Production Scheduling Method for Radio Frequency Identification Enabled Ubiquitous Shopfloor Environment. *Computers & Industrial Engineering*, 76, 89–97.

Zhong, R. Y., Dai, Q. Y., Qu, T., Hu, G. J., & Huang, G. Q. (2013). RFID-Enabled Real-Time Manufacturing Execution System for Mass-Customization Production. *Robotics and Computer-Integrated Manufacturing*, 29(2), 283–292.

Mehmet N. Aydın* and Ebru Dilan

Issues Regarding Deployment of IPv6 and Business Model Canvas for IPv6

1. Introduction

Today, importance of data is indisputable reality. The main goal in all new or planned technological studies is to collect, store and evaluate this data correctly. The concept of IoT (Internet of Things) that entered into our lives for the past few years has been very popular and taken technological studies to a whole new level. Briefly, IoT is the inter-networking of physical devices, which are permanently connected and communicated with each other or with larger systems. According to the American Federal Trade Commission, IoT refers to the ability of every-day objects to connect to the Internet and to send and receive data. All kinds of monitoring devices, sensors, biochips or access mechanisms are considered as Things. In other words, these Things need to have a network interface card (NIC) to communicate with each other. This applies not only to Things but also to all devices that communicate over the Internet.

IPv4 is the address that devices using the TCP/IP protocol use to exchange data over the network. The 32 bits wide address provides about 4.5 billion different addresses. This infrastructure, which was completed of its development in 1980, started to be inadequate due to both misallocation of resources and the rapid increase of the internet-using device population. IPv6's new addresses and its addressing structure presented in 1998 were expected to provide a solution to this problem. With a 128 bits wide structure, 3.4e+14 unique addresses and many new features that are not available in IPv4, this protocol could not make the progress that has been expected in terms of usage in the past 20 years although there were broad repercussions throughout the world.

There is a little progress from IPv4 to IPv6 in the world and particularly in Turkey (Bektaş, Soysal & Orcan, 2011). This study explores standardization efforts underlying the transition from IPv4 to IPv6 and aims to propose a model for a successful transition.

---

*   Corresponding author: Mehmet N. Aydın, mehmet.aydin@khas.edu.tr, Kadir Has University, Kadir Has Caddesi, Cibali/İstanbul, 34083, +90 212 533 6532 (ext. 1439).

In today's network sector, the biggest challenge faced by stakeholders is scantiness of available IP address (Thaler, Draves, Matsumoto & Chown, 2012). The current version of internet protocol has not changed since it was published in 1981. So, with the rapid growth of population and tremendous development of technology, the demand for IPv4 addresses becomes higher day by day running out available resources. Users in North America, who were early users of the Internet, were allocated over 60 % of all IPv4 addresses. Europe, a second mover, owns approximately 20 %. The rest of the world owns less than 20 %. So this historical experience shows us IPv4 addresses allocated disproportionately across global geopolitical regions because IPv4 addresses were allocated on a first come, first served basis, and not on the basis of need (Bouras, Karaliotas & Ganos, 2003). But we can see the difference since 2005, as IPv4 address allocation changed.

2. Method

In this study, IPv6 efforts around the world are examined and compared on a country basis (Grossetete, Popoviciu & Wettling, 2004). The major countries' studies, official progress reports, road maps, investments on IPv6 are reviewed and Turkey's position and progress in line with these global trends is explored.

The data is collected from a number of experts. Network specialists from three large-scale Turkish corporates were reached out and face-to-face interviews were conducted. Interviewee-1 is network and cybersecurity specialist of the vendor (Vendor1) which is one of the biggest white goods producers in Europe. Interviewee-2 is the former chief technology officer (CTO) of ISP and Telco operator in Turkey. Interviewee-3 is the system administrator of vendor (Vendor2) which is an e-commerce website.

The data is evaluated with inferences made from these interviews. At this stage, negative opinions about IPv6 and common concerns out of these three interviews were prioritized according to the experts. As a result, the most important one out of common problems was selected and model proposal started from scratch. It reached a desired point as a result of a collaborative work with an experienced network specialist after the model was proposed. Consequently, the obstacles to IPv6 have been identified, a *Business Model Canvas* has been prepared and presented to eliminate concerns and satisfy all stakeholders (Choudrie, Papazafeueioulou & Lee, 2003) resulting in socio-economic benefits (Dell, 2010).

The Business Model Canvas is a visual representation in one-page overview of current or new business models, generally used by strategic managers (Osterwalder and Pigneur, 2010). The Canvas provides a holistic view of the business as a whole and is especially useful in running a comparative analysis on the impact of an increase investment may have on any of the contributing factors (Amoss and Minoli, 2007).

Open-ended interview questions are as follows:

1. What are the weaknesses you faced about IPv4?
2. Do/Did you have a plan to change your system to IPv6?
3. Which factors prompt your company to start transformation process of their systems to IPv6?
4. What are your expectations about IPv6?
5. What type of difficulties you faced after implementation started?

The summary of interviews is displayed in Tab. 1.

3. Findings and Discussion

Based on the expert opinion we identify four major issues that may cause slowness of IPv6 Progress, which are competing standards worldwide, lack of business case, limited end-user perspective and varying transition techniques.

It was emphasized by the experts that IPv6 is not a technological issue; it is a value chain issue; pushing it up is important in every way. There is a need for a worldwide constitution of something that leaves IPv4 and switches to IPv6, creates a safe, value for people, brings convenience to automate everything but keeps it safe. We are not in a conjuncture where it could be. There are two solutions; the first one will both change the world conjuncture again and become a structure that all players will appreciate in globalization, and this constitution will be written accordingly. The second is that when two of the main-drive players (which may be the best match for America and China) are the ones that will make the best use of it, dictate the method, determine standards exactly and adopt it to the whole world, they will be literally switched to IPv6.

Standardization: The most important issue to deployment of IPv6 is standardization differences. Many countries have their own plan and roadmap for IP deployment. Some governments adjust their budget and their schedule properly but some of them are not. As explained before, IPv6 is working with end-to-end connectivity so there is no way to use IPv6 with standardization differences, all nodes have to use same parameters and metrics. If regulators want to deploy IPv6 as an agreed solution, stable standards are required to encourage large/medium enterprises to develop or invest in IPv6.

Business Case: Determining all components of business case can be useful for new released technology. However, IPv6 released at 1998; so when we look at taken time, IPv6 did not improve as expected. Demonstrating clear list of economic advantages of IPv6 can be helpful for enterprises to know what they will invest for. Also, key partners and activities must be identified; value proposition and revenue models have to be built.

End-User Perspective: To encourage end-users to change their systems to IPv6, all benefits like economical and technical enhancements have to articulate clearly. IoT and Internet of Everything (IoE) is the newest and popular technologies so IPv6 have critical position for IoT systems because of IPv4 address exhaustion. New features of IPv6 such as mobility, multicasting, P2P connectivity, plug and play and NAT avoidance features need to be understood by customers.

Transition: There are 3 types of transition techniques of IPv6; these techniques provide interoperability mechanism with IPv4 so users must be informed by internet service providers (ISPs) or Telco operators about transition process (Tadayoni and Henten, 2012). Without any clear information about process it is natural to have concerns and prejudice about transition. Also, Telco operators have to give affordable and reliable service to implement IPv6 systems in customer's network.

Tab. 1: Summary of Interviews.

	Vendor1	ISP	Vendor2
What are the weaknesses you faced about IPv4?	IP address shortage— for using IoT, IPv6 is necessity.	IPv4 is not suitable for IoT systems and P2P connectivity.	Network Address Translation (NAT) is the cause of security weakness and decreases scalability of networks.
Do/Did you have a plan to change your system to IPv6?	Hardware is ready for implementation but configuration is not.	Both hardware and configuration is ready for implementation.	Hardware is ready for implementation but configuration is not.
What are your expectations about IPv6?	Build unmanned factories, enhance R&D process of IoT products.	Provide IPv6 services to end-users and transform network systems in enterprise level.	Reliable and secure internal/external network systems provide larger address space.
What type of difficulties you faced after implementation started?	Absence of an agreement/ standardization about IPv6 between global actors (ISPs, Telco operators, application developers).	Unsettled value chain issue, underrated IoT systems, standardization differences of countries.	Transition process challenges, concerns about security, lack of information about IPv6 on end-user side.

Tab. 2 summarizes differences in country-based standardization efforts with respect to time, budget, planned projects, key approach, phases, partners, key activities and value propositions.

Tab. 2: Country-Based IPv6 Deployment Efforts.

	America (Frankel et al., 2010)	**China** (Wu et al., 2011)	**Europe** (Dhamdhere et al., 2012)	**Turkey** (Bolat and Tozer, 2009; Bektas et al., 2011)
Year	2005	2003	2001	2003
Budget	Over $150 billion	Over 1.6 billion RMB	Over $18.4 billion	Over $620,000
IPv6 Projects	3com, 6tap, ADC Communications, ESnet, Internet2, NY6IX, UNH Interop, Virginia Tech	China Next-Generation Internet(CNGI)	6INIT, COAIS, 6NET, GTPv6, NGNI, EURO6IX	ULAK6NET
Key Approach	The key point is to prepare organization for growth by expanding innovation and productivity. Government and Cisco play a very important role. Internet change major operations in industry and major network device vendors in America are providers of IPv6 products.	The major thing is that it should not only be the biggest but also the strongest. The key to CNGI is whether they can make some special application.	The main purpose of the project is to support the rapid introduction of IPv6 adoption in Europe.	The key point is the spread of broadband access, transition to e-government, closure of technological information gap and creation of IPv6 infrastructure.

	America (Frankel et al., 2010)	**China** (Wu et al., 2011)	**Europe** (Dhamdhere et al., 2012)	**Turkey** (Bolat and Tozer, 2009; Bektas et al., 2011)
Phases	The first phase accelerating IPv6 deployment in provisions mandates, NIST and ITL by US Government. The second phase is concerned with test lab. Final phase is about to identify business driven use cases.	The first phase is for technical challenges. The second and final phase is for achieving China Next-Generation Internet (CNGI) IPv6 program.	The first phase is a three-year European project. Second phase is concerned with security model. Final phase is a project to define a security framework particular for an IPv6 environment.	The first phase is inventory work, and planning and budgeting of purchases. The second phase is to gain support for IPv6 devices and purchase of new devices, and for making a pilot application. The third phase is for deployment.
Key Partners	US Federal and State Organizations (OMB, SAIC, ISP)	Government and non-government organizations including end-user device producers, and Telco operators.	Government and non-government organizations across EU countries, including application developers, regulators, networks and telcos.	Government led coordination including nongovernmental organizations.

	America (Frankel et al., 2010)	China (Wu et al., 2011)	Europe (Dhamdhere et al., 2012)	Turkey (Bolat and Tozer, 2009; Bektas et al., 2011)
Key Activities	Application development and certification processes have been underway to support IPv6.. Also, there were some models that improved security in a positive way such as security model changes to IT security policy and privacy considerations.	To build an information-based country supported by an IPv6 infrastructure, publicity of IoT and application development of next-generation Internet equipment and software played an important role in standards-setting for IPv6.	The implications of standardization on data protection resulting from the implementation and use of IPv6 must be a top priority, Europe had tested and developed software and equipment related to IPv6 multicast technologies.	Government policy, IPv6 Forum Turkey and Network deployment in public institutions, IPv6 test building, software development for IPv6 video conference. Hardware/ Software upgrade and job schedule need to be considered.
Value Proposition	Integrating systems along with the agencies' Enterprise Architecture, the Infrastructure Reference Model, and Security Reference Model. There is value in sharing best business practices among Federal agencies for IPv6.	Exploring technologies, cultivated personnel, innovated applications with commercial value.	Supporting changes to the directives and processing personal data in relation to the new services of electronic communications. communications.	Reducing transition costs and increasing domestic contributions along with secure, innovative, high quality services.

3.1 Business Model Canvas

The experts' common suggestion was that examining stakeholder perspective along with business model will be valuable. In line with this suggestion, we propose a business model using business model canvas method/technique (Fig. 1). Business model canvas is considered as a useful technique to accelerate the process of IPv6 deployment and articulate stakeholder aspect and value in successful deployment of IPv6.

Key Partners: Telecommunication operators transformed their hardware systems to support IPv6 features and solve IP address exhaustion problem. However, in 2017, IPv6 architecture evaluation is not completed and globally standardized yet. So it reveals worries about connection with other operators or customers. Even if manufacturers producing IPv6 devices are available for many years, when IPv6 end-to-end connectivity starts globally security problems will surely appear. Also there is no mechanism or standardization developed about this situation yet. With IPv6, P2P connections were used easily. Also with multicast feature, video streaming portals can develop easily with no server load but without enough audit, security and dependency of hosting service provider/operator, it is impossible to use this type of features of IPv6. Also, it affects IPv6's value chain negatively.

Government and regulation authorities have a critical role to regulate new procedures or rules about audit and security of IPv6 but still regulators are not giving enough importance except for a few developed countries. Many countries have not developed any plan yet about the global rising trend because of globalization. Globalization is easier than developing standards or protocols. Over the years, many technologies are developed and standardized by the developed countries' top producers. If countries are concerned about reliability of IPv6, they need to develop their own standards. Also, many countries have to change their strategies with building new organizational structure and stakeholder engagement to adapt IPv6.

Application developers directly get benefit from improving IPv6. They can work on enhanced and low-cost P2P communication and video streaming application. But developers did not realize possible opportunities yet because of their limited knowledge about network abilities. At the same time, developers do not get enough demand from customers to develop applications with P2P feature so it decelerates IPv6 progress.

End-user device producers have to manufacture devices which support IPV6 and they need to provide special support for customers when considerable problems occur.

Network device producers generally produce devices with IPv6 support. Also they need to give more support to governments or decision makers to develop new security and transfer protocols or standards which are available for IPv6 systems.

Government, enterprise and ordinary end user, education actors, application developers, hosting operators, international standard bodies, regulators, Telco vendors are all part of the ecosystem.

Key Activities: Low cost administrative applications need to be developed by application developers. So, these applications have to provide flawless service to get extra benefits of IPv6. Also, regulators must enable developers to improve security features and value chain of IPv6 without inhibitory attitude. As a result, it will help to change perception about IPv6 reliability.

Essential key activities include detailed dependency analysis and strategic planning. Interrelationships with other infrastructure programs ensure that the transition strategy drives IT investment decisions. Another critical activity is concerned with establishment of quarterly performance milestones. Furthermore one should take into account cybersecurity standards development for IPv6 based devices, equipment and applications.

Value Proposition: Besides providing more addresses to customer, many local network operators or ISPs still could not perceive the business value proposition of IPv6.

IPv6 is mandatory infrastructure standard for long run (>10 years) to solve IP address block availability problems. There are alternatives like CGNAT for short-term solutions.

Another value proposition is to have better and easier application development capabilities for P2P communication, which is not a motivation for major Telco operators.

Further emphasize on value proposition would be serverless video portals like de-centralized application potentials, which are not a motivation for major Telco or hosting operators.

Finally one can argue possibilities for improved cybersecurity in case of security centric architecture, hardware, firmware and software design and protocol improvements. There is no sufficient cooperation between governments, vendors and major operators. There is a very high risk of cybersecurity with IPv6 compatibility application without serious, main design based extra measures. Usage and infrastructure costs which have not increased with the increase in the number of users, thus, cost per user and per unit will occur.

Customer Relationship: CRM would need to support IPv6-related products and services, even if these systems are still operating in an IPv4 environment. As IPv6 adoption progresses, the CRM can also start to operate in an IPv6 environment.

Equipment, device and applications have to support seamless usage of benefits of improvements with IPv6. Customers cannot manage to handle that level of technical instantiations, setup details, security requirements with CRM support, etc. Applications and systems have to provide support by design.

Additional benefits of new applications along with complete, secure, advanced support of devices and applications with CRM are needed.

Customer Segment: IPv6 support is proving to be mandatory for the management of network equipment. IPv6 support in the devices purchased is a prerequisite for a significant customer segment. New customer segments can be added to portfolio with improved security measures. User centric seamless multi-device connectivity, device independent applications, inter-device communication will improve customer base.

Possible segments include early stage professionals (18–25 years old) mid-age professional end users (25–40 years old), remote technical support service providers, content producers such as journalists, device sellers and system integrators, vertical industries leveraging IoT driven opportunities.

Key Resources: All stakeholders need to be in sustained and have intensive, stable communication.

Cost Structure: Network infrastructure (new equipment and/or upgrades), architectural design, configuring, testing and managing the new IPv6 enabled elements of the network, staff training, upgrade of any customized software, IPv6 upgrade budget plans should assess the size and priority of each and allocate funds accordingly.

Revenue Streams: To improve the usage of IPv6, you need to encourage customers so that the revenue system meets the customer request. Also without profit, it is impossible to think about deployment because ISPs will not be willing to take on new technology soon. So, the new revenue system of IPv6 needs to encourage both sides. In Turkey value chain, the biggest barrier for IPv6 is clearly understandable because it is forbidden to get any profit from the customer when synchronizing networks to IPv6. As a result, ISPs are not ready for deployment and so it blocks IPv6 improvement in Turkey. New profit system has to be affordable for the customer without regard to the number of usage or user number.

Possible venues that are associated with revenue sources include an extention of the product life cycle of existing products (porting), innovative products that leverage aspects of the protocol (new product), innovative new services/features into existing product lines (product extension), service creation (product support).

Channel: It will be the best result to take the road with successful implementation. Also, it is not possible to mobilize related rituals without the possibility of a sample.

4. Conclusion

The need for a successful transition from IPv4 to IPv6 is inevitable for the premise of IoT, which is essential to make the dream of Industry 4 realizable.

IPv6 is the necessary standard for successful industry related practice. Most of the companies and ISPs are making IPv6 available permanently for their users and services. There is an increase in the use of Internet in the world and also in Turkey. This increase led to some changes in the business world. Internet protocol became a hot topic, as well as IoT. This study aims at finding the best business method of transition from IPv4 to IPv6 for Turkey. This study analyzed experiences of several ISP and vendors that had deployed IPv6.

This research points out appropriate standardization approach and value generation among stakeholders as the most critical aspects for successful transformation to IPv6. We provide a concise yet comprehensive analysis for representative standardization efforts all over the world. We furthermore articulate value generation aspect by using the business model canvas method. The contribution of this study to the literature and to the industry is to provide a reference model that can facilitate in finding out appropriate standardization approach and business model. In this regard, decision makers in governmental organization and key players in the industry may use it as a guiding model to determine a roadmap for successful transition.

Fig. 1: IPv6 Business Model Canvas for Turkey.

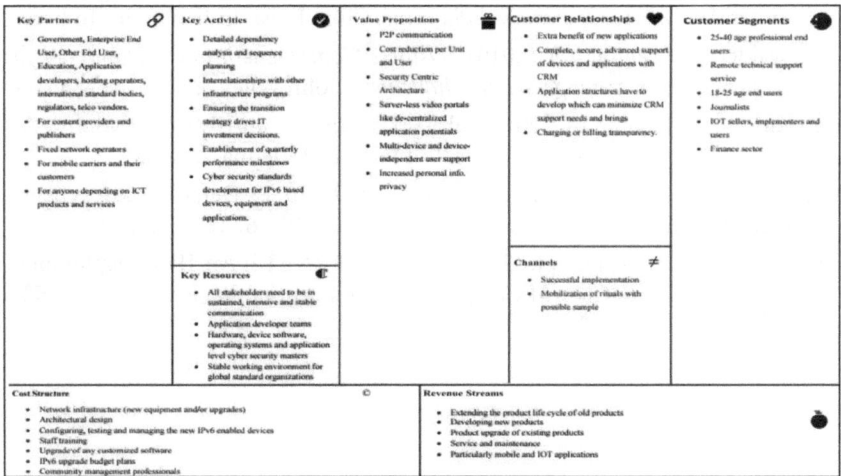

References

Amoss, J. J., & Minoli, D. (2007). *Handbook of IPv4 to IPv6 Transition: Methodologies for Institutional and Corporate Networks*. CRC Press. Florida.

Bolat, A., & Tözer, A. (2009). *IPv6 ve Türkiye*. Akademik Bilişim, Şanlıurfa.

Bektaş, O., Soysal, M., & Orcan, S. (2011). Türkiye İçin IPv6 Geçişi Zaman/Aşama PlanıÖnerisi. *IPv6 Konferansı*, 12, 11–17.

Bouras, C., Karaliotas, A., & Ganos, P. (2003). The Deployment of IPv6 in an IPv4 World and Transition Strategies. *Internet Research*, 13(2), 86–93.

Choudrie, J., Papazafeueioulou, A., & Lee, H. (2003). Applying Stakeholder Theory to Analyze the Diffusion of Broadband in South Korea: The Importance of the Government's Role. *ECIS 2003 Proceedings*, 46 Naples, Italy, 16–21 June 2003, AIS Electronic Library.

Dell, P. (2010). Two Economic Perspectives on the IPv6 Transition.info: Journal of Policy, Regulation and Strategy for Telecommunications, Information and Media, 12(4), 3–14.

Dhamdhere, A., Luckie, M., Huffaker, B., Elmokashfi, A., & Aben, E. (2012, November). Measuring the Deployment of IPv6: Topology, Routing and Performance. In *Proceedings of the 2012 ACM Conference on Internet Measurement Conference* (pp. 537–550). 14–16 November 2012, Boston, ACM Digital Library, ACM.

Frankel, S., Graveman, R., Pearce, J., & Rooks, M. (2010). Guidelines for the Secure Deployment of IPv6. *NIST Special Publication*, 800, 119. Gaithersburg, MD.

Grossetete, P., Popoviciu, C. P., & Wettling, F. (2004). *Global IPv6 Strategies: From Business Analysis to Operational Planning*. Cisco Press. Indianapolis, Indiana.

Osterwalder, A., & Pigneur, Y. (2010). *Business Model Generation: A Handbook for Visionaries, Game Changers, and Challengers*. John Wiley & Sons. New Jersey.

Tadayoni, R., & Henten, A. (2012, July). Transition from IPv4 to IPv6. In *23rd European Regional ITS Conference*, 1–4 July 2012 Vienna, Austria.

Thaler, D., Draves, R., Matsumoto, A., & Chown, T. (2012). Default Address Selection for Internet Protocol Version 6 (IPv6) (No. RFC 6724). California.

Wu, J., Wang, J. H., & Yang, J. (2011). CNGI-CERNET2: An IPv6 Deployment in China. *ACM SIGCOMM Computer Communication Review*, 41(2), 48–52.

Şebnem Akal[*]

Big Problem in Health 4.0: Access and Protection of Electronic Health Records

1. Introduction

Since Industry 4.0 has been announced as the financial, political and scientific future vision of Germany in Hannover Exhibition in 2011, its importance has gradually increased (Kagermann & Lukas, 2011). Germany's persistent policy on Industry 4.0 is supported by Chancellor Angela Merkel on the highest level. In one of the speeches which she gave in 2016, she emphasized that data protection and data processing are two competing elements, and in order to be a leader in automobile industry, it is important to utilize digital data instead of using protective law (Kollmann & Schmidt, 2016).

Industry 4.0 refers to three previous major changes. While the first three breakthrough could be explained with terms like mechanization, automation and computerizing, the last switchback can be translated in Turkish as "informatization" (Thuemmler & Bai, 2017). Informatics is related with software just like computerizing stage. While computerizing focuses on increasing activity and efficiency, informatics aims to increase system functionality by using the influence of real-world creatures on real world who were transformed into digital world. Its main theme is to provide integrated, digital and optimum production flow.

In recent years, terms like big data, Internet of Things (IoT), cloud informatics, mobile devices, digital and big data concepts have become prominent. However, big amount of data is not enough to make a difference and to be successful. With Industry 4.0, new terms like "Internet of Everything", Information o f Everything—a concept which was revealed by IBM—and cyber-physical systems have been put forward. Without a doubt, when these concepts are put into practice, Industry 4.0 will cause dramatic changes in every field including the health sector. In some studies, conducted in recent years, these changes are defined as Health 4.0 (Thuemmler & Bai, 2017).

---

[*]   Corresponding author: Şebnem Akal, sebnemakal@marmara.edu.tr, Marmara University, İstanbul Turkey.

Within the scope of 2013–2017 strategic action plan, digital hospital studies are being conducted in Turkey in order to collect health data and establish the integration of health institutions. With Health 4.0. (T.R. Ministry of Health, 2012), it is ensured that all stakeholders are included in new health eco-system. Health 4.0 vision is beyond digital data transformation and integration between hospitals. With this vision, hospital and health information systems and public health implementations are reestablished with fundamental changes.

With the effect of the aging world population and prolonging lifetime, health expenses increase all over the world. In 2018, average lifetime is expected to be 73.7 years. In this case, share of the population over 65 years reaches 10 % of total population and creates pressure on health expenses (Deloitte, 2014). In the last 50 years, the increase in total health expenses in OECD (Organisation for Economic Co-operation and Development) countries exceeds the increase in GDP. Unless there is a reform concerning health expenses, total health expenses are estimated to increase up to 100 % all over the world by 2040 (World Economic Forum, 2013).

The most important step of this transformation should be taken in creation, collection, storing, exchanging and protection stages used by all stakeholders of healthcare system.

In this study, current health implementations in Turkey are reviewed and necessary developments for Health 4.0 are emphasized briefly. The aim of this study is to discuss the regulation of electronic health records (EHR)—the smallest processor unit of Health 4.0 which have serious importance—and make suggestions.

2. Method

In order to make suggestions about changes which should be made and how systems should be created with Health 4.0, present condition should be identified and described. Within this scope, interviews were made with health sector employees in Turkey. As a result of the interviews, implementation of the present healthcare system is explained by means of resources.

The study method is based on secondary resources and literature scanning. While the resources are being observed, statistics and implementations in the United States and European Union countries have been researched and compared with the implementations in Turkey.

2.1 Present Condition of Healthcare System and Mobile Health Applications in Turkey

Health 4.0 concept did not merge suddenly in 2017 or it did not come out of nothing. It is not possible to separate old and new concepts with a sharp line. In the last 10 years, hospital information systems gained generally accepted standards. With these standards hospitals came out of being closed circuit systems and significant progress have been achieved in data sharing.

Although there are still some deficiencies concerning the subject, thanks to health data standards like HL7 and DICOM, most of the data can be stored and shared in digital environment. With this system, all stakeholders can reach required data nearly in real time. However, with the common usage of mobile devices a rapid increase has been observed in mobile health applications. Major progress has been achieved in digital data amount which has been created, stored and shared by the health sector stakeholders. Because of this situation, health data is created separately and remain fragmental in different resources. For the utilization of health data, it should be complete, chronologic and accessible.

Fig. 1: Stakeholders in Healthcare System (Specific Table). (Author's own diagram)

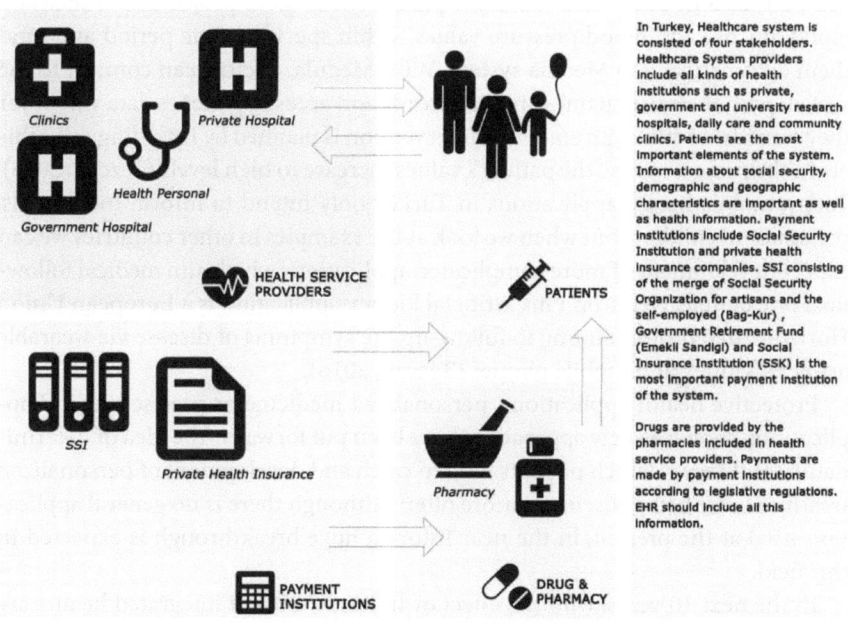

In Turkey, Healthcare system is consisted of four stakeholders. Healthcare System providers include all kinds of health institutions such as private, government and university research hospitals, daily care and community clinics. Patients are the most important elements of the system. Information about social security, demographic and geographic characteristics are important as well as health information. Payment institutions include Social Security Institution and private health insurance companies. SSI consisting of the merge of Social Security Organization for artisans and the self-employed (Bag-Kur) , Government Retirement Fund (Emekli Sandigi) and Social Insurance Institution (SSK) is the most important payment institution of the system.

Drugs are provided by the pharmacies included in health service providers. Payments are made by payment institutions according to legislative regulations. EHR should include all this information.

Between 2016 and 2022, mobile health market is predicted to reach up to 102.43 billion dollars with 32.5 % increase (Zion Market Research, 2016). Health market includes medical and mobile devices which have wireless data transferring specifications like glucose meters (blood glucose counters), "Bluetooth" thermometers, scales, blood pressure monitors, pulse oximeters, neurological monitoring systems, cardiac monitors like EKG/ ECG, apnea and sleeping monitors, wearable activity tracking devices, (bracelets with censors, smart watches and eyeglasses) along with the applications and services enabling data exchange between stakeholders. As it is understood from the predictions, in the future, observation of data stack and data sharing between stakeholders will be obligatory.

In short, mobile health, which means utilization of mobile technologies in accessing and sharing health services, forms only a small part of Health 4.0. Mobile health refers to the utilization of cell phones, tablets, wearable technologies and censor technologies in health data sharing between people, hospitals and other institutions (stakeholders). Health 4.0 system does not only include mobile health but also integrates all healthcare systems transparently enabling the perfection of interoperation of diverse system components and data flow.

There is a large number of current mobile health projects in Turkey. Its samples can be found in 2013–2017 strategic plan. One of these projects is to measure blood glucose and blood pressure values within specified time period and send them automatically to Medula system. With Medula, doctors can connect to the system with their user names and passwords and access a patient's data whenever they want. In addition, an emergent intervention is planned by including call centers in the project in case the patient's values increase to high levels (Tezcan, 2016). Today, mobile health applications in Turkey only intend to inform individuals, patients and children, but when we look at the examples in other countries we can expect development of more complicated applications which aim medical follow-up. For example, Nephron Plus artificial kidney application is a European Union Horizon 2020 project aiming to follow-up the symptoms of disease via wearable artificial kidney and mobile phones (Tezcan, 2016).

Protective health applications, personalized medicine or precise medical applications known as new approaches have been put forward. The idea of determination of diseases which patients tend to catch and development of personalized treatments are being discussed more often. Although there is no general application used at the present, in the near future a huge breakthrough is expected in this field.

In the next 10 years, with the effect of Industry 4.0 and integrated healthcare systems, interoperation of Health 4.0 and mobile devices via invisible integration and development of preventive medicine, patient-oriented treatment and precise medical applications will create a major difference.

A standard model named EMRAM (Electronic Medical Record Adoption Model) measuring health institutions' information technology utilization level with a rating system defined between 0–7 has been developed by HIMSS[2] in 2005. On the 7th—the highest digital level—nothing is saved in paper environment. All processes are executed via information technologies. On the 6th level, almost all processes are executed in digital environment, and only a small part of the processes is executed in paper environment. In Turkey, 18 health institutions (one of them is on the 7th level, the remaining 17 are on the 6th level) meet the 6th and 7th level EMRAM standard which could be described as digital hospital (TR Ministry of Health, 2017). In the United States, 35 % of health institutions are digital hospitals. In the last quarter of 2016, approximately 1,934 of 5,478 institutions were digital hospitals (HIMSS Analytics, 2017).

Although Turkey's EMRAM score is close to the average score of Europe, two values pointing the 3rd level are far from the requirements of Health 4.0 target. One of the details which attracts attention on Fig. 2 is that none of the European countries come close to 6th or 7th EMRAM level which shows digitally supported health processes.

Fig. 2: EMRAM Score According to Countries. Source: HIMSS Analytics Database, 2017.: (HIMSS Analytics Database. (2017). Emr adoption in europe q4/2016 Web address http:// www.himss.eu/sites/himsseu/files/HE_EMRAM_Score_Distribution_Q4_2016.pdf)

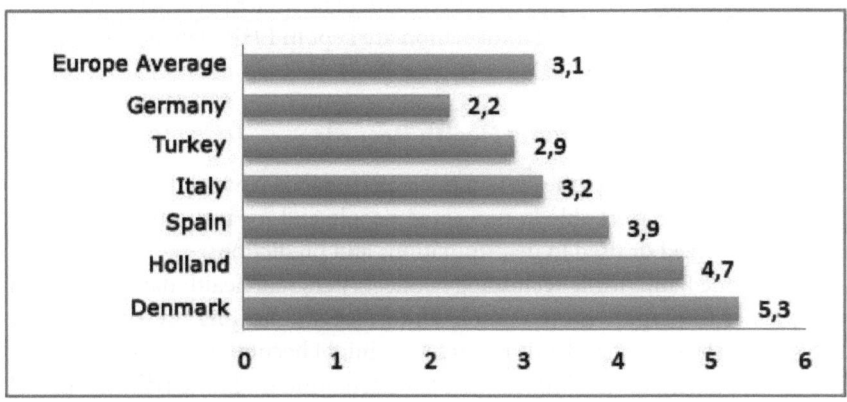

---

2   HIMSS (Healthcare Information and Management Systems Society), established in 1961, is a non-commercial worldwide organization aiming the utilization of IT in healthcare services (HIMMS Org, 2017).

2.2 Electronic Health Records

Electronic health records (EHR) is the basic data enabling the sharing of patients' medical data by all stakeholders between diversified systems. It can be described as the information concerning patients' symptoms which is used, exchanged and shared by different units during medical treatment. Information is either exchanged within hospital information system or it is exchanged between diversified hospital information systems according to necessity.

EHR contains personal information including blood pressure, blood sugar, various blood values, allergy test results, information concerning anesthesia, past diseases, along with the drugs taken during treatment.

Since its establishment until today, EHR is referred with different names such as medical patient records, electronic patient records and personal health records. In time, discrepancies have been formed between these names. EHR refers to all kinds of medical and chronological health data belonging to a person collected from different data resources. Its content has enriched with mobile and wearable technologies.

3. Findings

In the previous chapter, we mentioned about present developments and emphasized that the demand on instant accessing, evaluation and sharing of personal data has increased with these developments. It has been remarked that this data which becomes increasingly personalized should be under personal responsibility. Since all kinds of personal information are kept in EHR, data becomes more sensitive.

3.1 Sensitivity of Electronic Health Records

One of the major risks of the system is the personal data which has vital importance can become accessible to unwanted people and /or institutions; it can be changed, analyzed or used in determinations and predictions.

On the one hand, while technology collects personal health data since birth and enables to prepare more reliable and successful healing prescriptions for diseases, on the other hand, this information might become a huge security concern in the wrong hands. This extensive information might be risky in several ways including treatment turning into expensive and hard to reach by profit-oriented, greedy people and biological terror attacks effecting people who have certain specifications.

Because of this reason, there are significant problems in Health 4.0 implementation like protection of patient confidentiality, possession of cost, law assurance and computer skills of the users.

The major problem of EHR data is the healthcare providers' bearing cost while all the benefits are utilized by patients, payers (e.g., insurance companies) and the health system.

From patients' perspective, their personal data being under the initiative of other people might be a problem. Other important problems concerning EHR are (Yuan, Lin & McDonnell 2017):

- EHR may not be shared between stakeholders on the grounds of confidentiality, integration or information system diversity.
- Even if the data is shared by well-intended institutions or people, integration might not be accomplished due to the diversity between the systems.
- Sharing of various and large amount of data becomes hard and costly due to the incompatibility between diversified data and database systems.
- Due to inadequate organization, patients' data is kept in diversified database remaining missing and separate. Because of this missing data, tests and examinations have to be repeated and this causes time, cost and labor loss. In some cases when patients' medical conditions change, anamnesis remains incomplete due to the inaccessibility of vital chronological health data.
- Because paper is still commonly used as a sharing method, health data which was recorded in electronic environment is carried by the patients as printouts between healthcare providers. This causes deficient and/or wrong information which has a negative effect on diagnose and treatment process.
- Data storing systems require high-cost technological investment and it does not have the standards which ensure compatibility. This causes three major problems, including data integration, data security and interoperability.

3.2 Using Blockchain as a Solution in Protecting Patient Data and Determination of EHR Possession

Although Blockchain is not recognized by many people, the situation changes when the name Bitcoin is pronounced. Bitcoin was created in 2008 by an unknown man using the nickname Satoshi Nakamoto. Bitcoin is a virtual or digital currency which enables to transfer money between different countries without contacting any official organization or bank, without paying fee or tax, without dealing with intermediary institutions. Thanks to the currency which e-trade needed for a long time, many payments including food orders, clothing purchases along with money transfers between companies and banks are made reliably and error free. Since the identity of money senders and receivers is hidden, it can be used in illegal e-trade called dark web in money laundering or trading of illegal goods such as drugs. Due to this situation, many criticisms are being made. On the other hand, some people see Bitcoin as currency of the future and anticipate that it would entirely change financial operations.

In short, Blockchain is a distributed database which is the general name of working principle on which Bitcoin currency is based. This structure derived from the name given to "record chains" being linked to each other with time tags called blocks protects data against future changes and enables to share it reliably. "Cryptographic hash" protects data against changes and intervention. This safe design provides high error tolerance.

Fig. 3: Working Principle of Bitcoin.(Author's own diagram based on bitcoin.org website)

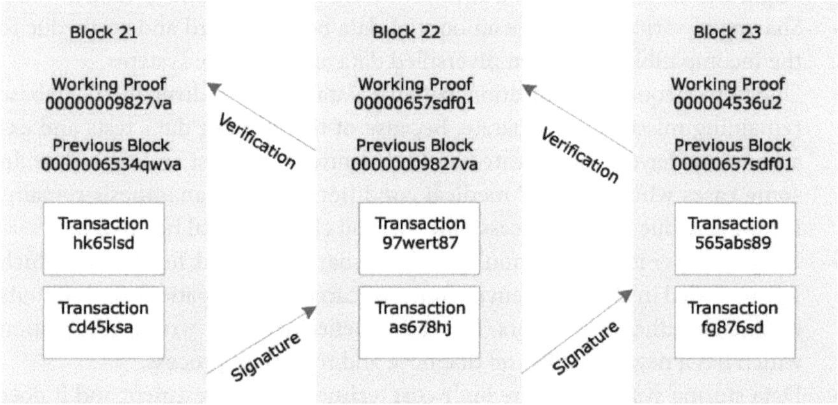

On Blockchain, transactions are shown with groups called "block". Each "block" is connected to its previous block. A "block" is valid only if all previous upper blocks are valid (Fig. 3). Blocks lined in chronological order are accepted as valid after the transaction's accuracy is verified by block-miners. (Various Blockchain variations are possible where miners can be state institutions, private individuals and institutions.) If the number of miners verifying accuracy of the transaction is more than the number of wrong miners, transaction is accepted as correct and valid. The first miner (in some cases it can be miners) who verifies the accuracy of transaction is given incentive. For example, in Bitcoin, approximately 25 Bitcoins are given as incentive to the first miner who verifies within 10 minutes.

Although it is easy to access transaction books in Blockchain (ledger), it is hard to change them. Some or all of the transactions cannot be deleted or changed by one stakeholder (Yuan, Lin & McDonnell, 2017). The safe structure of Blockchain against error and changes is ideal for sharing health records in institutions such as laboratories, hospitals and pharmacies (Halamka & Lippman, 2017). For example, a system was designed by using Blockverify Blockchain technology in order to prevent forgery and fraud in cases where there is more than one stakeholder involved. This structure which can be used in supply chain (e.g., drug supply chain) enables to record changes which will be made between stakeholders in time, in an unchangeable way with the help of a tag which will be attached on the product. In this way, it will be possible to avoid luxury goods and insurance forgery (BlockVerify Inc., 2017).

With this system, it is possible to follow drug usage of the patient along with drug duration and its contribution in the healing process from a single source. (Present healthcare systems enable to access this information but since the process require so many inquiries and comparison of database stored in different places, it becomes hard and complicated.) With the help of this distributed system, it is also possible to create automatic records without human intervention and share data without destroying integrity or making changes (Halamka 2017).

Utilization of Blockchain can be a solution for record sharing and data possession problems which have been mentioned before. Patient data exchange between institutions can be problematic due to the diversity in data systems. With this structure, a decentralized system can be actualized and data sharing problems which occur due to the diversity of institution standards can be solved. In Blockchain architecture, data is shared between stakeholders but none of them can have possession (Yuan, 2017).

4. Discussion and Conclusion

As a result, since the fragmental data is stored in several independent servers, Blockchain can be used as accessible, unchangeable, reliable and verifiable decentralized health database in tracking changes.

With data safety and interoperation specifications, data integration ensures to overcome the problems of present information systems. In traditional closed or open information systems, these three specifications cannot be obtained at the same time.

However, the situation concerning the structure set up is still open to discussion. It could either be set up as an open database containing miners consisting of independent individuals or it could be utilized as a system used only by government employees to make accuracy controls. It could be set up as heterogeneous control mechanism, partly controlled by independent individuals, government employees or other employees such as insurance companies.

At first, this innovation might be hard to accept since it will cause changes in stakeholder's behavior and usage habits. In this respect, creation of business models and development of financial approaches require time and specific work. Since the present structure is more flexible, this precise control mechanism might be disturbing for some institutions and individuals since it would prevent malicious usage of healthcare system.

Protection of rapidly growing data could be accomplished by keeping health records but there are many technological and legal obstacles concerning this subject. For the implementation of such a comprehensive change, existing protective and regulatory laws have to be revised. But this will be a subject of another study.

References

BlockVerify Inc. (2017). *BlockVerify Application Fields*. Retrieved 26.04.2017 from http://www.blockverify.io/

Deloitte. (2014). *2015 Global Life Sciences Outlook Adapting in an Era of Transformation*. Retrieved 20.04.2017 from https://www2.deloitte.com/content/dam/Deloitte/global/Documents/Life-Sciences-Health-Care/dttl-lshc-2014-global-life-sciences-sector-report.pdf

Halamka J. D., Lippman M. D. (2017). *The Potential for Blockchain to Transform Electronic Health Records*. Retrieved from https://hbr.org/2017/03/the-potential-for-blockchain-to-transform-electronic-health-records

HIMMS Org. (2017). *About HIMMS*. Retrieved from http://www.himss.org/about-himss

HIMSS Analytics. (2017). *EMRAM: A Strategic Roadmap for Effective EMR Adoption and Maturity*. Retrieved from http://www.himssanalytics.org/emram

HIMSS Analytics Database. (2017). *EMR Adoption in Europe Q4/2016*. Retrieved from http://www.himss.eu/sites/himsseu/files/HE_EMRAM_Score_Distribution_Q4_2016.pdf

Kagermann, V. H., & Lukas, W.-D. (2011, 2011-04-01). *Industrie 4.0: Mit dem internet der dinge auf dem weg zur 4. Industriellen revolution*. Retrieved from http://www.vdi-nachrichten.com/Technik-Gesellschaft/Industrie-40-Mit-Internet-Dinge-Weg-4-industriellen-Revolution

Kollmann, T., & Schmidt, H. (2016). *Deutschland 4.0 wie die digitale transformation gelingt* (Baskı ed.): Gabler Verlag, Springer Gabler, Wiesbaden.

T. R. Ministry of Health. (2012). *T.R. Ministry of Health Strategic Plan 2013–2017*. Retrieved 10.04.2017 from https://sgb.saglik.gov.tr/Dkmanlar/Strategic%20 Plan%202013-2017.pdf

T. R. Ministry of Health. (2017). *Digital Hospital 6th and 7th Level Hospitals*. Retrieved from http://dijitalhastane.saglik.gov.tr/TR,4971/emram-6-ve-7-se viye-hastanelerimiz.html

Thuemmler, C., & Bai, C. (2017). *Health 4.0 How Virtualization and Big Data Are Revolutionizing Healthcare*. C. Thuemmler & C. Bai (Eds) Edition ed.: Springer International Publishing Switzerland.

Tezcan, C. (2016). *Turkish Industrialists' and Businessmen's Association: Innovative Perspective on Healthcare: Mobile Healthcare* (TIBA-T/2016-03/575). Retrieved from http://www.tusiad.org/tr/yayinlar/raporlar/item/8676-dunya-ornekleri-isiginda-turkiyede-mobil-saglik

World Economic Forum. (2013). *Sustainable Health Systems Report Visions, Strategies, Critical Uncertainties and Scenarios* (REF 1501113). Retrieved 20.04.2017 from http://www3.weforum.org/docs/WEF_SustainableHealthSystems_Re port_2013.pdf

Yuan B., Lin W., & McDonnell C. (2017). *Academic Projects: Blockchains and Electronic Health Records*. Retrieved 20.04.2017 from https://www.fer.unizg. hr/_download/repository/blockchain_ehr.pdf

Zion Market Research. (2016). *Global mHealth Market: By Device Type, Stakeholder, Size, Share, Segments, Analysis and Forecast 2015–2021*. Retrieved from https://www.zionmarketresearch.com/report/mhealth-market

Tuğba Koç*, Alptekin Erkollar and Birgit Oberer

Industry 4.0 and Turkey: A Chance or a Thread?

1. Introduction

After the first industrial revolution (Industry 1.0), followed by mass production (Industry 2.0), also known as Fordism, and Industry 3.0, which is still retaining its update, finally a new paradigm called "Industry 4.0" has occurred. Robeco company has published a report in 2015 and has pointed out that although there are different concepts such as Internet of Things (IoT), Industrial Internet of Things (IIoT), cyber-physical systems (CPS), they all serve the same purpose with Industry 4.0 (Koychev, 2015). To avoid ambiguity and become more explicit, we prefer to use the term Industry 4.0 in our study. The term Industry 4.0 was first coined in 2011 at the Hannover Fair in Germany. Starting with the manufacturing area, the term has expanded in all sectors rapidly under the guidance of some leading companies, such as Intel, Siemens AG and General Electric Company (Industry ARC, 2018, Lydon, 2014).

There is another confliction in the literature if the term "Industry 4.0" is a revolution or is just an evolution. When we call something as developing, it means gradually changing, carrying out a job in a different way or evaluating something with different perspectives; on the other hand, revolution represents a radical change. A strategic consulting company has asserted (Roland Berger, 2014) this transformation has some disruptive and fundamental processes and nations have no chance to return from this path. With the same viewpoint, Schumacher, Erol and Sihn (2016) have emphasized the necessity of the integration of companies both vertically and horizontally to cope with this development and have advocated that Industry 4.0 is indeed a revolution. Schumacher et al. (2016) have also stated that the triggers of Industry 4.0 revolutions are entirely different than the other three industrial developments, so they assume that guidance for countries and companies is a definite need throughout this transformation process. A study, conducted in Germany in 2014, shows that

---

* Corresponding author: Tuğba Koç, tcekici@sakarya.edu.tr, Sakarya Üniversitesi İşletme Fakültesi Dekanlık Zemin Kat D12, +90 264 295 70 96.

at least 41 % of German firms are aware of the importance of the subject and they have launched some initiativs in their organizations to keep up with these developments (Weiss, Zilch & Schmeiler, 2014).

Therefore, the countries that are currently in the top level of automation will take the fastest pace and benefit the most from this new transformation process (Deloitte, 2015). The fact that Germany has allocated about 20 billion Euro into a year with developing new smart and real-time systems and it is considered that adaptation will become faster and more accessible in some leader countries that already have some attempts in the field of embedded systems designing and security/commercial software develop (LaValle et al., 2011).

As a developing country, in Turkey, until now there is no serious attempt to adopt Industry 4.0, neither in practice nor academic research, since the term is relatively new. In addition to shortcomings in the application of Industry 4.0, there exists no doctoral thesis in the Turkish Council of Higher Education database. We have encountered only one master's thesis investigating the current status of the manufacturing sector in Turkey via a survey. The most noteworthy study is a report published by TÜSIAD in 2016, where it is analyzed why smart companies are essential for Turkey and which strengths of the country are significant regarding Industry 4.0 transition. Within this framework, productivity, growth, investment and employment are evaluated as the primary fields during the conversion to Industry 4.0 for Turkey. In this report, TÜSIAD has focused on four main opportunities, which are efficiency, growth, investment and workforce, a notably young population with an internalized technology and is emphasized that big data analytics, smart robots, simulation, horizontal/vertical integration, IoT, cybersecurity, cloud, augmented reality and additive manufacturing are the nine triggered factors. However, the report is a roadmap for Turkey; there is lack of detailed information about challenges about Industry 4.0 and lack of recommendations that developing countries can be traced.

There are some maturity models developed to measure readiness level of countries and organizations toward Industry 4.0 (Schumacher, Erol & Sihn, 2016; PricewaterhouseCoopers, 2014). When these models are examined, countries described as "ready" had easy adaptation skills and located in a leading position. The most critical sub-criteria for being ready to change are qualification of information infrastructure, adaptation at the national level, qualified labor, horizontal/vertical integration in the value chain of the organizations in the country and the integrated solutions presented as a result.

This study seeks to highlight the challenges and threads that Turkey will face and its opportunities and strengths that will take it a step forward during this transformation process. In addition to intensely reviewed literature, an expert interview was conducted, and a Strengths, Weaknesses, Opportunites, and Threats (SWOT) table was acquired (Tab. 1). Also, some recommended strategies were presented in Fig. 2, and finally in Fig. 3 the "Turkey and Industry 4.0 Matrix" was obtained. To put it briefly, Turkey is in the critical situation. If it can succeed in evaluating this (r)evolution as an opportunity, it will be one step closer to the level of developed countries; otherwise, it will be excluded from the globalization race and will remain just as a spectator.

2. Method

In this study, the SWOT analysis was used to reveal the current status of Turkey and be a guide for future strategies during the Industry 4.0 transition process. SWOT analysis is one of the most important methods used to determine the internal state of any event, problem or situation and also external factors affecting it. This research method, in which internal dynamics (strengths-weaknesses) and external dynamics (opportunities-threats) are evaluated together to be able to reflect the general point of view, is frequently used in academic research (Valentin, 2005). The success of this method comes from its simplicity, flexibility and pellucidity (Cadle, Paul & Turner, 2010). After a correctly applied SWOT, it can be much easier to understand the market position for a company or the mission of the country depending on the aim of your research, thus taking future precautions against the competitors effectively. To increase the accuracy of our SWOT analysis, an interview with an expert whose research areas are mainly focused on new technologies and Industry 4.0 was conducted. The chosen expert has had more than 30 years of sector experience besides his academic career. When assessed in this context, our research can be called as a descriptive and qualitative study which aims to offer a framework for Turkey's current situation. In addition to this general framework, possible future strategies and initiatives are also discussed.

3. Results

After conducting a detailed literature review and expert interview, SWOT analysis results are presented in Tab. 1. The young population and their potential are considered as the most powerful side of Turkey which offers the advantages of low-cost production. According to the BCG Global Manufacturing Cost-Competitiveness Index 2016, the average unit cost is 98 in Turkey while the cost for the United States is 100 and for Germany is 121. To ensure the sustainability of this low unit price advantage, additional initiatives should be planned in the period of fourth industrial revolution. Although Turkey has the advantage of its young population, another report's results have shown that unfortunately, the average of senior high school graduates per year is about 36 % in Turkey (OECD, 2016) which proved that sufficient importance is not given to the education of youths. Turkey Information Society has also emphasized in their report (2016) that especially in the field of information and communication technologies, there is lack of qualified labor. During our interview, the expert has also stated that "When we have a look at the current structure of Turkey both economical and socio-demographical sides, it can be seen that the qualification of education is insufficient The expected benefit from technology may not be provided." For all these reasons, the strengths of the young population would be a thread if it is not well structured and evaluated.

Tab. 1: Industry 4.0 and SWOT Analysis for Turkey.

Strengths	Weaknesses
• Young population and their potential • Government's driving force and its incentives • Organizations support SMEs' development and their entrepreneurship such as KOSGEB • Techno-parks located at universities	• Low quality in education and unskilled labor • Excess number of SMEs and the lack of capital • Inadequacy of information systems infrastructure and challenges for process integration • Resistance of employees
Opportunities	**Threads**
• Critical importance of country location, both export, and import • Increasing global competition and reaching country development level • Expected fall in inflation • Expected rise in productivity	• Just watching the leaders and never achieve to develop new products • Full dominance of the market by big companies • Unsatisfied customers because of cultural differences • Resistance of citizens • Declining employment

Another critical issue is about cultural differences since this transformation process deals with the whole country. Since the concept of Industry 4.0 has its origins in Germany, any country has to evaluate its culture and the expected adaptation level toward this new revolution to avert any potential conflict. There have been many studies in the literature investigating the adaptation of new technologies at the national level (Montealegre, 1998; Davison and Martinsons, 2003; Kenneth,

Teng & Cheon, 2010); almost all come to agree that culture is one of the most critical dimension. During interviewing, culture also becomes the main topic of our conversation and expert has stated that "Companies and nations struggle to adopt new technologies to compete with other competitors. However, their large investments will go down the drain if the same benefits cannot be obtained from the customers willing to buy the products". Expenditures for Industry 4.0 technologies and applications have exceeded 650 million Euro just for Germany by the year of 2015 (Erol, Schumacher & Sihn, 2016). When it comes to Turkey, while the most of the companies are small and medium enterprises (SMEs) and the capital shortage is the main problem of them, financial issues also have to be discussed in terms of how capital sources will be separated into sectors and how SMEs will get a benefit from this financial crisis. Deloitte (2015) emphasized that SMEs abstain investing their sources to new technologies because of their limited budget and lack of bank credit support. There is a positive relationship between the adaptation process, the infrastructure capability of the company, and the company size. Countries with the inadequate technological infrastructure also have a slight chance to transform since they cannot foresee the benefits of their future investments and perceive this revolution as a thread (Pricewater-houseCoopers, 2014).

Fig. 1: SWOT Analysis—Main Issues.

EXTERNAL DYNAMICS / INTERNAL DYNAMICS	OPPORTUNITIES	THREADS
STRENGTHS	S/O strategies: Which strategies make these opportunities utilize?	S/T strategies: Which strengths can reduce the threads at minimum level?
WEAKNESSES	W/O strategies: Because of which weaknesses, some opportunities get lost?	W/T strategies: What kind of risks can be emerged in our weak areas?

Another study about transformation to Industry 4.0 has strongly agreed that this transitional process should not be regarded as only related with the organization's itself, but it also captures all the stakeholders and the other factors of its value chain (Baur & Wee, 2015). From this point of view, it seems struggling to integrate all value chain dimensions, especially in developing countries, as most companies still prefer to use traditional methods in their operational processes and have no common standards or rules in their way of doing the job. If so, the integration problem will disappear; the real-time data and the integration of all processes will be provided.

In Fig. 1, detailed information about the meaning of dynamics and SWOT analysis factors can be seen. Fig. 2 contains the sample strategies which may be implemented for the actualization of the determined criteria. Lastly, Fig. 3 is a dynamic matrix which shows the purposes, aims and initiatives that have to be fulfilled to determine strategies.

Fig. 2: Industry 4.0 Strategies for Turkey.

EXTERNAL DYNAMICS → INTERNAL DYNAMICS ↓	OPPORTUNITIES O1 Critical importance of country location, both export, and import O2 Increasing global competition power and reaching country development level O3 Expected fall in inflation O4 Expected rise in productivity	THREADS T1 Just watching the leaders and never achieve to develop new products T2 Fully dominance of the market by big companies T3 Unsatisfied customers because of cultural differences T4 Resistance of citizens T5 Declining employment
STRENGTHS S1 Young population and their potential S2 Government's driving force and its incentives S3 Organizations support SMEs' development and their entrepreneurship such as KOSGEB S4 Technoparks located at universities	S/O strategies • Positioning of SMEs in global markets • Revitalization of trade via decreasing customs duty	S/T strategies • Reducing customer satisfaction because of provided rapid incentives to SMEs
WEAKNESSES W1 Low quality in education and unskilled labor W2 Excess number of SMEs and lack of capital W3 Inadequacy of information systems infrastructure and the challenges for process integration W4 Resistance of employees	W/O strategies • Enhancing the variety of products and improving the productivity of SMEs	W/T strategies • Declining employment for SMEs • Lack of country-level adaptation • Loss of the country reputation

Fig. 3: Turkey and Industry 4.0 Matrix.

S/O strategies		
S#	O#	Purpose(P) / Aim(A) / Initiative(I)
3-4	2	Strategically sustainable development to achieve a better positioning of SMEs in the global market (P). Obtaining products and facilitating their transmission to customers (A). Investigating products and customers (I).
3-4	1	Revitalizing trade (P). Providing the most benefit from Turkey's strategic location (A). Remove the obstacles that decelerate import and export through the government(I).
S/T strategies		
S#	T#	Purpose(P) / Aim(A) / Initiative(I)
3-4	3	Increasing customer satisfaction (P). Detailed analysis for revealing the current status of the market and the expectations of customers (A). Establishing standards to get SMEs ready for a possible transition process (I).
W/O strategies		
W#	O#	Purpose(P) / Aim(A) / Initiative(I)
2	4-3	Making SMEs more competitive (P). Minimum fault and maximum productivity (A). Increasing the variety of products, create potential niche markets, encouraging entrepreneurship by providing financial support (I).
W/T strategies		
W#	T#	Purpose(P) / Aim(A) / Initiative(I)
1	5	Increased employment through new technologies (P). Reducing blue collar working class and make more investment on information workers (A). Examine the education system and start initiatives to improve quality (I).
1	4	Be a part of Industry 4.0 and make it a part of our culture (P). Remove prejudice towards Industry 4.0 (A). Make it interesting with advertising and other public relations activities (I).
3	1	Be a developed country (P). Strengthen the information infrastructure and having a well-planned integration process (A). Detailed analysis of all process and apply business process re-engineering for which it is required (I).

4. Conclusion

High labor costs in Europe is the main reason why Industry 4.0 has come to exist, and it is spreading all around the world starting from the manufacturing sector and expanding to other ones rapidly. While the developed countries have already begun to open smart companies, the developing ones are regarded as they are in an adaptation process. Because every country has its unique conditions, there is no standard rule or a specific guide that will help them during their conversion. Even the companies located in the same country but carry on their businesses in a different sector would have different strategies and distinctive milestones to reach their Industry 4.0 goals.

In this study, a SWOT analysis was carried out that uncovers the possibilities and opportunities for Turkey during the Industry 4.0 transition phase. Additionally, to become more descriptive and informative, a dynamic matrix was also presented. Thanks to this matrix, it has been possible to investigate the positive and negative factors for the concept and to decide the right initiatives. For further studies, it is intended to develop a maturity model that suits Turkey's conditions and meets the requirements of the country. For doing this, case studies, different scenario techniques and technology scouting can be used as different methods.

References

Barnes, B. (2015). *Manufacturing and the Internet of Everything*. International Manufacturing Technology Show.

Baur, C & Wee, D. (2015) manufacturing's next act.McKinsey. Retreived October 11,2018 from https://www.mckinsey.com/business-functions/operations/our-insights/manufacturings-next-act

Cadle, J., Paul, D., & Turner, P. (2010). *Business Analysis Techniques: 72 Essential Tools for Success*. BCS Chartered Institute for IT.

Davison, R., & Martinsons, M. (2002). Empowerment or Enslavement: A Case of a Process-Based Organizational Change in Hong Kong. *Information Technology & People*, 15(1), 42–59.

Deloitte. (2015). *Industry 4.0 Challenges and Solutions for the Digital Transformation and Use of Exponential Technologies*. Zurich. Retrieved October 10, 2018 from https://www2.deloitte.com/content/dam/Deloitte/ch/Documents/manufacturing/ch-en-manufacturing-industry-4-0-24102014.pdf

Erol, S., Schumacher, A., & Sihn, W. (2016). Strategic Guidance towards Industry 4.0 – a Three-Stage Process Model.. *In D. Dimitrov & T. OosthuizenIn (Eds.), International Conference on Competitive Manufacturing (pp. 495–501). Stellenbosch, South Africa: The International Academy for Production Engineering, January.*

Global Management Consulting. (2014). *The BCG Global Manufacturing Cost-Competitiveness Index*. Retrieved October 10,2018 from https://www.bcg.com/publications/interactives/bcg-global-manufacturing-cost-competitiveness-index.aspx

IndustryARC. (2018). *Industrial Internet of Things (IIoT) Market: By Component (Transmitter, Memory, Others); By Industry Verticals (Energy, Healthcare, Transportation, Others); By Connectivity (Wired, Wireless, Cellular, others), by Geography – Forecast (2018–2023), ESR 0027,* Retrieved October 10, 2018 from http://industryarc.com/Report/7385/industrial-internet-of-things-(IIoT)-market-report.html

Kenneth, J., Teng, J., & Cheon, M. (2010). Impact of National Culture on Information Technology Usage Behavior: An Exploratory Study of Decision Making in Korea and the USA. *Behavior & Information Technology*, 21(4), 293–302.

Koychev, D. (2015). *The Industrial Internet of Things—Invest in a More Connected World.* ROBECO.

LaValle, S., Lesser, E., Shockley, R., Hopkins, M., & Kruschwitz, N. (2011). Big Data, Analytics and the Path from Insights to Value. *MIT Sloan Management Review*, 52, 21–32.

Lydon, B. (2014). *The 4th Industrial Revolution, Industry 4.0, Unfolding at Hannover Messe.* Retrieved from https://www.automation.com/automation-news/article/the-4th-industrial-revolution-industry-40-unfolding-at-hannover-messe-2014

Montealegre, R. (1998). Guest Editorial for a Special Issue on IT in Latin America. *Information Technology & People*, 11(3), 169–172.

OECD. (2016). *Education at a Glance 2016: OECD Indicators.* Paris: OECD Publishing.

Pfohl, H.-C., Yahsi, B., & Kurnaz, T. (2015). The Impact of Industry 4.0 on the Supply Chain. *In W. Kersten, T. Blecker, & C. M. Ringle (Eds.), Hamburg International Conference of Logistic (HICL)-20 (pp. 32–58). Hamburg: epubli, August.*

PricewaterhouseCoopers. (2014). *Opportunities and Challenges of the Industrial Internet.* Booz & Company Group.

Roland Berger. (2014). *The New Industrial Revolution: How Europe Will Succeed.* Retrieved October 10, 2018 from https://www.rolandberger.com/publications/publication_pdf/roland_berger_tab_industry_4_0_20140403.pdf

Schumacher, A., Erol, S., & Sihn, W. (2016). A Maturity Model for Assessing Industry 4.0 Readiness and Maturity of Manufacturing Enterprises. *In A. Nassehi & S. Newrman (Eds.), The Sixth International Conference on Changeable, Agile, Recınfigurable and Virtual Production (Vol. 52, pp. 161–166). Bath, United Kingdom: The International Academy for Production Engineering, September.*

Türkiye Bilişim Derneği. (2016). *Türkiye Bilişim Derneği 2015 Yılı Değerlendirme Raporu. Retrieved October 10, 2018 from* http://www.tbd.org.tr/wp-content/uploads/2016/07/2015_degerlendirme_raporu.pdf

TÜSİAD. (2016). *Türkiye'nin Küresel Rekabetçiliği İçin Bir Gereklilik Olarak Sanayi 4.0.* Gelişmekte Olan Ekonomi Perspektifi. (N. Numanoğlu, M. E. Eynehan, G. Morkoç-Nikelay, & E. Aksoy, Eds.) Boston Consulting Group. İstanbul. Retrieved October 10, 2018 from https://tusiad.org/tr/tum/item/download/749 1_853576e56b585a09c02f06fb22d22e46

Valentin, E. (2005). Away with SWOT Analysis: Use Defensive/Offensive Evaluation Instead. *The Journal of Applied Business Research*, 21(2), 91–105.

Weiss, M., Zilch, A., & Schmeiler, F. (2014). *Industrie 4.0 Status Quo und Entwicklungen in Deutschland.* Eine Analyze der Experton Group.

Sevinç Gülseçen* and Doğan Aydın

The Rising Fundamental Skills of IT Field in Industry 4.0 Age

1. Introduction

Industry is the process of operationalizing the labor or the materials by processing them through various phases (Ertin, 1998). Industry not only is responsible for meeting demands but also is a significant, integral part for the development of countries. Urbanization and industrialization related to it have been consistently progressing in parallel to the growing population of the world. In fact, the innovations made on behalf of the industry have been described as the industrial revolution by means of these developments (Wilson, 2014). In the 18th century, the second industrial revolution had taken a new dimension due to the industrial and technological developments (Hackett, 1992), resulting in the start of using steam engines in production and the developments in textile industry (Deane, 1979). Especially the invention of electricity during the second industrial revolution allowed the industrial production to become mass produced (Atkeson & Kehoe, 2001). Even today, data transmission with tools that we have discovered, thanks to this industrial revolution, such as diesel engines (Diesel, 1912), internal combustion engines (Hendrickson III, 2014) and radio waves, is an indispensable part of many sectors. Moreover, it played a role in the development of the sectors which have the greatest market share in the world, such as the automotive, steel and energy sectors (Mokyr, 1998). The third industrial revolution, which occurred at the end of the Second World War, introduced to the world with notions such as nuclear energy, computer technologies and biogenetics, Wi-Fi, telecommunication, nanotechnology and holography. The classical production methods of the third industrial revolution have still maintained their validity. However, as the competition in the nature of the global market increases, it becomes necessary for the enterprises to explore several innovations in order to increase their competitive capacities. Therefore, recently, the new technological discoveries have frequently taken place in the agenda of the world.

---

* Corresponding author: Sevinç Gülseçen, gulsecen@istanbul.edu.tr, 16 Mart Şehitleri ave. Dr. Şevket Apt. No: 8, Postal code: 34134 Fatih/Istanbul/Turkey, +90 212 440 0000 – 10037.

1.1 Toward the New Generation of the Industry Revolution

According to the report published by "We Are Social", which is a digital platform marketing agency that operates worldwide, 3,419 billion people actively used the Internet in 2015 (Kemp, 2016). If non-human things are also included, then this number would increase even more. The "Internet of Things" (IoT), which is a step of the new industrial revolution, has emerged along with the inclusion of non-human things into this interaction on the Internet (Rubmann et al., 2015). Thanks to this new trend, both people and objects can make contact with each other. Various technological developments such as the Li-Fi (Light Fidelity) wireless communication technology, which is almost 100 times faster and safer than the Wi-Fi technology, IoT, cloud computing, augmented reality caused changes in the methods of production and service (Kahraman, 2016; Rani, Chauhan & Tripathi, 2012). In addition to these developments, due to the fact that the production has been transferred to Asian countries, where the labor force is much cheaper, the leading countries of the industry were forced to take some precautions (Şuman, 2017). Such developments become the footsteps to radical changes in production and industry.

The concept of Industry 4.0 has arisen in Germany (Heiner, Fettke, Thomas & Hoffmann, 2014) which aims to increase its competitive capacity with as the Asian countries became effective in production (Zhou, Zhou & Liu, 2015), thanks to their cheap labor force.

Industry 4.0, in other words smart factory, as Lee (2015) defined in his work, basically states that workflows are monitored over wireless networks, necessary information is composed and analyzed to be decided by the people on the production and applied to the production (Marr, 2016).

1.2 Industry 4.0 in Turkey

It is predicted that Industry 4.0 Revolution will contribute to the development of global competitive power and labor force quality in Turkey (TÜSİAD, 2016). Industry 4.0 not only manifests itself in mass producing enterprises but also in many other fields as well. Turkey has entered the Industry 4.0 period with the automotive sector, which is one of its most dynamic sectors (Ersoy, 2016). It is considered that this technological leap will also come into effect in other basic sectors as well (Şuman, 2017). In the event that this transition is made, the production capacity will increase between 5 % and 15 % with Industry 4.0, which has low cost in itself and high efficiency (Aksoy, 2017). Thus, it is stated that it will make a great contribution to the economy (Yazıcı & Düzkaya, 2016). With Industry 4.0, which means more digitalization in the production and everything being online, it is believed that many changes will occur in the qualifications/skills of many fields (BRICS Skill Development Working Group, 2016).

Today, Industry 4.0, which changed the production logic in the world, brought new profession areas (World Economic Forum, 2016) and caused the specific changes in the current professions (Brown & Buntz, 2017). These developments also bring along the problem of competent staff. It is considered that in the new period, the enterprises will face new problems such as the adaptation of current competent human resource to the new system in their body and the inability to find proper human resources for the new systems.

1.3 Literature Review

When the literature is reviewed, it is seen that there are similar studies about the determination of the basic skills expected from the candidates who will apply for the post of information technology (IT) specialist in several sectors. Todd, McKeen and Gallupe (1995) examined the job advertisements for the positions of programmers, system analyst and information system managers between the years of 1974–1990 in their study. At the end of the study, they have concluded that only the system analysts' technical skills were increased. Gallivan, Truex and Kvasny (2004) made a research in their study, which evaluates whether tendencies within the last 17 years in the informatics sector support the forecasts made in the previous researches. They examined various job advertisements in a specific period of years in order to determine this. Surakka (2005) examined the job advertisements to determine what kinds of changes have happened for the technical skills of software developers between the years of 1990–2004 in his study, and he concluded that multiple changes have occurred for technical skills of software developers. Another study, made by Nelson, Ahmad, Martin and Litecky (2007), compared job descriptions and job skills for the IT field and focused on the analysis program (job advertisement search engines) used to analyze the content of job advertisements. Huang et al. (2009) examined academic articles, practitioner literature and online job advertisements in their studies on job positions in 19 advertisements related to IT field. They have stated that the online job advertisements powerfully represent the other data collection tools and they have also reached the conclusion, and they have state that business acumen is also important as well as technical skills.

1.4 Aim of This Study

In this study, the answer to the question "what are the basic skills for information systems career candidates in various sectors?" is searched based on questions such as "what kind of qualifications do the enterprises expect from the employees to be employed in this new industry period?", "which departments can be opened to meet this demand or what kind of changes can be done in the current departments?". The aim is to determine and analyze the basic skills for the new production logic in the information systems sector, which directly interacts with Industry 4.0. In addition, coming up with recommendations related to the subject for the sectors in Turkey is one of the sub-goals of this study.

2. Method

This research is a qualitative study and content analysis has been used as a data collection tool.

The skills that were described in the job advertisements have been examined during the stage of content analysis. For the job advertisements, 68 advertisements have been examined between the dates April 17 and May 29, 2017, by using the keywords "IT", "developer" and "industry 4.0" among the career sites shown in Tab. 1.

Tab. 1: Career Sites.

Career Sites	Country
http://www.stepstone.de	Germany
http://www.jobsinmunich.com	Germany
https://www.monster.de/	Germany
http://www.infineon.com	Germany
https://de.indeed.com	Germany
http://www.thelocal.de/	Germany
https://krb-sjobs.brassring.com	Switzerland
https://www.totaljobs.com	United Kingdom
http://www.jobg8.com	United Kingdom
https://www.glassdoor.com	United States
https://jobs.dell.com	United States
https://www.indeed.com	United States
https://jobs.cvshealth.com	United States
https://jobs.veritude.com	United States
https://www2.jobdiva.com	United States
http://jobs.verizon.com	United States
https://lennar.taleo.net	United States
https://www.usajobs.gov	United States
http://jobs.myvoca.com	United States

Career Sites	Country
https://careers-vencore.icims.com	United States
https://jobs.boeing.com	United States
https://theapplicantmanager.com	United States
https://www.learn4good.com	United States

The positions, job descriptions and the competence information of the advertisements have been included in the examination.

2.1 Restrictions and Limitations

In the scope of this study, it has been attempted to propose determinations and suggestions about which skills would start to rise, which new skills would emerge with Industry 4.0 in Turkey. When determining the skills, it has been focused on only "automotive and IT" sectors by taking into consideration the width of impact area of Industry 4.0. It has been attempted to determine basic skills demanded for job description related to IT field in these sectors with Industry 4.0.

For this purpose,

– Restrictions for data collection tools: Data used in the research is restricted with 68 job advertisements received from the career sites between the dates of April 17 and May 29, 2017. At the data collection stage of the study, job advertisements have been examined in Germany, which is the leader of the Industry 4.0 revolution (Lee, Kao & Yang, 2014), and the United States (Itasse, 2016), which has entered the Industry 4.0 in 2014 with the "Industrial Internet" notion. Due to their pioneering roles in the Industry 4.0 revolution, data collection has been restricted with Germany and the United States.
– Limitations on sectoral dimensions: Sector selection is restricted with job positions only in IT and automotive sectors in the research.

3. Findings

The findings have been examined in two stages. In the first stage, as shown in Fig. 1, the advertisements within the job searching sites determined in the introduction section have been classified for the positions of "Information Security Expert", "Software/System Developer", "Software Engineer", "IT Specialist/Architect", "Manager", "Analyst/Data Scientist–Expert", "Consultant" and "Technician" according to their titles, and examined according to their prominence. When Fig. 1 is examined, the mainly demanded position is the Software/System Developer with the rate of 38 %, followed by Analyst/Data Scientist–Expert with 26 %, IT

Specialist/Architect with 13 %, Engineer with 10 % and other positions (Director, Information Security, Consultant and Technician) with 15 %.

Fig. 1: Distribution of Advertisements According to Job Position.

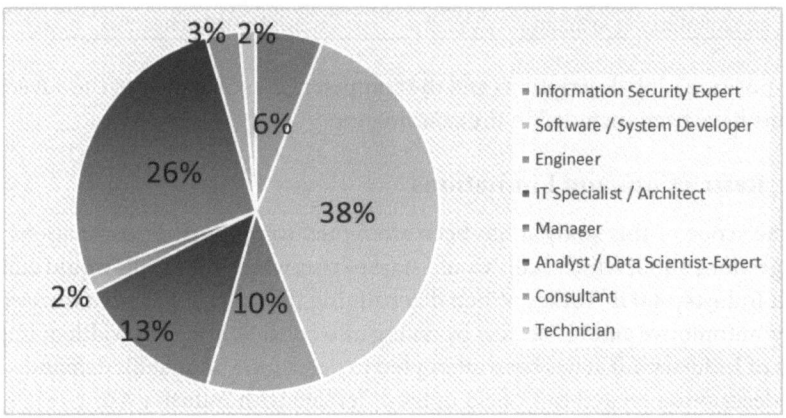

In the second stage made to examine the findings, the advertisements seen in Fig. 2 have been classified as "Analysis", "Software Development", "Design", "Management" and "Technical Support" according to technical competences required from the candidates for the positions needed. When Fig. 2 is examined, it is seen that, Analysis (39 %) and Software Development (36 %) competences are mainly demanded from the candidates for the positions in the IT field, followed respectively by Design with 10 %, Management with 9 % and Technical Support with 6 %.

Fig. 2: Distribution of Job Advertisements According to Their Contents.

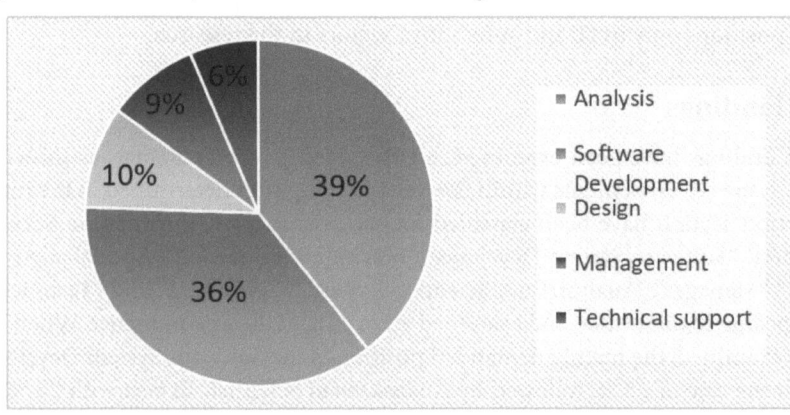

When job advertisements are examined in detail based on their areas of competence, it has been observed that software and design competence areas (Fig. 2) have been separated into sub-branches among themselves.

3.1 Software Competence Area

As shown in Fig. 3, software competence area is separated into three branches as "Middle Level Programming Languages" (Visual Basic, C, C++, Delphi, .NET), "High-Level Programming Languages" (Java, C#, Pascal) and "Web Programming Languages" (HTML, HTML5, PHP, All Script Languages). When Fig. 3 is examined, percentiles of these branches belonging to the "Software Development" competence area are seen. The competence of High-Level Programming Language with 40 % is mainly demanded in the field of software, followed by Web Programming Language with 33 % and finally Middle Level Programming Languages with 27 %.

Fig. 3: Distribution of Software Competence Area According to Its Sub-Branches.

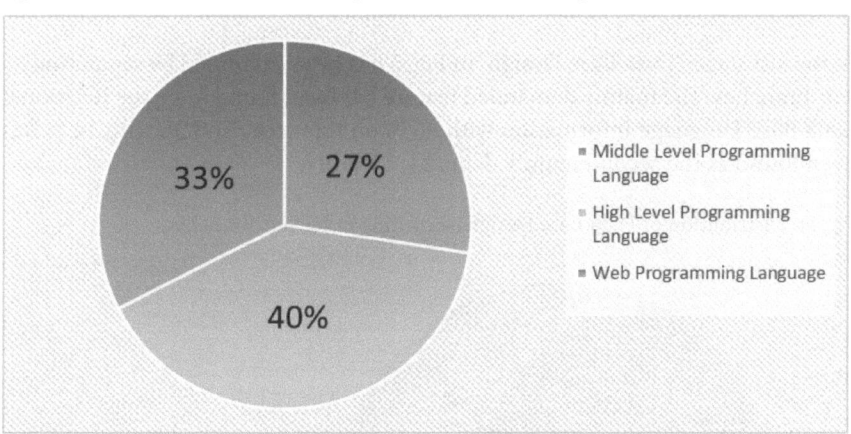

3.2 Design Competence Area

"Design" competence area is separated into three branches: "System (Network) Design", "Software Architecture Design" and "Database Design" (Fig. 4). After that, Data Base Design is separated as "Relational Data Base Languages" and "Non-Relational Data Base Language" (NoSQL) in itself as shown in Fig. 5. When Fig. 4 is examined, the mainly demanded feature on design competence area belongs to Data Base Design with 71 %, followed by Software Architecture Design with 24 % and System (Network) Design with 5 %.

Fig. 4: Distribution of Design Competence Area According to Its Sub-Branches.

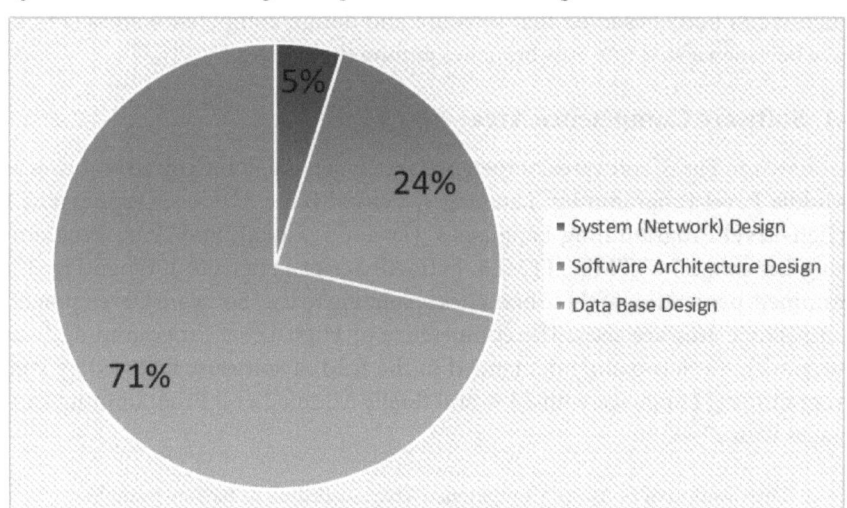

In the last stage, "Data Base Design" in Fig. 5 has been examined by separating its sub-branches. The mainly demanded feature has been found to be the Relational Data Base Language information with 86 % on this area. NoSQL with 14 % has been found as the second mainly demanded feature.

Fig. 5: Distribution of Data Base Design According to Its Sub-Branches.

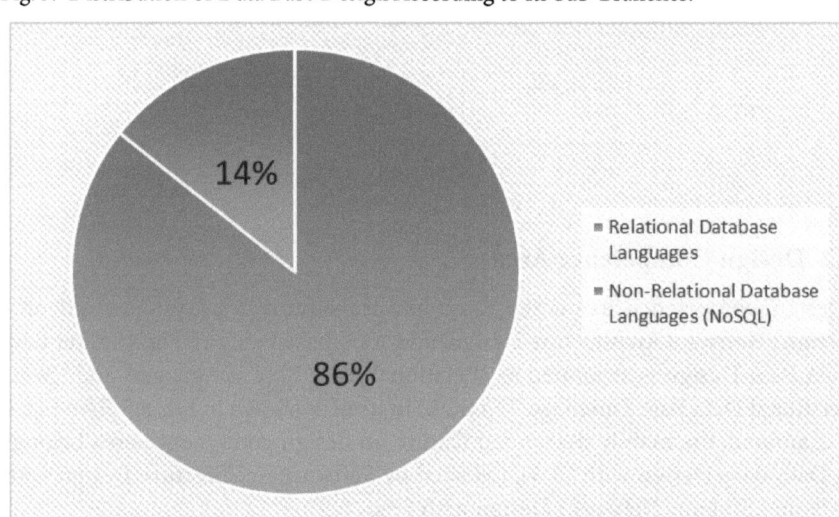

4. Discussion and Conclusion

In this study, the content of job advertisements for the IT positions has been examined within the career sites in Germany and the United States. Based on these examinations, it is aimed to determine which competences/skills will be demanded for job positions in IT sector in this new period named as the Industry 4.0. As a result of this study, it is considered that the outcomes coming in view will cast a light on educational institutions, which train the competent labor force.

When we examine job vacancy advertisements, it is seen that most of the advertisement holders requires the labor force for "Software or System Developer" (38 %) position at the highest rate (Fig. 1). "Analyst" with 28 %, "IT Specialist or Architect" with 13 % and "Engineer" with 10 % have been also found as the mainly demanded other positions. Besides, it is determined that the most sought labor force for the IT position are the software/system developer and analysts.

When the content in the advertisements is examined, the competences on "Analyst" with 39 % and "Software" with 36 % came into prominence at most for the candidates sought by the employers. It is considered that it would be beneficial for the programs related to the field of computer sciences in the current universities to focus on the software development and also especially on the courses related to the data analysis, which has come under the spotlight in the recent years. It is considered that it could be a topic for future researches to determine how much the expectations of the sector reflects on the curriculums of the departments in the field of IT.

When the competences are deeply examined, it can be seen that each field of competence is different within itself. Competence of "High-level Programming Languages" is the most sought-after (40 %) knowledge in the field of "IT" competence by the job holders. "Web Programming Languages" with 33 % and "Midlevel Programming Languages" with 27 % became the other highly sought software knowledge.

As compared with other programming languages, it is believed that in order to increase job and career opportunities in the sector software developers should use the software languages in the category of high-level programming languages.

"Design", which is another competence area, is separated into the branches in itself and the mainly demanded sub-competence became "Data Base Design" with 71 %. Other sought-after sub-competences are respectively "Software Architecture Design" with 24 % and "System (Network) Design". Finally, it has been concluded that the most sought-after competence in the field of "Data

Base Design" is "Relational Data Base Language" with 84 %. Apart from this, it has been seen that knowledge of "Non-Relational Data Base Languages" is sought-after for only 14 % in the relevant areas. This means that the large part of the sector prefers the current relational data base systems for big data, which is one of the Industry 4.0 dynamics. In Fig. 5, it is seen that the knowledge on non-relational databases is also demanded besides the relational database knowledge. In universities, it is thought that it would be beneficial to teach on the non-relational data base systems alongside with the relational data bases in the current program curriculums.

One of the main tasks of the universities is to bring competent human resource to the sector. Universities are among the important representatives of the business chain that provides real applicable knowledge along with a sectoral, entrepreneurial and research-oriented culture to their alumni. University–industry cooperation is an overemphasized subject especially in today's world (Edmondson et al., 2012). This subject is often discussed in recent years also in Turkey (Erdil, Pamukçu, Akçomak & Erden, 2013). In 2014, Turkish Union of Chambers and Exchange Commodities asked about the situation of cooperation with the universities in the studies they have made with 500 firms. In the reports published, they have determined as a result of the study that 76 % of the participant firms did not cooperate with the universities. It is considered that the decrease of this rate and an effective Research and Development relationship between the industry and the university are primary subjects on the way of becoming a technology producing country, instead of just a technology using one. It is considered that it could be a topic for future researches to determine the level and subjects of the university–industry cooperation.

References

Aksoy, S. (2017). Değişen teknolojiler ve Endüstri 4.0: Endüstri 4.0'ı anlamaya dair bir giriş. *Teknoloji*, (4), 34–44.

Atkeson, A., & Kehoe, P. J. (2001). *The Transition to a New Economy after the Second Industrial Revolution* (Vol. w8676). Cambridge, England: National Bureau of Economic Research.

Aulbur, W., Arvind C., J., & Bigghe, R. (2016). Skill Development for Industry 4.0. New Deplhi: Roland Berger GMBH.

Brown, D., & Buntz, B. (2017). *How Industry 4.0 Will Change Your Job*. Retrieved April 13, 2017, from Internet of Things Institute: https://www.iotworldtoday.com/2017/01/30/how-industry-40-will-change-your-job/

Deane, P. M. (1979). *The First Industrial Revolution*. United Kingdom: Cambridge University Press.

Diesel, R. (1912). The Diesel Oil-Engine, and Its Industrial Importance, Particularly for Great Britain. *Proceedings of the Institution of Mechanical Engineers*, 82(1), 179–280.

Edmondson, G., Valigra, L., Kenward, M., Hudson, R. L., & Belfield, H. (2012). *Making Industry-University Partnerships Work: Lessons from Successful Collaborations*. London: Science Business Innovation Board.

Erdil, E., Pamukçu, M. T., Akçomak, İ. S., & Erden, Y. (2013). Değişen Üniversite-Sanayi İşbirliğinde Ünivresite Örgütlenmesi. *Ankara Üniversitesi Siyasal Bilgiler Fakültesi Dergisi*, 68(2), 95–127.

Ersoy, A. R. (2016). *Endüstri 4.0 Sürecinde Neredeyiz?* Retrieved April 10, 2017, from Endüstri 4.0: http://www.endustri40.com/endustri-4-0-surecinde-nere deyiz/

Ertin, G. (1998). Türkiye'de Sanayi. In N. Serter, *Türkiye Coğrafyası* (pp. 165–182). Eskişehir: Anadolu University.

Gallivan, M. J., Truex, D. P., & Kvasny, L. (2004). Changing Patterns in IT Skill Sets 1988–2003: A Content Analysis of Classified Advertising. *The Data Base for Advances in Information Systems*, 35(3), 64–87.

Hackett, L. (1992). *Industrial Revolution*. Retrieved April 07, 2017, from International World History Project: http://history-world.org/Industrial%20 Intro.htm

Heiner, L., Fettke, P., Thomas, F., & Hoffmann, M. (2014). Industry 4.0. *Business & Information Systems Engineering*, 6 (4), 239–242.

Hendrickson III, K. E. (2014). *The Encyclopedia of the Industrial Revolution in World History* (Vol. 3). Lanham, Maryland, USA: Rowman & Littlefield.

Huang, H., Kvasny, L., Joshi, K., Trauth, E., & Mahar, J. (2009). Synthesizing IT Job Skills Identified in Academic Studies, Practitioner Publications and Job Ads. In *Proceedings of the Special Interest Group on Management Information System's 47th Annual Conference on Computer Personnel Research* (pp. 121–128). May 28 – 30, 2009, Limerick, Ireland: ACM Press. doi:10.1145/1542130.1542154.

Itasse, S. (2016). *USA: Industry 4.0 the American Way*. Retrieved June 10, 2017, from Process Worldwide: http://www.process-worldwide.com/usa-industry-40-the-american-way-a-536602/

Kahraman, H. (2016). *4. Sanayi Devrimi'ne Geçişin 9 Ayağı*. Retrieved December 20, 2017 from Endüstri 4.0: http://www.endustri40.com/4-sanayi-devrimine-industry-4-0-gecisin-9-ayagi/

Kemp, S. (2016). *Digital in 2016*. London: We Are Social.

Lee, J. (2015). Smart Factory Systems. *Informatik Spektrum*, 38(3), 230–235. doi:10.1007/s00287-015-0891-z

Lee, J., Kao, H.-A., & Yang, S. (2014). Service Innovation and Smart Analytics for Industry 4.0 and Big Data Environment. *Procedia CIRP*, 16, 3–8.

Marr, B. (2016). *What Everyone Must Know about Industry 4.0.* Retrieved April 7, 2017, from Forbes: https://www.forbes.com/sites/bernardmarr/2016/06/20/what-everyone-must-know-about-industry-4-0/#71231490795f

Mokyr, J. (1998). *The Second Industrial Revolution, 1870–1914.* Retrieved April 07, 2017, from U.S. History Scene: http://ushistoryscene.com/article/second-industrial-revolution/

Nelson, H. J., Ahmad, A., Martin, N. L., & Litecky, C. R. (2007). A Comparative Study of IT/IS Job skills and Job Definitions. *Proceedings of the 2007 ACM SIGMIS CPR Conference on Computer Personnel Research* (pp. 168–170). April 19–21, 2007, St. Louis: The Global Information Technology Workforce. doi:10.1145/1235000.1235038.

Rani, J., Chauhan, P., & Tripathi, R. (2012). Li-Fi (Light Fidelity)—The future technology in Wireless communication. *International Journal of Applied Engineering Research*, 7(11) 1517–1520.

Rubmann, M., Lorenz, M., Gerbert, P., Waldner, M., Justus, J., Engel, P., & Harnish, M. (2015). *Industry 4.0: The Future of Productivity and Growth in Manufacturing Industries.* Boston: The Boston Consulting Group.

Surakka, S. (2005). Analysis of Technical Skills in Job Advertisements Targeted at Software Developers. *Informatics in Education: An International Journal*, 4(1), 101–122.

Şuman, N. (2017). *Akıllıüretim çağı: Endüstri 4.0.* Retrieved April 10, 2017, from Fortune Turkey: http://www.fortuneturkey.com/akilli-uretim-cagi-endustri-40-42841

Todd, P. A., McKeen, J. D., & Gallupe, R. B. (1995). The Evolution of IS Job Skills: A Content Analysis of IS Job Advertisements from 1970 to 1990. *MIS Quarterly*, 19(1), 1–27.

TÜSİAD. (2016). *Türkiye'nin Küresel Rekabetçiliği için Bir Gereklilik Olarak Sanayi 4.0: Gelişmekte Olan Ekonomi Perspektifi.* İstanbul: TÜSİAD.

Wilson, D. C. (2014). Arnold Toynbee and the Industrial Revolution: The Science of History, Political Economy and the Machine Past. *History And Memory*, 26(2), pp. 133–161.

World Economic Forum. (2016). *The Future of Jobs: Employment, Skills and Workforce Strategy for the Fourth Industrial Revolution.* Cologny: World Economic Forum.

Yazıcı, E., & Düzkaya, H. (2016). Endüstri Devriminde Dördüncü Dalga ve Eğitim: Türkiye Dördüncü Dalga Endüstri Devrimine Hazır mı? *Eğitim Ve İnsani Bilimler Dergisi: Teori Ve Uygulama*, 713, 49–88.

Zhou, K., Zhou, L., & Liu, T. (2015). Industry 4.0: Towards Future Industrial Opportunities and Challenges. *12th International Conference on Fuzzy Systems and Knowledge Discovery*, (pp. 2147–2152). 15–17 Aug. 2015, Zhangjiajie, China: IEEE, doi:https://doi.org/10.1109/FSKD.2015.7382284.

Umut Kaya, Atınç Yılmaz* and Kadir Keskin

Vehicle Sales Prediction Using Neural Fuzzy Logic Method in Industry 4.0

1. Introduction

There have been three main revolutions in the industrial sector since the 18th century. The first one is the usage of steam-powered machines, the second one is the mass production of Ford in automobile industry and the last one is the development of Internet technologies and usage of computer-controlled systems. Every revolution has changed the whole business world structures in that time, respectively (URL1).

With the occurrence of rapid technological developments and the industrial revolutions, the human factors are tried to be eliminated or it is aimed to make these work machines to remove the mistakes originating from the people.

The greatest example of this is the industrial revolution, Industry 4.0. The interaction of machine with Industry 4.0 is evident, and the next generation of intelligent machines is manifesting in the whole realm of human life.

The automotive sector is a large industrial sector that deals with many industries in the field of economy and industry all over the world, and it is a continuously growing sector. For countries, the automotive sector is one of the parameters for development. Wang and colleagues, using neural fuzzy logic and artificial neural networks for predicting automobile sales in Taiwan, have shown that the neural fuzzy logic model yields better results (Wang et al., 2011). According to the research, it is said that the number of motor vehicles per capita in Turkey is much lower than the EU average (Karatlı, Helvacıoğlu, Ömürbek & Tokgöz, 2012). Automotive sector in Turkey for the first quarter of 2017 decreased by 8.65 % to 237,717 units, production increased by 22.13 % to 573,239 units and exports increased by 31.49 % to 472,632 units (URL 2) (Fig. 1).

---

* Corresponding author: Atınç Yılmaz, atincyilmaz@beykent.edu.tr, Beykent University, 444 1997 (5146).

Fig. 1: Automobile Market Graph. Source: Association of Automobile Distributors, http://www.odd.org.tr/folders/2837/categorial1docs/1832/Sekt%C3%B6rel%20 De%C4%9Ferlendirme%20Nisan%202017.pdf

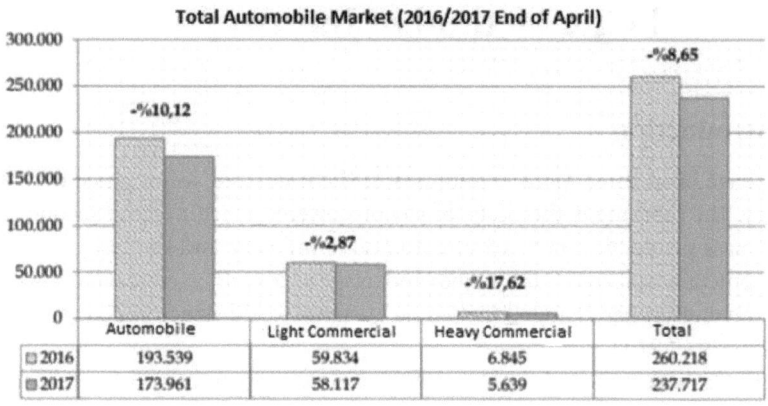

	Automobile	Light Commercial	Heavy Commercial	Total
2016	193.539	59.834	6.845	260.218
2017	173.961	58.117	5.639	237.717

In the study, the neural fuzzy logic method of artificial intelligence is proposed to provide a sales forecasting model in Industry 4.0, which will contribute to the production of automobiles.

1.1 Industry 4.0

Three industrial revolutions took place until the age of information and technology. The productivity of the three objective industrial revolutions is the common goal according to Pan et al. (2015, p. 1). Intelligent systems developed with Industry 4.0 will be the most important factor in user management and control. In addition, intelligent systems are thought to maximize production by maximizing resources.

Today, it has become a very ordinary event for people to communicate and interact with each other at any time and place in the Internet environment with information technology. This interaction with the imagination of the fourth revolution of the world has become the interaction of objects with the Internet and objects through the interaction of people. In line with these results, Industry 4.0 aims at industrial systems where people communicate with each other, machines and systems communicate with each other and produce in this way (URL3).

Industry 4.0, by the simplest definition, will bring together new technologies (programmed machine technology) and new generation (intelligent) production techniques to bring new technologies to market and add industry to business techniques and production lines.

1.2 The Role of Machines in Production

With Industry 4.0, machines use distributed object links to reduce the amount of human error, intelligent connections and distributions. By means of integrated communication capabilities, objects can have the right decision-making ability, thanks to artificial intelligence decision mechanisms, but also they can intervene in every stage where the product will come into constant interaction with environmental equipment.

Thanks to the ability to organize machines and equipment in-house, they will be able to develop very complex production scenarios that people will find difficult to solve.

Even those special products that need to be produced separately by this machine organization and communication can be produced and presented to the market without human influence (URL 2).

1.3 The Role of People in Production

Contrary to the computer-integrated systems developed by the 3rd-generation revolution, the industry did not intend to remove the beliefs altogether during production. With Industry 4.0, though it is aimed at passing through intelligent systems and minimizing the role of the human being, the human being is actively involved in controlling systems and solving problems. The fourth revolution is a system that aims to move people and intelligent machine technologies together. The "Diversity of Nature" approach presents a different approach, even if social–technological problems arise between the ability of these systems of systemic self-determination and action by themselves and human relations. A different intelligent system is needed to eliminate problems that arise in intelligent decision-making systems and to keep the system running smoothly. In Industry 4.0, this system was considered as human and the opportunity to work actively was presented to the human. Here, people can keep production and system control from beginning to end, if necessary (URL2).

1.4 Interaction of Industry 4.0 with Other Disciplines

In industrial revolution, the Industry 4.0 is the first cross-interaction strategy between engineering sciences and information technology. Especially open and changeable systematic production concept is one of the elements to be understood.

However, the technology and technology that will make Industry 4.0 live to make this industrial revolution productive must be well known (URL2).

Another important factor is how to ensure the interaction between intelligent machines and people. Simple and easy to understand interface applications are required for this. Practices that clearly convey all the steps of production are therefore one of the characteristics of this paradigm.

Interdisciplinary work will be mandatory as Industry 4.0 depends on the condition that communication between human and machine must operate at maximum efficiency. So, it is a human–computer interaction. Some of the disciplines involved in this area are (Preece, 1994; Shneiderman, 1998):

- Human behavior
- Psychology, cognitive sciences
- Computer technologies
- Software engineering
- Ergonomics
- Graphic and industrial design
- Sociology
- Anthropology
- Education science

1.5 Intended for Industry 4.0

With Industry 4.0, which we can call smart production cycle, the international competition element is also increasing rapidly. The most important element in this competitive environment is the provision of an efficient production environment, which is the most important goal of Industry 4.0. When productive production is carried out, it is necessary to determine the demands correctly.

It would be much more efficient to have this process performed on machines as it would be too much trouble to make demand forecasts manually in smart systems like Industry 4.0.

In this study, demand forecasts are made for the future stages of the automotive sector using historical data. With Industry 4.0, which we can call smart production cycle, the international competition element is also increasing rapidly. The most important element in this competitive environment is the provision of an efficient production environment, which is the most important goal of Industry 4.0. When productive production is carried out, it is necessary to determine the demands correctly. It would be much more efficient to have this process performed on machines as it would be too much trouble to make demand forecasts manually in smart systems like Industry 4.0.

In this study, demand forecasts are made for the future stages of the automotive sector using historical data. Thus, overproduction will be avoided, and supply demand balance will be ensured and efficiency will be increased.

1.6 Demand Forecasting

According to the definition of Tekin (1996), consumers are demanding quantities that they are ready to buy a product or service at a certain price level.

There are many factors that affect the demand for a product, especially the type and quality of the product. The most important of these is the type of product, price, other service or product prices (complementary products and services), consumer income level, consumer pleasure and preferences (Case & Fair, 1999).

In case the product is available, demand will be used at the same time as sales demand. In the literature, sales forecasting is used instead of demand forecasting.

2. Method

2.1 Fuzzy Logic and Artificial Neural Networks

Fuzzy logic and artificial neural networks are the areas of application of artificial intelligence. Artificial intelligence is that human beings can be imitated by machines. Definition of artificial intelligence according to Konar (Konar, 2000: 3) is to define the problem from the characteristics of human beings and to solve the problem by finding the necessary information for the problem.

Artificial neural networks are one of the artificial intelligence applications that are used to imitate new information learned by the human brain and to be imitated by machines modeled based on linking this information. Even the quantum computers that are currently being used cannot reach the storage capacity of the human brain. On the positive side, studies on the brain and modeling of the human brain have been developed to model artificial neural networks by modeling machines based on the ability of biological neural networks to process complex information and to relate such information to events. Artificial neural networks have been established to acquire the properties of biological neural networks with machines (Samer & Schlenkoff, 2009).

Blurred selection of the entire nerve cells, blurring measurement to remove the network instability, blurring of the neural network by changing the basic properties of the artificial neural network cells and methods proposed by Baykal and Beyan in the combination of artificial network and fuzzy logic (Baykal & Beyan, 2004a, 2004b). Hybrid systems are constructed using methods of blurring the neural network and changing the basic properties of artificial neural network cells. Because of the integration of neural networks with fuzzy logic, neural fuzzy methods have emerged. ANFIS, FALCON, RuleNet, NEFCLASS, NEFCON are the systems created using neural fuzzy methods. The aim of the study is to estimate the demand using ANFIS (Adaptive Neuro Fuzzy Inference System) method.

2.2 Artificial Neural Fuzzy Inference System (ANFIS)

Neural fuzzy logic is a hybrid learning model that is derived by combining the strengths of both artificial intelligence methods because of combining the fuzzy logic method with the decision-making ability by using the learning ability of artificial neural networks (Yilmaz, 2017).

The Takagi-Sugeno-Kang (TSK) neural fuzzy method, which is formed by combining the Kang neural network method with the Takagi-Sugeno fuzzy logic method, forms the basis for the ANFIS method. The ANFIS method for identifying nonlinear functions, estimating chaotic time series and identifying nonlinear elements in control systems was developed by Jang in 1993. The Fuzzy Logic Module in the Matlab software user interface allows the use of the method developed by Jang (Yücel, 2010). In the following Fig. 2, "Layer" refers to the ANFIS architecture, which is expressed by Jang in 1993, indicating the layers of the statement.

Fig. 2: ANFIS Architecture. (Source: Jang, 1993).

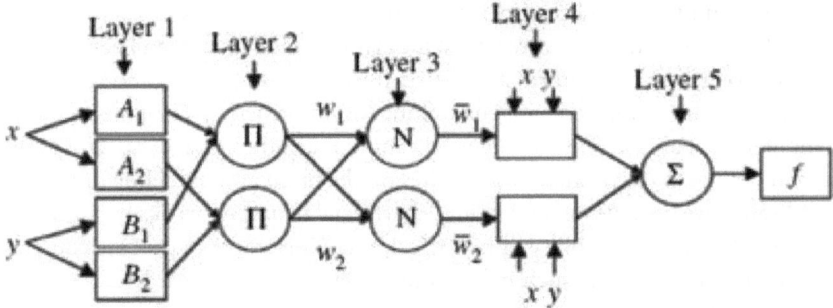

The structure formed by the values of A and B in the first layer defines the verbal variables. These variables are transferred to the second layer after determining the membership grades generated by the membership function. In the second layer, the inputs are multiplied and transferred to a single node in the third layer. The input products transferred to the layer are defined as the ignition power value.

These values are proportional to the net total ignition power value for normalization. The TSK method is applied to the values after the fourth layer. Thus, it is ensured that the output values are obtained as a constant number or polynomial function (Şen, 2004). The average weight method is used to calculate output values in the TSK model (Babuska, 2001). This also clarifies the output values. The sum of the output values is obtained after the fifth layer.

$$x_0 = \frac{\sum a_i . x_i}{\sum a_i} \qquad (1)$$

In addition, TSK method output values are clear values. If the extraction is done by this method, no further clarification is necessary. The total output value resulting from the fifth layer is derived from the model. According to the rule set below, the model is explained by Doğan (Doğan, 2016):

IF x is A_i AND y is B_i THEN $f_i = p_i.x + q_i.y + r_i$ (2)

Layer 1: The membership level of the verbal variables is determined by the membership function ($\mu(x)$), which indicates the level of ignition ($\mu_{Ai}(x), {}_{Bi}(y)$).

Layer 2: $w_i = \mu_{Ai}(x).\mu_{Bi}(y)$.

Layer 3: $= \overline{w_i} = w_i / \sum w_i$ refers to the level of normalized ignition.

Layer 4: $\overline{w_i} f_i$ refers to layer output.

Layer 5: $x_0 = \sum w_i . \dfrac{f_i}{\sum w_i}$.

2.3 Application

The data set required for modeling is taken from www.odd.org.tr which is the web page of the Automotive Distributors Department. The data set includes the sales quantities of the cars produced for 2014, 2015 and 2016. In addition, monthly PPI, YUI, S & P and Moody's credit rating agencies have been integrated into the Turkish credit memo data to directly affect automobile sales. Of the 3199 data in the data set, 2952 were in the training set; 246 were used in the test set. The 2952 data set used for the training set is defined as the output of the system, the credit ratings of Moody's credit rating agencies, vehicle type and vehicle brand entry values for 2014–2016 and vehicle sales between 2014 and 2016. There are 7 inputs and 1 output in the modeled structure. Entries were determined as vehicle brand, vehicle type, monthly PPI, monthly salary, Moody's credit rating on a quarterly basis, S & P credit rating on a quarterly basis and sales date monthly. The output of the system gives the name of automobile sales. The architectural structure of the method used is shown in Fig. 3 and Fig. 4.

Fig. 3: The Architectural Structure of Method.

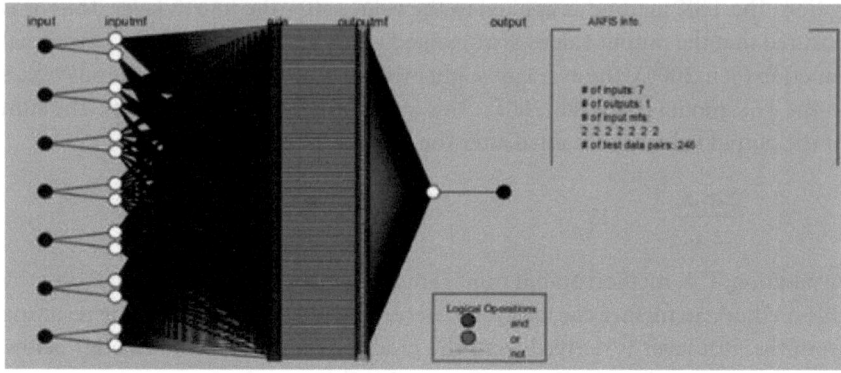

Fig. 4: Anfis Structure.

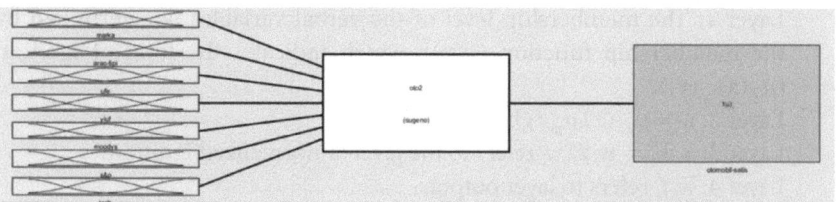

The training set is presented to the system to learn the problem of the modeled system, to form membership functions of the fuzzy clusters and to determine the rules. In the learning process, 200 epochs eventually reached the 0.1 error threshold (Fig. 5).

Fig. 5: The Learning Process.

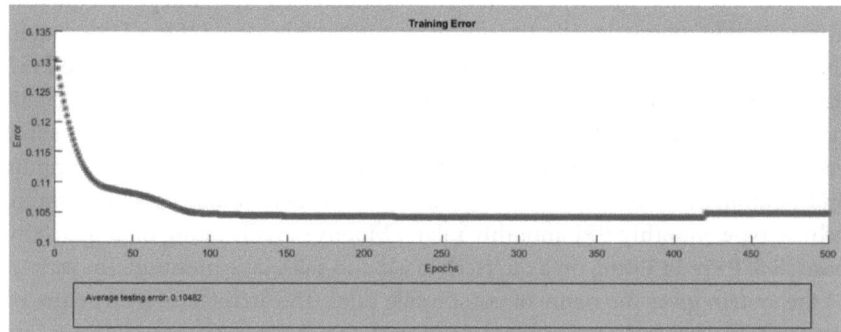

The system that learned the problem has been created with the help of the membership functions of the fuzzy clusters (Fig. 6) and 33 fuzzy rules training (Fig. 7).

Fig. 6: The Membership Functions of the Fuzzy Clusters.

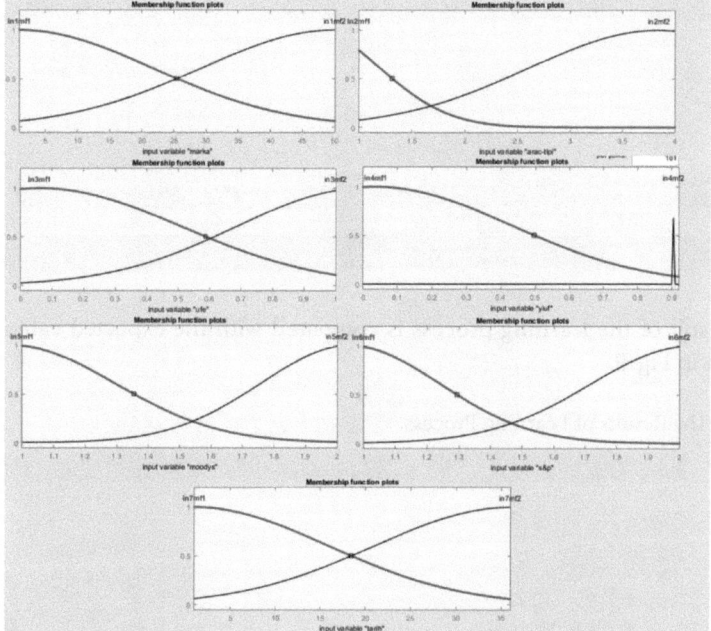

Fig. 7: Rules.

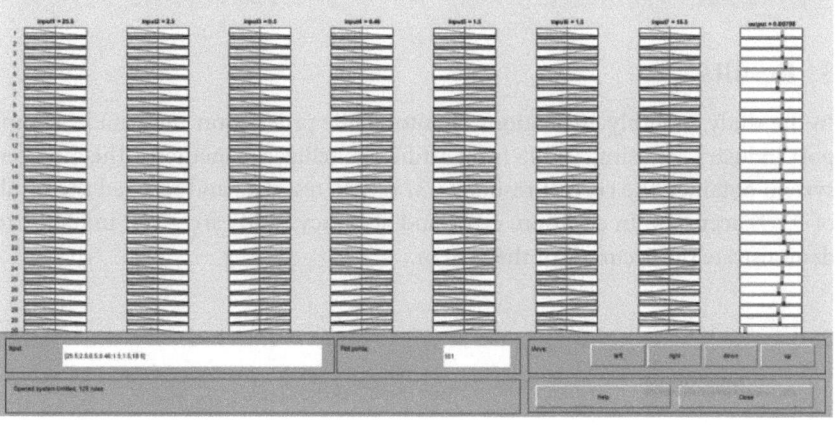

The comparison of the outputs of the learning system with the expected values in the test set is given in Fig. 8.

Fig. 8: Outputs of the Learning System with Expected Values in the Test Set.

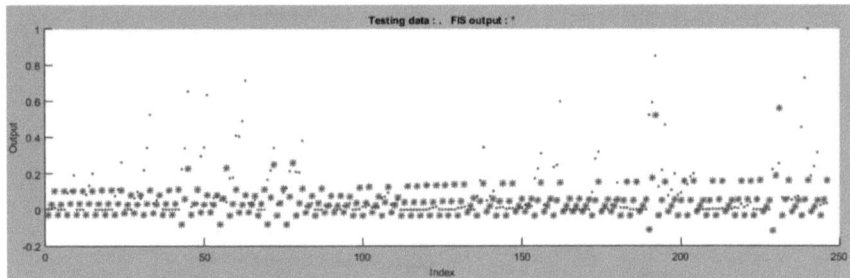

The result of the learning process is compared with the expected value of the system in Fig. 9.

Fig. 9: The Results of Learning Process.

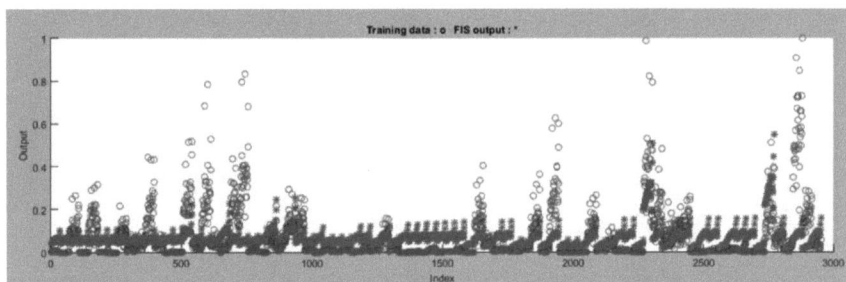

3. Results

In the study, monthly estimations of automobile production were made to support Industry 4.0 using ANFIS from artificial intelligence methods. The modeled system obtained the correct result at 232 of 246 test data and reached the result of 93 % accuracy. In addition, error and accuracy ratios are given in Tab. 1 to demonstrate the accuracy of the system.

Tab. 1: The Error and Accuracy Ratios of the System.

Total Error $\sum_{i=1}^{n}(y_i-d_i)$	Mean Error Sum, $\sum_{i=1}^{n}(y_i-d_i)^2$	Accuracy Ratio
23.58	8.49	93%

In order to strengthen the accuracy of the model, a box-whisker graph of Fig. 10 is drawn on the error values.

Fig. 10: Box-Whisker Graph.

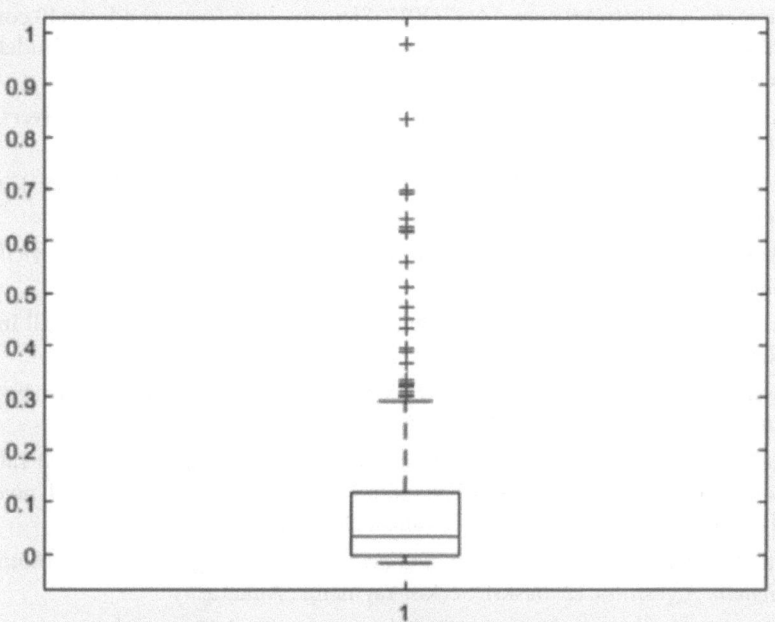

With these values obtained, it is revealed that the model will contribute to automobile production on Industry 4.0. With such demand forecasts, the lowest number of vehicles that can be produced and sold can be reduced. This will contribute to the reduction of production costs and to the total production efficiency.

4. Conclusion

This work is an example to illustrate the benefits of using artificial intelligence in the industry, and further work is aimed at using more advanced artificial intelligence methods for Industry 4.0.

References

URL1: Üretimin Yeni Çağı "Endüstri 4.0" | BİLEŞİM Yayıncılık, Fuarcılık ve Tanıtım Hizmetleri A.Ş., İstanbul. http://www.otomasyondergisi.com.tr/arsiv/ yazi/93-uretimin-yeni-cagi-endustri-40#sthash.g9rqmgrk.dpuf Erişim Tarihi: 14/05/2017 10:05.

URL2: Otomotiv Distribütörleri Derneği, Nisan 2017 http://www.odd.org.tr/fold ers/2837/categorial1docs/1832/sekt%c3%b6rel%20de%c4%9ferlendirme%20 nisan%202017.pdf)

URL3: Kahraman, H. İnsan Makine Etkileşimi, Osnabrücker Zeitung, T-Systems, Siemens, endüstri4.0.com, 14/05/2017, Siemens, http://www.endustri40.com/ insan-makine-etkilesimi-1-endustri-4-0la-insan-ve-makinenin-uretimdeki-yeri/.

Babuska, R. (2001). *Fuzzy and Neural Control* (Lecture Notes). Delft University of Technology, Delft.

Baykal, N., & Beyan, T. (2004a). *Bulanık Mantık İlke Ve Temelleri.* Bıçaklar Kitabevi, Ankara.

Baykal, N., & Beyan, T. (2004b). *Bulanık Mantık, Uzman Sistem Denetleyiciler.* Bıçaklar Kitabevi, Ankara.

Case, K. E., & Faır R. C. (1999). *Principles of Mikroeconomica.* Prentice-Hall Inc., New Jersey.

Chase, R. B., & Aquilano, N. J. (1981). *Production and Operations Management: A Life Cycle Approach.* (Third Edition). Richard D. Irwin Inc., Illinois.

Doğan, O., 2016. "Uyarlamalı Sinirsel Bulanık Çıkarım Sisteminin (ANFIS) Talep Tahmini İçin Kullanımı ve Bir Uygulama" Doktora Tezi, Dokuz Eylül Üniversitesi İktisadi ve İdari Bilimler Fakültesi Dergisi, 31(1), pp. 257–288,

Elmas, Ç. (2011). *Yapay Zeka Uygulamaları (Yapay Sinir Ağları-Bulanık Mantık-Genetik Algoritma.* (2. Baskı). Seçkin Yayıncılık, Ankara.

Ergün, M. E. (2008). *Sermaye Bütçelemesi Ve Türk Sanayi İsletmelerinde Uygulaması.* Yüksek Lisans Tezi. Çukurova Üniversitesi Sosyal Bilimler Enstitüsü, Adana.

Halaç, O. (1995). *Kantitatif Karar Verme Teknikleri.* (4. Baskı). Alfa Basım Yayım Dagıtım, İstanbul.

İpek, M. (1995). *Tam Zamanında Üretim Sistemi Ve Bir Simülasyon Uygulaması.* Yüksek Lisans Tezi. İstanbul Teknik Üniversitesi Fen Bilimleri Enstitüsü, İstanbul.

Jang, Jyh-Shing. (1993). ANFIS Adaptive-Network-based Fuzzy Inference System. Systems, Man and Cybernetics, IEEE Transactions on. 23. 665–685. 10.1109/21.256541.

Karatlı, M., Helvacıoğlu, Ö.C., Ömürbek, N., & Tokgöz, G. (2012). Yapay Sinir Ağları Yöntemi İle Otomobil Satış Tahmini. *Uluslararası Yönetim İktisat Ve İşletme Dergisi*, 8(17).

Konar, A. (2000). *Artificial Intelligence and Soft Computing-Behavioral and Cognitive Modeling*. CRC Press LLC, Florida.

Lee, K. H. (2005). *First Course on Fuzzy Theory and Applications*. Springer, Berlin.

Pan, M., Sikorski, J., Kastner, C., Akroyd, J., Mosbach, S., Lau, R., & Craft, M. (2015). Applying Industry 4.0 to Jurong Island Eco Industrial Park. *The 7th International Conference on Applied Energy—ICAE2015* (pp. 1536–1541). during March 28–31, 2015, in Abu Dhabi, United Arab Emirates, New York , US, Elsevier.

Preece, J. (1994). Interaction Style. In Preece, Jennifer, et al (Ed), *Human-Computer Interaction*. Addison-Wesley Wokingham. UK (United Kingdom).

Samer, E., & Schlenkoff, A. (2009). ANFIS and BP Neural Network for Travel Time Prediction. *World Academy of Science, Engineering and Technology*, 57, 116–121.

Schuh, G., Potente, T., Wesch-Potente, C., Weber, A. R., & Prote, J. P. (2014). Collaboration Mechanisms to Increase Productivity in the Content of Industrie 4.0. *Robust Manufacturing Conference* (pp. 51–56). Elsevier B:V. 7–9 July 2014, Bremen, Germany, publisher location : Amsterdam, Netherlands.

Shneiderman, B. (1998). Human Factors of Interactive Software. In *Desiging the User Interface: Strategagies for Effective Human-Computer Interaction*. Addison-Wesley. Indiana.

Şen, Z. (2004). *Mühendislikte Bulanık Mantık (Fuzzy) İle Modelleme Prensipleri*. Su Vakfı Yayınları, İstanbul.

Tekin, M. (1996). *Üretim Yönetimi*. Arı Ofset Matbaacılık, Konya.

Tenkorang, F. A. (2006). *Projecting World Fertilizer Demand in 2015 and 2030*. Doktora Tezi. Purdue University, Indiana.

Yılmaz, A. (2017). *Yapay Zeka*. Kodlab Yayıncılık, İstanbul.

Yücel, A. (2010). *Tedarikçi Seçimi Probleminde Bütünleşik Sinirsel Bulanık Mantık Yaklaşımı*. Doktora Tezi. İstanbul Üniversitesi Endüstri Mühendisliği Bölümü, İstanbul.

Wang, F. K., Chang, K. K., & Zeng, C. W. (August 2011) Using Adaptive Network-Based Fuzzy Inference System to Forecast Automobile Sales. *Expert Systems with Applications*, 38(8), 10587–10593, Elsevier.

Kadir Alpaslan Demir* and Halil Cicibaş

The Next Industrial Revolution: Industry 5.0 and Discussions on Industry 4.0

1. Introduction

Various advancements in technology help us to achieve a significant increase in industrial output. This significant increase in industrial output creates a chain reaction in industries and businesses. New products in large quantities become available. Supply and demand relations in many products and services change. New businesses are born. Existing businesses adapt. New industries are born and some become obsolete. Changes in inter- and intra-industrial relations are observed. Science and technology research receive funding for further research. All these changes affect the society in many ways. Therefore, we call this an industrial revolution.

There were three industrial revolutions in history. According to many, we are at the dawn of a new industrial revolution. This fourth revolution is called "Industry 4.0". The concept was recently introduced in 2011. "Smart Manufacturing" or "Smart Factories" are at the core of Industry 4.0. Since the introduction, there were many talks, discussions, conferences, seminars and scientific research related to Industry 4.0. However, while there is a remarkable support for Industry 4.0, there is also some criticism from various experts and scholars. One of the basic critique related to Industry 4.0 is that it is not a revolution but the same old information technology (IT)-supported manufacturing. Furthermore, we have yet to see the changes previously observed in earlier industrial revolutions. So far, there is much talk but little change.

Even though it has been only a few years since the introduction of Industry 4.0, some intellectuals (such as Gotfredsen, 2016; Østergaard, 2016; Rada, 2015; Rendall, 2017; Sachsenmeier, 2016) started a discussion for Industry 5.0. Industry 4.0 is conceptualized and defined by a panel of experts and scholars with support from the German government. Therefore, Industry 4.0 is actually a predefined concept. The problem with a predefined concept is its difficulty in changing the definition. Therefore, when experts have ideas related to an industrial revolution

---

* Corresponding author: Kadir Alpaslan Demir, kadiralpaslandemir@gmail.com, Turkish Naval Research Center Command, Pendik, İstanbul, Turkey, +90 532 333 3988.

and these are outside its currently defined scope, it is natural that some come up with a new definition and a new version.

In this chapter, we first overview industrial revolutions. Then we provide a detailed critique of Industry 4.0. In the following section, we present the current discussions and our view on Industry 5.0. In addition, we provide a brief discussion on the topic of collaboration between humans and robots in organizations, since it is the center of many Industry 5.0 visions. Finally, we conclude the chapter with our arguments for what the next industrial revolution -regardless of its name and version- should be about.

2. Methods

In this section, we briefly overview the industrial revolutions.

2.1 Industrial Revolutions

According to the Oxford dictionary, an industry is the "economic activity concerned with the processing of raw materials and manufacture of goods in factories". The origin of the word dates back to the 15th century. However, the first industrial revolution started towards the end of the 18th century. Naturally, all industrial revolutions are fueled by technological advances in manufacturing processes. Fig. 1 depicts the industrial revolutions on a timeline. Note the shortening of periods between each industrial revolution. There are 100 years between the revolutions up to the third revolution. However, there are only 40 years between the third and the fourth industrial revolution.

Fig. 1: From Industry 1.0 to Industry 4.0.

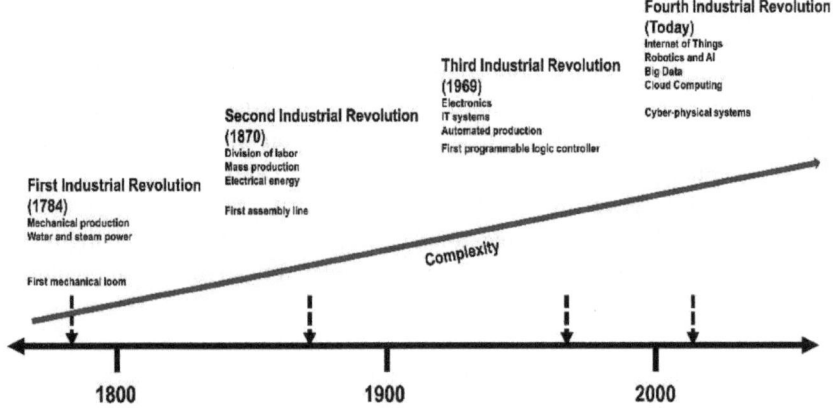

Industry 1.0: At the end of 17th century, we started to observe manufacturing facilities equipped with water- or steam-powered mechanical manufacturing equipment. We observed a significant boost in productivity with the help of industrial machines and new power technologies.

Industry 2.0: Second industrial revolution started in the 1870s with the introduction of a division of labor and mass production. Electrical energy became the main power source during this industrial revolution. Towards the end of the 18th century, we started to put more science into manufacturing and management of industries. In addition to technical advances in manufacturing and mass production, we also observed advances in management. During these times, the influential classical management theory was born. In the beginning of 1900s, Frederick Winslow Taylor published "The Principles of Scientific Management" (Taylor, 1914). Henri Fayol published the "14 Principles of Management" in his book titled as "Administration Industrielle et Générale" based on his work and experiences as a mining company director (Fayol, 1918).

Industry 3.0: Third industrial revolution started in 1969 with the introduction of electronics and information technology (IT) systems use in manufacturing. While electricity is the main power source, computing is the conceptual power in the Industry 3.0. The computing power enabled by electronics and IT was the key to this industrial revolution.

Industry 4.0: According to many industrialists, experts and scholars, we are in the beginning of the fourth industrial revolution. As in the previous industrial revolutions, various technological developments fuel the fourth revolution. We discuss Industry 4.0 in detail in the next section.

Industry 5.0: The next industrial revolution is already being discussed in various forums and blogs. It is actually quite interesting to see that various futurists are talking about the next revolution while we are still discussing the beginning of an industrial revolution. Actually, this may be due to the shortcomings of Industry 4.0. We devote a separate section to Industry 5.0.

According to World Robotics 2016 Report, in the beginning of 2016, the total worldwide stock of operational industrial robots is about 1.6 million units. In addition, between 2016 and 2019, an annual growth of more than 13 % is estimated in industrial robot installations all around the world. In the next few years, a double-digit growth is expected. According to the report (International Federation of Robotics, 2016), this high growth is the result of various developments some of which are presented in Tab. 1. The increase in supply and demand in industrial robots signals the steps of an industrial revolution.

Tab. 1: Reasons for High Growth in Industrial Robot Installations. Source: International Federation of Robotics, 2016.

Industry 4.0, linking the real-life factory with virtual reality, will play more and more important role in worldwide manufacturing.
The human-robot team-effort will have a breakthrough in this period.
Compact and easy-to-use collaborative robots will drive the market in the coming years.
Worldwide competition needs continued modernization of production facilities.
Energy-efficiency and using new materials, for example carbon-composites, require continued retooling of production.
Growing consumer markets require expansion of production capacities.
A decline in products' life cycle and an increase in the variety of products require flexible automation.
Continuous quality improvement needs sophisticated high-tech robot systems.
Robots improve the quality of work by taking over dangerous, tedious and dirty jobs that are not possible or safe for humans to complete.
Small and medium-sized companies will increasingly use industrial robots.

3. Findings

In this section, we present the findings based on our analysis of the discussions related to Industry 4.0 and Industry 5.0.

3.1 Industry 4.0

The goal of Industry 4.0 is not different from the previous industrial revolutions. It is basically achieving mass production with the help of new technologies. Obviously, in this sense, technology is the fuel for an industrial revolution. This is, in fact, a cyclic development. Technology helps industries and industries create a need for technology. Basically, the need for achieving efficiency and effectiveness in industries creates a necessity for technology.

Smart production is at the core of Industry 4.0 (Erkollar and Oberer, 2016). The motto set by Germany Trade and Invest (GTAI) for Industry 4.0 is "Smart Manufacturing for the Future" (Germany Trade and Invest, 2014). According to the policy document (Industrie 4.0 – Smart Manufacturing for the Future), smart industry or "Industrie 4.0" is the technological evolution of embedded systems to cyber-physical systems. It also makes way for a paradigm shift from centralized to decentralized production.

Naturally, Industry 4.0 benefits from various trend technologies. These trend technologies are in fact information technologies (IT) not originally developed and intended for Industry 4.0. These technologies are:

➤ Internet of Things (IoT)
➤ Cloud Computing
➤ Big Data
➤ Robotics and Artificial Intelligence (AI)

These technologies are commonly accepted as the core technologies supporting Industry 4.0. There are also various other technologies such as 3D printing.

European Union and especially Germany is pushing forward the Industry 4.0. Germany is one the most important manufacturers of the world and leader in many aspects related to manufacturing. Much has been written about Industry 4.0 and its promises. Therefore, we focus on the shortcomings of Industry 4.0 and discussions related to Industry 5.0.

3.2 Discussions on Industry 4.0

The discussions about Industry 5.0 before the widespread realization of Industry 4.0 clearly indicate the inadequacies of current Industry 4.0 or similar industrial concepts. The definition and the concept of Industry 4.0 are set forth by a set of visionaries including industry experts and academics. The concept was developed in a series of meetings and discussions. In this aspect, there is a top-down approach in the fourth industrial revolution. It is possible to argue that a bottom-up approach is also required. For example, a study indicates that currently small to medium enterprises (SMEs) find Industry 4.0 irrelevant (Maier and Student, 2015).

Various scholars also approach Industry 4.0 with skepticism (Hirsch-Kreinsen, Weyer & Wilkesmann, 2016; Sachsenmeier, 2016). First, they question the feasibility of the revolution. The cost to deploy the mentioned technologies such as IoT, robotics and big data is high for many businesses in many industries. SMEs are at a disadvantage compared to big corporations. Note that SMEs have a crucial role in many industries. There is also the fear of negative social consequences (Hirsch-Kreinsen, Weyer & Wilkesmann, 2016). Many people may be out of jobs due to increased automation. On the other hand, industrial innovations create new job possibilities. The number of new jobs should be higher than the number of job losses. Otherwise, the public perception towards the industrial revolution will be negative.

According to some, Industry 4.0 concept lacks innovation. They argue that it is the same stuff with a different packaging. The concept of IT-supported production was already the basis for the previous industrial revolution. Therefore, they claim that there is nothing new about supporting the industry with new information technologies.

The increased use of information technology in the industry will create a heavy dependency on the IT sector. IT companies will get richer, bigger and powerful. This creates a certain level of fear among some. On the other hand, IT companies were becoming powerful regardless of the industrial development. Because people use IT for many purposes, especially in their daily routines. As various industries and IT evolve, implications for each other will not be predictable. The fear towards unknown is natural.

There are also concerns related to the increased autonomy of robotic systems. Experimental autonomous vehicles are on the roads. Unmanned aerial systems are occupying the skies. While such autonomy in robotic systems is exciting, it also raises a variety of concerns among the public. Safety and invasion of privacy are at the top of the list.

IT companies gather enormous behavioral data for various purposes. Even when the intention is good and economic, the result may not be so harmless for the people. Zuboff (2016) criticizes such business behavior and calls it "Surveillance Capitalism".

Sachsenmeier (2016) states that the definitions related to Industry 4.0 and similar concepts are incomplete. These concepts clearly need more development before they become an actual industrial reality. Sachsenmeier (2016) claims that these types of concepts and their implementations follow a well-known path. First, they are born with limited scope. Then, these concepts are bought and enlarged by various stakeholders such as political, academic, consulting and business players. The concept is shaped according to the agendas of these players. The increased use of the word "transformation" is observed during related discussions (Sachsenmeier, 2016). However, these concepts have not yet impacted the actual businesses and society. According to April 2016 findings by the German Allensbach Institute, the public attitude towards Industry 4.0 has shifted from indifference to mistrust (Sachsenmeier, 2016).

An investigation into the 64 papers presented at the 10th Asia Pacific International Conference on Information Science and Technology (APIC-IST 2015) held in July 2015, at Da Nang, Vietnam, showed that most researchers focused on IoT and wireless sensor networks in their studies (Chung and Kim, 2016). 3D printing, sharing economy, driverless cars, nanotechnology and biotechnology did not get enough attention from researchers. Chung and Kim (2016) suggest that we need to focus on a diverse research agenda for achieving the fourth industrial revolution.

Increase in mass production seems to be the center and goal of the Industry 4.0. With its current definition, one of the main inadequacies of Industry 4.0 is the lack of environmental considerations. According to Rada (2017), sustainability and waste prevention should be at the center of the vision for Industry 5.0. Sustainability is also at the heart of the bioeconomy vision of the European Commission. Bioeconomy is expected to have a deep impact on businesses and industries.

3.3 Industry 5.0

There are various visions for Industry 5.0. Some futurists argue that while Industry 4.0 is essentially about connecting devices together, Industry 5.0 is about collaboration between humans and machines on the factory floors (Johansson, 2017). Gotfredsen (2016) lists the benefits of a collaborative man and machine workforce. There will be a creative human touch on the production instead of a standard robotic production. New jobs will be created. Human workers will assume better roles on the factory floor. According to Østergaard (2016), Industry 5.0 is the return of the human touch on the factory floors. Rendall (2017) argues that while Germany leads the fourth industrial revolution, North America is uniquely positioned to lead the next industrial revolution – Industry 5.0. Rendall (2017) and many others share the vision of man-machine collaboration for the Industry 5.0. There are many discussions and posts on the Internet related to Industry 5.0. However, Johansson (2017) argues that two industrial revolutions so close to each other may actually be treated as one. Therefore, Industry 4.0 may incorporate both connectivity and man-machine collaboration (Johansson, 2017). We also share the view of Johansson (2017). One of the pillars of Industry 4.0 is robotics and AI. A natural extension of robotics and AI is man-machine collaboration. Just the introduction of human-machine collaboration on the factory floors falls short of supporting the case for a new industrial revolution.

According to Sachsenmeier (2016), Industry 5.0 is related to bionics and synthetic bionics. Bionics is "the imitation or abstraction of the inventions of nature" (Sachsenmeier, 2016). According to European Commission (2012), bioeconomy is

> the production of renewable biological resources and the conversion of these resources and waste streams into value-added products, such as food, feed, bio-based products, and bioenergy. It includes agriculture, forestry, fisheries, food and pulp and paper production, as well as parts of chemical, biotechnological and energy industries. Its sectors have a strong innovation potential due to their use of a wide range of sciences (life sciences, agronomy, ecology, food science and social sciences), enabling and industrial technologies (biotechnology, nanotechnology, information and communication technologies (ICT), and engineering), and local and tacit knowledge.

Bioeconomy is crucial in achieving a sustainable economy (Schütte, 2018). Smart use of biological resources for industrial purposes will help to achieve a balance between ecology, industry and economy. German federal government gave a priority to bioeconomy in the national research policy agenda when the government adopted the "National Research Strategy BioEconomy 2030" in 2011 (German Federal Ministry of Education and Research, 2011). In the research agenda, five fields of action are given priority. These are:

➢ Securing global nutrition.
➢ Ensuring sustainable agricultural production.
➢ Producing healthy and safe foods.
➢ Using renewable resources for industry.
➢ Developing biomass-based energy carriers.

Bioeconomy is important for Europe. It has an annual turnover of 2 trillion Euros and 22 million people underemployment. According to Schütte (2018), biologization, the guiding principle of the bioeconomy, has the potential to create a fundamental change in the industry. Therefore, bioeconomy should be a key part of next industrial revolution.

Another vision for Industry 5.0 is set forth by Michael Rada (Rada, 2015; Rada, 2017). Rada states that the priority of Industry 5.0 is "to utilize an efficient workforce of machines and people, in synergy with the environment". He also provided a definition for Industry 5.0 (Rada, 2017). The theme of this vision is Industrial Upcycling. This vision focuses on waste prevention. Furthermore, Rada points out that we need to turn to the human element in the manufacturing process. He criticizes the current digitization trend that is the effort to embed 1s and 0s into any living organism (Rada, 2015).

According to Rada, Industry 5.0 includes 6R methodology and LED principles. The 6R is:

➢ Recognize: First, we need to recognize the opportunities offered by Industrial Upcycling. An awareness is the first required step.
➢ Reconsider: We need to evaluate and reconsider our business and manufacturing processes. A redesign of processes to realize the benefits of Industrial Upcycling is an essential step.
➢ Realize: After recognizing the opportunities and reconsidering business processes, we need to realize the business process improvement or innovation.
➢ Reduce: Reducing the use of resources to achieve efficient outcomes is the essence of the methodology.

➢ Reuse: Reusing the materials considered as useable prior to process improvement is also at the center of the methodology.
➢ Recycle: Recycling as much as possible is one of the main expected outcomes of the Industrial Upcycling effort. Naturally, the ideal is the zero waste.

6R methodology defines a business improvement model. Depending on the specific case, it can be considered as a business process improvement or a business process innovation. Therefore, the 6R methodology is subject to the rules, assumptions and dynamics of process improvement efforts.

LED stands for Logistics Efficiency Design. It is designed for global supply chain efficiency improvements. Its goal is to eliminate the waste created by the current modern standard buyer-supplier business relations. LED is the concurrent application of transparency, profit sharing and efficiency in the supply chain (Rada, 2017).

Four types of waste are identified in Industrial Upcycling:

➢ Physical Waste: The actual physical waste introduced during and after the production. It is basically the trash.
➢ Social Waste: It is the unused potential of the manpower. People unemployed is at the heart of social waste.
➢ Urban Waste: This type of waste includes brownfields, empty spaces and inadequate infrastructure.
➢ Process Waste: Overproduction, overstocking, empty transport vehicles on the roads are among the process waste.

According to Rada, Industry 4.0 focuses on quantity and mass production. However, the focus of Industry 5.0 is a higher life standard and creativity with high-quality custom-made products. The theme of Industry 5.0 is simply sustainability. Note that in recent years many companies started programs for green manufacturing and production. Furthermore, they focus on social responsibility projects. The awareness for environmental protection is increasing among people. Customers begin choosing products developed by companies promoting green production.

In Tab. 2, we provide a comparison of Industry 4.0 and Industry 5.0. Note that both industrial revolutions have yet to occur. This comparison is based on the current discussions. The actual revolutions may be quite different than what is discussed. A quick analysis of the comparison shows that Industry 5.0 will have a wider and deep impact on society.

Tab. 2: A Comparison of Industry 4.0 and Industry 5.0.

	Industry 4.0	Industry 5.0
Motto	Smart Factory	Bioeconomy
Motivation	Mass Production	Sustainability
Power Source	Electrical Power – Fossil-Based Fuels – Renewable Power Sources	Electrical Power – Renewable Power Sources
Involved Technologies	– Internet of Things (IoT) – Cloud Computing – Big Data – Robotics and Artificial Intelligence (AI)	– Sustainable Agricultural Production – Bionics – Renewable Resources – Human–Robot Collaboration
Involved Research Areas	– Organizational Research – Process Innovation and Improvement – Business Administration	– Agriculture – Biology – Waste Prevention – Organizational Research – Process Innovation and Improvement – Business Administration

3.4 Integrating Robots into Organizations

Some of the visionaries (Gotfredsen, 2016; Johansson, 2017; Østergaard, 2016; Rendall, 2017) strongly argue that Industry 5.0 will be about increased collaboration between humans and robots in factories and organizations. Therefore, we devote a brief section to the subject. With the rapid advancements in the robotics field, robots are already becoming a part of our lives (Demir, 2017). Both autonomous and remote-controlled unmanned aerial vehicles (UAVs) are being used for various purposes (Demir, Cicibas & Arica, 2015). Autonomous cars are being tested on roads. Medical robots help surgeons in complex medical procedures. The price of educational and entertainment robots has become affordable in recent years. There are many other types of robotic applications in development (Demir, 2017). The robot named Pepper is designed as a day-to-day companion. According to its developers, Pepper is "the first humanoid robot capable of recognizing the principal human emotions and adapting his behavior to the mood of his interlocutor" (Softbank Robotics, 2017). In Japan and France, Pepper is being used to greet customers in stores and train stations. Currently, the cost of Pepper is less than 25,000 Euros. Therefore, it is affordable for many businesses. In the near future, we should not be surprised to see robots as administrative assistants or in some other roles in offices and other workplaces. A collaborative human and robot workforce will create many changes in our organizational and business environments. Therefore, increased coordination between humans and robots in the workplaces will be an important part of Industry 5.0. In a prior study, we identified certain issues in integrating robots into organizations. Note that in Tab. 2, we list organizational research, process innovation and improvement and business administration as important research areas for Industry 5.0. Humans and robots working together will lead to significant changes in the organizations. We may observe evolutions in some of the management and organization research areas such as organizational behavior, organizational structures and workflows and work ethics. Furthermore, acceptance of robots in the workplace and discrimination against robots or humans will be among the important issues in Industry 5.0. In Tab. 3, we list some of the important issues in integrating robots into organizations.

Tab. 3: Issues in Integrating Robots into Organizations. Source: Demir et al., 2017.

Evolution in organizational behavior
Acceptance of robots in the workplace
Evolution of organizational structures and workflows
Evolution in work ethics
Discrimination against robots or people
Privacy and trust in a human-robot collaborative work environment
Education and training
Redesign of workplaces for robots

4. Discussion and Conclusions

Industry 4.0 is still in its early stages. It is officially introduced in the beginning of 2010s. In only a few years, various visionaries started discussing Industry 5.0. Moreover, these visionaries also point out the inadequacies of Industry 4.0 and propose Industry 5.0 to overcome the shortcomings of Industry 4.0. One logical conclusion is that Industry 4.0 was introduced without adequate vision. Previous industrial revolutions occurred naturally unlike Industry 4.0, which is formally defined and forced upon the industry. It is possible to argue that this artificial revolution start is premature and proposed without adequate maturity. To call a concept an industrial revolution, we need to observe a widespread change in industries, businesses and society. Currently, with its current definition, Industry 4.0 should actually be a proposal. It is clear that Industry 4.0 is still under development.

Smart mass production seems to be the goal of Industry 4.0. Sustainability is the main theme in Industry 5.0 proposals. Actually, they are both inadequate by themselves. Note that sustainability and mass production are not mutually exclusive. Therefore, combining these two goals or themes and redefining the next industrial revolution may be a better approach. As a result, the motto of the next industrial should at least be "sustainable smart production". Furthermore, the next industrial revolution – regardless of its name, version and definition – should encompass the following technologies and research areas:

➢ IoT
➢ Cloud Computing
➢ Big Data
➢ Robotics and AI
➢ Sustainability and Environmental Protection
➢ Bioeconomy
➢ Waste Prevention
➢ Business Administration and Organizational Research

5. Acknowledgments and Disclaimers

The views and conclusions contained herein are those of the authors and should not be interpreted as necessarily representing the official policies or endorsements, either expressed or implied, of any affiliated organization or government. This chapter is an extended version of the research (Demir and Cicibas, 2017) presented in the 4th International Management Information Systems Conference, Istanbul, Turkey, on October 17–20, 2017.

References

Chung, M., & Kim, J. (2016). The Internet Information and Technology Research Directions based on the Fourth Industrial Revolution. *KSII Transactions on Internet & Information Systems*, 10(3). 1311–1320.

Demir, K. A. (2017). Research Questions in Roboethics. *Mugla Journal of Science and Technology*, 3(2), 160–165.

Demir, K. A., Caymaz, E., & Elçi, M. (2017). Issues in Integrating Robots into Organizations. In Proceedings of the 12th International Scientific Conference on Defense Resources Management in the 21st Century, Brasov, Romania, November 9–10, 2017.

Demir, K. A. & Cicibas, H. (2017). Industry 5.0 and a Critique of Industry 4.0, 4th International Management Information Systems Conference, Istanbul, Turkey, October 17–20, 2017.

Demir, K. A., Cicibas, H., & Arica, N. (2015). Unmanned Aerial Vehicle Domain: Areas of Research. *Defence Science Journal*, 65(4), 319–329.

Erkollar, E., & Oberer, B. (2016). Endüstri 4.0 AkıllıÜretim İçin Politika ve Programlara Ait Bir Örnek: Alman AkıllıÇözümleri, In *Smart Technology & Smart Management*. Edited by Vahap Tecim, Çiğdem Tarhan, Can Aydın, İzmir, Turkey.

European Commission (2012). Innovating for Sustainable Growth—A Bioeconomy for Europe. Retrieved from https://ec.europa.eu/research/bioeconomy/pdf/official-strategy_en.pdf , Accessed on 10 October 2018.

Fayol, H. (1918). *Administration industrielle et générale* (Vol. 2). H. Dunod et E. Pinat, Edituers, Paris.

German Federal Ministry of Education and Research (2011). National Research Strategy BioEconomy 2030—Our Route Towards a Biobased Economy.

Germany Trade and Invest (2014). Industrie 4.0. Smart Manufacturing for the Future.

Gotfredsen, S. (2016). Bringing Back the Human Touch: Industry 5.0 Concept Creating Factories of the Future. Retrieved 10 October 2018 from http://www.manmonthly.com.au/features/bringing-back-the-human-touch-industry-5-0-concept-creating-factories-of-the-future/

Hirsch-Kreinsen, H., Weyer, J., & Wilkesmann, J. D. M. (2016). *"Industry 4.0" as Promising Technology: Emergence, Semantics and Ambivalent Character*. Soziologisches Arbeitspapier Nr. 48/2016, Technische Universität Dortmund, Germany.

International Federation of Robotics (IFR) (2016). Executive Summary. World Robotics 2016 Industrial Robots, Retrieved 10 October 2018 from https://ifr.org/img/uploads/Executive_Summary_WR_Industrial_Robots_20161.pdf

Johansson, H. (2017). *Profinet Industrial Internet of Things Gateway for the Smart Factory*. Master's Thesis in Embedded Electronic System Design, Department of Computer Science and Engineering, Chalmers University Of Technology, University Of Gothenburg Gothenburg, Sweden.

Maier, A. & Student. D. (2015) Industrie 4.0 – der große Selbstbetrug, Manager Magazin. Retrieved 10 October 2018. from http://www.manager-magazin.de/magazin/artikel/digitale-revolutionindustrie-4-0-ueberfordert-deutschen-mittelstand-a-1015724.html

Østergaard, E. H. (2016). Industry 5.0 – Return of the Human Touch. Retrieved 10 October 2018. from https://blog.universal-robots.com/industry-50-return-of-the-human-touch

Pfeiffer, S. (2017). The vision of "Industrie 4.0" in the making—a case of future told, tamed, and traded. *NanoEthics*, 11(1), 107–121.

Rada, M. (2015, December 1). INDUSTRY 5.0—From Virtual to Physical, Retrieved 10 October 2018 from https://www.linkedin.com/pulse/industry-50-from-virtual-physical-michael-rada

Rada, M. (2017, February 3). INDUSTRY 5.0 definition, Retrieved 10 October 2018 from https://www.linkedin.com/pulse/industrial-upcycling-definition-michael-rada

Rendall, M. (2017). The New Terminology: CRO and Industry 5.0. Retrieved 10 October 2018 from https://www.automation.com/automation-news/article/the-new-terminology-cro-and-industry-50

Sachsenmeier, P. (2016). Industry 5.0—The Relevance and Implications of Bionics and Synthetic Biology. *Engineering*, 2(2), 225–229.

Schütte, G. (2018). What Kind of Innovation Policy Does the Bioeconomy Need? *New Biotechnology*, 40 (Part A), 82–86.

Softbank Robotics (2017). Retrieved 10 October 2018 from https://www.ald.softbankrobotics.com/en/robots/pepper

Taylor, F. W. (1914). *The Principles of Scientific Management*. Harper. New York, USA.

Zuboff, S. (2016). The Secrets of Surveillance Capitalism, FAZ.NET. Retrieved 10 October 2018 from http://www.faz.net/aktuell/feuilleton/debatten/the-digital-debate/shoshana-zuboff-secrets-of-surveillance-capitalism-14103616.html?printPagedArticle=true#pageIndex_2

Uğur Keleş* and Zümrüt Ecevit Satı

Industry 4.0 as High Technology and Evaluation of Turkey

1. Introduction

From the beginning of industrialization which started during the 18th century, technological developments have triggered significant transformations in industries and societies. These paradigm shifts, which are termed as industrial revolutions, are divided into four eras. Mechanization and steam power led to the first industrial revolution wave. Development of electrical power made way for assembly lines and mass production for the second revolution. Automation of production with the help of computers and electronics started the third age. Nowadays, a new visionary idea of modular and efficient manufacturing system (Lasi et al., 2014) is creating the new paradigm shift which is established as the fourth industrial revolution or "Industry 4.0".

Industry 4.0 represents a digitalized value chain that includes a self-controlled and smart manufacturing process. To achieve this objective, the whole business model must comprise two dimensions: integration (Oesterreich & Teuteberg, 2016; Qin, Liu & Grosvenor, 2016; Roblek, Mesko & Krapez, 2016) and intelligence (Martin & Juergen, 2016; Qin, Liu & Grosvenor, 2016). Machines and humans are integrated through digital networks to communicate and transfer information while machines and logistics systems manufacture and transfer products by optimizing themselves with the network information simultaneously.

Most companies and universities have already started working on different aspects of Industry 4.0 (Drath & Horch, 2014), but many companies and organizations have also reluctant attitude to develop corporate strategies because of the uncertainty of technology (Schmidt et al., 2015). Regarding the dilemma, the purpose of this chapter is to provide an overview of the researches about Industry 4.0 and high technology and present generic strategies for the management of Industry 4.0 transformation.

---

* Corresponding author: Ugur Keles, ugur.keles@ogr.iu.edu.tr, Istanbul University, +90 538 245 2246 (ext.)

2. Method

In this study, the literature about Industry 4.0 and high technology is comprehensively reviewed and revealed and Industry 4.0 concepts are compared with high technology trends. According to Industry 4.0 and high technology relationship, strategies for several business operations which are valid for both small-sized companies and international giants are developed. In the first section, the principals of Industry 4.0 are discussed and related high technologies are introduced. The importance of high-tech management for the Industry 4.0 transformation is presented in the second section. Also, high technology definition and classification is reviewed to apply Industry 4.0 technologies. In the third section, the current state of high technology is reviewed and the future of Industry 4.0 in Turkey is discussed regarding Turkey's high technology competency. Finally, generic high technology strategies are developed to manage business operations in the fourth industrial age in the fourth section.

3. Findings

3.1 The Future of Technology: Industry 4.0

Applications of technologies which are related with fourth industrial revolution have been practiced for several years but the term of Industry 4.0 has been introduced as a holistic concept during Hannover Fair event in 2011 (Qin, Liu & Grosvenor, 2016). German government presented Industry 4.0 idea as a vision for future manufacturing process. This futuristic prediction has enough impact to make lots of companies aim on technological advances while small- and medium-sized companies have still uncertain attitude (Schumacher, Erol & Sihn, 2016) on developments related with Industry 4.0 concept.

Social, economic, strategic and technological developments have triggered Industry 4.0 idea not only by creating pressure on companies to use their resources in most efficient way possible but also by lending them impetus to understand, demand and answer it in the most flexible and fastest way. New technological developments in the fields of internet, robotics, computers, etc., pushing digitalization, flexibility, individualization and decentralization; aging of population (United Nations, 2015), changing human resource (Erol et al., 2016; Hecklau, Galeitzke, Flachs & Kohl, 2016) concerning the need of high level of educated staff, changing demand, the need of resource efficiency (Marre, Beihofer & Haggenmueller, 2015), the effect of green management and sustainability (Gabriel & Pessl, 2016; Seliger & Stock, 2016) pulls industries to embrace Industry 4.0 vision.

Industry 4.0 is a multidisciplinary research area which is very complex and there are abundant researches to explore its dimensions, key features and technologies. The concepts that reached a consensus among researchers and industries will be explained here to construct a complete frame.

Digitalization is the integration of digital technologies into business models. Machines and humans are connected to each other by network systems, internet of things and smart products. In today's world, not only recent manufacturing systems but also the whole business processes engage by adapting digital developments directly into their value chains.

Individualization is tailoring a service or product for customers specified by their demands. Digitalization is one of the key triggers of the individualization trend, which leads companies to have the ability to exchange information between humans and machines via wireless sensors and send the data to smart services or factories (Roblek, Mesko & Krapez, 2016).

Flexibility in manufacturing is the basic element and the necessity of individualization. The ability of producing almost every product individually in the same factory can be achieved by the help of cyber-physical systems, 3D printers and robots and the integration of these machines and humans.

Decentralization of functions, departments, business units including manufacturing improves efficiency by reducing hierarchy. Decentralized autonomous systems allow processing of highly complex data (Brettel, Friederichsen & Kelle, 2014) and make decision making faster. It is essential to adopt flexible manufacturing to cope with individual conditions of demand (Lasi et al., 2014) for the Industry 4.0 vision.

Integration in all levels of business operations is the most critical step to achieve challenges of the new industrial age (Saucedo-Martínez, Pérez-Lara & Marmolejo-Saucedo, 2017). There are three types of integration: horizontal integration over the entire value chain, vertical integration of manufacturing systems and end-to-end integration across the product life cycle.

Industry 4.0 is related with many technologies that have several applications in individual cases. The phenomena will be materialized by the advancement of fundamental components and their common practice in one value chain. Nevertheless, smart technologies have already applied to many commercial products:

- Smart products have embedded sensors which collect real-time information. Also they can give feedback to customer by processing data and transferring the information to data center to improve product development (Qin, Liu & Grosvenor, 2016).

- Smart factory is the next-generation of manufacturing systems where they analyze, predict, manage, maintain and control the production process autonomously. In a smart factory, all machines and logistic systems are integrated to each other. The factory is very flexible and intelligent to optimize the production and resources.
- Cyber-physical systems are one of the key components of smart technologies. They are the integration of mechatronic components and computation (Hermann, Pentek & Otto, 2016). Physical processes are managed by computer-based algorithms while they are interacting with many other processes. Besides cyber-physical machines are not only embedded to each other, they are also linked to humans with networks via cloud in real time (Schumacher, Erol & Sihn, 2016).
- Internet of things connects objects to each other via mobile or fixed networks. In this respect, it is the fundamental technology behind the communication of machines. It enables the optimization of processes with the help of big data and computer algorithms.

3.2 Understanding the High Technology

According to many empirical researches, high technology markets grow faster than any other segment of international trade (Srholec, 2007). This reason alone is enough for companies to aim on research and development (R&D) of high technology products and/or apply the product itself into their manufacturing systems. All concepts related with the Industry 4.0 phenomenon are not actualized yet, even if a number of technologies are ready for industrial applications. There are already many sophisticated manufacturing companies and factories in Germany (Qin, Liu & Grosvenor, 2016); the German federal government was the one that declared Industry 4.0 as a policy based on high technology strategy of 2020 (Lasi et al., 2014). Today, the rest of the world has embraced the vision but it is crucial to discover the high technology basis of Industry 4.0 before being adapted to it in every aspect.

There are numerous difficulties to distinguish a high technology from low technology. The most important concern is about the content: what makes a technology high? Is it related with the output based on technical sophistication of final product or input based on complexity of technology and intensity of research to produce it (Baesu, Albulescu, Farkas & Drăghici, 2015)? The second constraint is the determination of distinctive criteria to identify the technology content of an industry (Hatzichronoglou, 1997). R&D investments, number of technical personals, patents, licenses, copyrights, know-how, collaborations, lifetime of technology, obsolescence rate and velocity of innovation are important factors. But it is extremely challenging to calculate the impact of some of them. Another discussion is the break point in the scale which technologies are separated to classes. Why are some technologies considered as high-tech after a certain level while some of them are accepted as low? The last debate is the segmentation of products or industries. Today's products and manufacturing systems, especially Industry 4.0 concept, are so complex that they are constructed by the combination of multiple technologies.

There is not a consensus as to how to classify high technology in the various researches on high-tech sectors. High-tech products are very complex goods which make companies to learn several key competences while designing, manufacturing and selling them (Wiechoczek, 2016). The intensity of R&D studies highlights how much effort is required to create them. Besides R&D intensity is the most inclusive criteria among other factors. According to OECD (Organisation for Economic Co-operation and Development) working papers (Hatzichronoglou, 1997) (OECD, ISIC Rev. 3 Technology Intensity Definition: Classification of Manufacturing Industries into Categories Based on R&D Intensities, 2011), both direct and indirect R&D intensity is taken into consideration. Additionally, technology-producer and technology-user aspects are presented, respectively. These two approaches resolve the content concern by combining technology of product and sector. As a result, a high technology product is generally associated with the R&D intensity (Hatzichronoglou, 1997) (Sandu & Ciocanel, 2014) to develop and/or produce it.

The high technology sectors (regarding product approach) and several products related with sector are listed here according to product approach of OECD classification (Hatzichronoglou, 1997):

- Aircraft and spacecraft: aircrafts, direction finding compasses, other navigational instruments
- Pharmaceuticals
- Office, accounting and computing machinery: computers, data processing machines
- Radio, TV and communications equipment: telecommunications equipments, optical media
- Medical, precision and optical instruments
- Scientific instruments
- Electrical machinery
- Chemistry
- Non-electrical machinery
- Armament

Classification of technologies needs to be updated with the new technologies as OECD and Eurostat do. Because existing technologies evolve or new technologies emerge thanks to scientific progress, technical development and innovation, so others become obsolete in time. These lists are helpful to understand the existing situation in market, however tracking technology cannot be an option for companies in a dynamic competitive environment. Companies need to pay close attention to R&D of new products and technologies. Because R&D investments are crucial for high technology enterprises (Schumacher, Erol & Sihn, 2016) as R&D intensity is a fundamental indicator of high technology export (Sandu & Ciocanel, 2014).

3.3 Strategies to Manage Industry 4.0 as a High Technology Transformation

Industry 4.0 concepts are formed by emerging high-tech products and/or manufacturing systems, therefore these concepts are the extension of high technology trends. Understanding these trends may create opportunities for competition in the Industry 4.0 era. Because technology intensive firms win new markets and create superior value by innovating more and using resources more effectively (Hatzichronoglou, 1997).

Business Models
Industry 4.0 will not only change how we manufacture but also the whole business models. Eventually, pressure on existing value chains will increase during this transformation (Kagermann, 2015). Customer satisfaction will be gained by adapting digital technologies into value chains. Therefore value chains need to be restructured and adapted to Industry 4.0 in all levels. Due to the horizontal integration of business operations and vertical integration of production systems, value will not be created by a step-by-step chain but will be created by an end-to-end circular process.

Product Development
Objects become smart which can collect and analyze user experience while service functions are tracked by embedded sensors in products. Giving feedbacks to customers will be a must to enhance satisfaction. Additionally products are improved during design phase with the help of feedbacks from smart products which also affect product development processes. The big data of past is processed to design future products. But more importantly, products are designed modular which can be specialized for unique needs of people. As a result, designing both

products and manufacturing systems modular is an essential product development strategy for Industry 4.0.

Complementary products that cooperate with many functions occur in various industries (Wiechoczek, 2016). This kind of complexity is a basic characteristic of high technology products. Thus flexibility and integration concepts of Industry 4.0 are derivative approaches of the high technology product trend. Complementary goods are developed with less cost by modular designs, so the flexibility concept is achieved. Also, integration is required to make these individual designs work in harmony.

Manufacturing
Smart factories are the heart of production. Individualized demands of customers are converted to smart goods in the fastest way. Due to increasing cumulative knowledge and faster product development time for changing demand, lifetime of technologies become shorter. In this respect, flexibility is indispensable for manufacturing as much as product development processes. In a smart factory, while autonomous systems manufacture products with the help of cyber-physical machines, they also optimize processes to reduce costs and resource consumption. Efficient resource management becomes more and more important regarding corporate responsibility with Industry 4.0 transformation.

Supply Chain
Logistics systems need to be intelligently designed to supply raw materials for production and deliver finished goods to customers immediately. A smart logistic management is possible only if suppliers are integrated to system. So, in addition to the integration of humans and machines via networks, integration between companies must be accomplished as well.

Networking
The importance of networking will emerge because of integration of business processes between companies to manage such complexity. Companies, suppliers, manufacturers, distributors, resellers, banks and even consumers must achieve strength cooperation due to increasing sophistication of high technology products to overcome the management of larger number of entity in their businesses (Wiechoczek, 2016). Networking will accelerate product development performance by increasing knowledge share and efficiency while shortening development duration (Wiechoczek, 2016).

Marketing
Despite marketing departments not having as powerful role in product development decisions as R&D in high-tech firms (Workman, 1998), the impact of

marketing will be more important with Industry 4.0 due to feedbacks from smart products. Marketing and R&D departments will need to cooperate to evaluate these feedbacks to improve product design. Moreover, marketing performance can be increased by affecting consumer behavior with high technologies such as social networks and display technologies (Huarng, Yu & Lai, 2015). Managing individual demands of customers is also a challenge for marketing as much as manufacturing and R&D.

Human Resources

Routine tasks on worker level will no longer exist with the increasing effect of automatization and digitalization in manufacturing (Schumacher, Erol & Sihn, 2016). But the need of qualified and well-educated staffs increases in all levels of companies. Management of human resources will become more important in a hyper information-laden work environment. As a result of all the advancement listed below, Industry 4.0 will change how people produce as well as how they live.

Resource Management and Sustainability

The effect of industrialization on environment regarding climate change and air pollution is disruptive which makes current production systems unsustainable (Wang et al., 2016). Growing ecological expectations on demand side, efficiency necessity and corporate citizenship of supply side makes self-optimized smart factories imperative. Developing resource efficient production systems and using renewable technologies with Industry 4.0 concepts is a beneficial way of value creation. Hence companies must implement Industry 4.0 technologies as a social responsibility strategy.

3.4 State of Art in Turkey

Despite a number of contradictive researches, it is widely accepted that high technology sectors have an important effect on the growth of new economies regarding productivity (Karahan, 2015). Thus supporting and regulating high technology researches and investments is critical for a country to enhance industrial development and competitiveness. So governments promote R&D investments for companies by providing financial aids and reducing taxes. Because high technology markets are the most growing trade in international markets and their dynamism is enhanced mostly by the development of knowledge (Wiechoczek, 2016).

According to the World Bank high technology export data, average high technology export rate of both OECD and EU member countries in total manufactured exports was 17.7% and 16.9%, respectively, in 2015. While the world average export value was 18.5%, these countries have started to close the gap between world and them in the last decade. An OECD member country, Turkey's export was only 2% in the same year and this rate has not improved significantly after 2000. Turkey's low rate of high-tech export in total manufactured exports is a significant point to start improving. In this section, Turkey's Industry 4.0 and high technology and competitiveness potential is discussed and the general strategies are applied to Turkey's position.

Structural change is necessary by increasing technology intensity and share of high-tech manufacturing in all sectors to achieve the economic growth by advancing high technology (Karahan, 2015). As an important point, high technology firms need to achieve both innovating constantly and turning these innovations into satisfactory products (Dutta, Narasimhan & Rajiv, 1999). Also the spillover effect of high technology manufacturing helps improving the expansion rate of international trade (Hatzichronoglou, 1997). Arising technology clusters can be effective to increase spillover effect (Ecevit Satı, 2014). So developing Industry 4.0 ready manufacturing systems can also affect the whole manufacturing markets positively in Turkey.

Fig. 1: Range of Revealed Technological Advantage in Economies by Field, 2010–13. Source: OECD, OECD Science, Technology and Industry Scoreboard 2015: Innovation for Growth and Society, 2015).

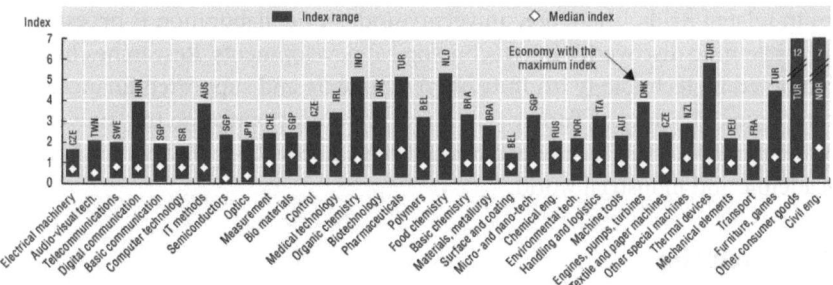

The high technology state of Turkey regarding to economical fields is very varied. According to OECD's Technology and Industry Scoreboard 2015 report, Turkey has the technological advantage in pharmaceuticals, thermal devices, furniture, games and other consumer goods markets (Fig. 1). On the other hand, Turkey has the least economical advantage in control, environmental, micro and nano technologies. The development level of these fields has the range of high to low technologies. But it is important to remember that an industry may be very technology intensive in one country and less in another (Hatzichronoglou, 1997). From this point, Turkey should reveal its own less developed technologies independent from the technologies general level in the world and invest them to improve these weak points to develop Industry 4.0 ready manufacturing systems and products.

International experience has positive influence on global marketing performance and marketing strategy affects significantly innovation performance of a firm (Wu, 2011). Therefore, commercialization of technology is important as its development for companies to create growing opportunity (Kim & Huarng, 2011). To achieve business success, high-tech products must be commercialized with a global marketing strategy. There are already several Turkish global brands but also high technology brands must be raised to improve competitiveness during Industry 4.0 age.

Turkey has much potential regarding its young population. According to OECD indicators, Turkey is the fourth youngest country among other OECD countries (OECD, Young Population, 2017). But it has also the highest share of inactive NEETs (15- to 29-year-olds not in employment, education or training) with Mexico (OECD, Society at a Glance 2016: OECD Social Indicators, 2016). Turkey needs to transform these unemployed young people to well-educated labor with a developing high industry.

Another important driver of Industry 4.0 accomplishment is scientific studies in related fields. Therefore, university–industry collaboration is necessary to turn scientific discoveries into technological developments. Researches must be prompted by technology institute of governments and supported financially by companies. The duty of Turkey is to direct its young population into universities or industries to strengthen researches and help improve the collaboration between academic and industrial world.

4. Conclusion

The main intention of this article is to focus on high technology trends and strategies to implement them to manage Industry 4.0 transformation. Business models need to be updated according to new end-to-end value chains that digital technologies get used in them because value creation process starts with the user experience which is gained by smart products, continues with the self-improvement of products by individual needs of consumers, development of advanced products with the collected data by embedded sensors and ends after manufacturing them by resource efficient autonomous systems and presenting them to consumers again. In this circular process that starts and ends with the consumer, humans and machines are integrated to whole steps of value creation. To achieve the agility to present every costumer individual products, not only manufacturing systems are decentralized and flexible but also all suppliers and distributers must be integrated to business networks.

A particular constraint of the chapter is that no survey has been made. But the primary purpose of the study is to review and reveal main elements of Industry 4.0 and discover their similarities with high technology trends to develop generic strategies. For further studies, researchers should focus on detailed investigation of strategies related with each business operation. Additionally case studies can be performed to enlighten the current state of Industry 4.0 and understand the trends of high technology. A valid scale of technological accomplishment for Industry 4.0 is also needed for both industry and academy.

References

Baesu, V., Albulescu, C. T., Farkas, Z.-B., & Drăghici, A. (2015). Determinants of the High-Tech Sector Innovation Performance in the European Union: A Review. *Procedia Technology*, 19, 371–378. doi:10.1016/j.protcy.2015.02.053

Brettel, M., Friederichsen, N., & Kelle, M. (2014). How Virtualization, Decentralization and Network Building Change the Manufacturing Landscape: An Industry 4.0 Perspective. *International Journal of Mechanical, Aerospace, Industrial, Mechatronic and Manufacturing Engineering*, 8(1), 37–44.

Drath, R., & Horch, A. (2014, 618). Industrie 4.0: Hit or Hype? *IEEE Industrial Electronics Magazine*, 8(2), 56–58. doi:10.1109/MIE.2014.2312079

Dutta, S., Narasimhan, O., & Rajiv, S. (1999). Success in High-Technology Markets: Is Marketing Capability Critical? *Marketing Science*, 18(4), 547–568. doi:10.1287/mksc.18.4.547

Ecevit Satı, Z. (2014). The Analysing of High-tech Clusters Potential of Turkey as an Emerging Economy. *IIB International Refereed Academic Social Sciences Journal*, 5(15), 112–132.

Erol, S., Jäger, A., Hold, P., Ott, K., & Sihn, W. (2016). Tangible Industry 4.0: A Scenario-based Approach to Learning for the Future of Production. *Procedia CIRP*, 1–6. doi:10.1016/j.procir.2016.03.162

Gabriel, M., & Pessl, E. (2016). Industry 4.0 and Sustainability Impacts: Critical Discussion of Sustainability Aspects with a Special Focus on Future of Work and Ecological Consequences. *Annals of the Faculty of Engineering Hunedoara*, 14(2), 131–136.

Hatzichronoglou, T. (1997). *Revision of the High-Technology Sector and Product Classification. OECD.* Paris: OECD Publishing. doi:10.1787/134337307632

Hecklau, F., Galeitzke, M., Flachs, S., & Kohl, H. (2016). Holistic Approach for Human Resource Management in Industry 4.0. *Procedia CIRP,* 1–6. doi:10.1016/j.procir.2016.05.102

Hermann, M., Pentek, T., & Otto, B. (2016). *Design Principles for Industrie 4.0 Scenarios.* 49th Hawaii International Conference on System Sciences (pp. 3928–3937). Hawaii, 2016 IEEE. doi:10.1109/HICSS.2016.488

Huarng, K.-H., Yu, T. H.-K., & Lai, W. (2015). Innovation and Diffusion of High-Tech Products, Services, and Systems. *Journal of Business Research,* 68, 2223–2226. doi:10.1016/j.jbusres.2015.06.001

Kagermann, H. (2015). Change through Digitization—Value Creation in the Age of Industry 4.0. In *Management of Permanent Change* (pp. 23–45). Horst Albach, Heribert Meffert, Andreas Pinkwart, Ralf Reichwald. Wiesbaden: Springer. doi:10.1007/978-3-658-05014-6_2

Karahan, Ö. (2015). Intensity of Business Enterprise R&D Expenditure and High-Tech Specification in European Manufacturing Sector. *Procedia—Social and Behavioral Sciences,* 195, 806–813. doi:10.1016/j.sbspro.2015.06.180

Kim, S.-H., & Huarng, K.-H. (2011). Winning Strategies for Innovation and High-Technology Products Management. *Journal of Business Research,* 64, 1147–1150. doi:10.1016/j.jbusres.2011.06.013

Lasi, H., Kemper, H.-G., Fettke, P., Feld, T., & Hoffmann, M. (2014). Industry 4.0. *Business & Information Systems Engineering,* 239–242. doi:10.1007/s12599-014-0334-4

Marre, M., Beihofer, D., & Haggenmueller, W. A. (2015). Forming for Resource-Efficient Industry 4.0. 5th International Conference on Accuracy in Forming Technology 2015, 88. Chemnitz.

Martin, P., & Juergen, W. (2016). Industry 4.0 and Object-Oriented Development: Incremental and Architectural Change. *Journal of Technology Management & Innovation,* 104–110. doi:10.4067/S0718-27242016000200010

OECD. (2011, 77). ISIC Rev. 3 Technology Intensity Definition: Classification of Manufacturing Industries into Categories Based on R&D Intensities. Retrieved from OECD: www.oecd.org/dataoecd/43/41/48350231.pdf

OECD. (2015). *OECD Science, Technology and Industry Scoreboard 2015: Innovation for Growth and Society.* Paris: OECD Publishing. doi:10.1787/sti_scoreboard-2015-en

OECD. (2016). *Society at a Glance 2016: OECD Social Indicators.* Paris: OECD Publishing. doi:10.1787/9789264261488-en

OECD. (2017, 829). Young Population. OECD iLibrary doi:10.1787/3d774f19-en

Oesterreich, T. D., & Teuteberg, F. (2016). Understanding the Implications of Digitisation and Automation in the Context of Industry 4.0: A Triangulation Approach and Elements of a Research Agenda for the Construction Industry. *Computers in Industry*, 121–139. doi:10.1016/j.compind.2016.09.006

Qin, J., Liu, Y., & Grosvenor, R. (2016). A Categorical Framework of Manufacturing for Industry 4.0 and Beyond. *Procedia CIRP*, 173–178. doi:10.1016/j.procir.2016.08.005

Roblek, V., Mesko, M., & Krapez, A. (2016). A Complex View of Industry 4.0. *SAGE Open*, 1–11. doi:10.1177/2158244016653987

Sandu, S., & Ciocanel, B. (2014). Impact of R&D and Innovation on High-Tech Export. *Procedia Economics and Finance*, 15, 80–90. doi:10.1016/S2212-5671(14)00450-X

Saucedo-Martínez, J. A., Pérez-Lara, M., & Marmolejo-Saucedo, J. A. (2017). Industry 4.0 Framework for Management and Operations: A Review. *Journal of Ambient Intelligence and Humanized Computing*, 9(3), 1–13.

Schmidt, R., Möhring, M., Härting, R.-C., Reichstein, C., Neumaier, P., & Jozinović, P. (2015). *Industry 4.0—Potentials for Creating Smart Products: Empirical Research Results. International Conference on Business Information Systems* (pp. 16–27). Poznan: Springer. doi:10.1007/978-3-319-19027-3_2

Schumacher, A., Erol, S., & Sihn, W. (2016). A Maturity Model for Assessing Industry 4.0 Readiness and Maturity of Manufacturing Enterprises. *Procedia CIRP*, 161–166. doi:10.1016/j.procir.2016.07.040

Seliger, G., & Stock, T. (2016). Opportunities of Sustainable Manufacturing in Industry 4.0. *Procedia CIRP*, 40, 536–541. doi:10.1016/j.procir.2016.01.129

Srholec, M. (2007). High-Tech Exports from Developing Countries: A Symptom of Technology Spurts or Statistical Illusion? *Review of World Economics*, 227–255. doi:10.1007/s10290-007-0106-z

United Nations. (2015). *World Population Ageing*. New York: Department of Economic and Social Affairs.

Wang, S., Wan, J., Zhang, D., Li, D., & Zhang, C. (2016). Towards Smart Factory for Industry 4.0: A Self-Organized Multi-Agent System with Big Data Based Feedback and Coordination. *Computer Networks*, 101, 158–168. doi:10.1016/j.comnet.2015.12.017

Wiechoczek, J. (2016). Creating Value for Customer in Business Networks of High-Tech Goods Manufacturers. *Journal of Economics and Management*, 23(1), 76–90.

Workman, J. P. (1998). Factors Contributing to Marketing's Limited Role in Product Development in Many High-Tech Firms. *Journal of Market Focused Management*, 2(3), 257–279.

World Bank. (2015). High-Technology Exports (% of Manufactured Exports). Retrieved 2017 from The World Bank: http://data.worldbank.org/indicator/TX.VAL.TECH.MF.ZS?view=map&year=2015

Wu, C.-W. (2011). Global Marketing Strategy Modeling of High Tech Products. *Journal of Business Research*, 64, 1229–1233. doi:10.1016/j.jbusres.2011.06.028.

Gülay Ekren*, Birgit Oberer and Alptekin Erkollar

Augmented Reality 4.0: Opportunities and Challenges for Smart Factories

1. Introduction

The recent industrial revolution named as Industry 4.0 shapes new digital industrial technologies that especially create disruptive changes in the process of industrial production. The philosophy of Industry 4.0 has produced the smart factory of the future, driven by cyber-physical systems (CPS). CPS refers to the physical production steps of Industry 4.0 that are accompanied by computer-based processes (Schmidt et al., 2015). In accordance with the Industry 4.0 approach, the consolidation of processes and network abilities equipped with latest technology in factory-set (Gorecky, Schmitt, Loskyll & Zühlke et al., 2014) as machines, robots, devices, applications, services, systems, products and people are all interconnected via Internet in a smart way.

The next-generation of the Internet is called as the "Internet of Things" (IoT). IoT represents a universal system that merging overall objects such as computer networks, sensors, actuators and devices via Internet within the bounds of possibility (Fantana et al., 2013). Thus, Industry 4.0 and IoT describe a paradigm shift in production technology (Valdeza et al., 2015). However, they are not only boosting manufacturing productivity, but also shifting economics, creating significant investments, fostering industrial growth and even revenue growth, modifying the profile of the workforce by building up technical capacity and abilities of employees in the development of software systems as well as information and communication systems (Rüßmann et al., 2015), and they also newly create supply-chain management systems, business models and innovation opportunities in a considerable extent (Schmidt et al., 2015). Nowadays, these developments put a new face on mass customization and provide several opportunities especially for manufacturers who are potential investors of technologies coming together with Industry 4.0 (Rüßmann et al., 2015). One of these technologies is called as augmented reality (AR), which provides new opportunities in manually driven manufacturing processes, warehouse picking tasks of logistics area, maintenance

---

* Corresponding author: Gülay Ekren, gekren@sinop.edu.tr; Sinop University, Ayancık Meslek Yüksekokulu, 57400 Ayancık/Sinop, Turkey; +90 368 613 3436 (ext. 6915).

scenarios for reducing complexity as well as training (Büttner et al., 2017). Additionally, AR particularly holds good for educational or training purposes, simulation and guidance of technicians via remote experts (Kans, Galar & Thaduri, 2016). AR supports various additional services, such as extracting relevant information from warehouse and sending them over to mobile devices (Bahrin, Othman, Nor & Azli, 2016). Accordingly, virtual reality (VR) technology can be used in manufacturing companies to carry out product development (Choi, Jung & Noh, 2015), as it provides enactive training for assembly or maintenance tasks or skills (Gavish et al., 2015). Büttner et al. (2017) offer a design space of mixed reality (AR blended with VR) applications to create an auxiliary environment in manufacturing processes and puts forward a taxonomy of existing industrial research projects related to mixed reality.

This study aims to present an overview of requirements for industrial AR ecosystem and to assess the opportunities and challenges of AR in the way to Industry 4.0 as well as the smart factory by introducing the main areas of industrial AR applications.

1.1 Which Fits More Closely with Industry 4.0, AR or VR?

VR, AR or mixed reality converges on the information in real and virtual world from different perspectives. Although VR emphasizes the virtual objects without real environment connection, AR puts real objects and virtual objects together by finding a middle ground (Quint, Sebastian & Gorecky, 2015). VR and AR technologies find way in many applications or activities in industrial environments on the way to Industry 4.0, such as industrial maintenance activities (Aschenbrenner et al., 2016; Rodriguez et al., 2015; Webel et al., 2013), assembly task training (Gavish et al., 2015), product assembly (Ong, Pang & Nee, 2007), manufacturing processes (Novak-Marcincin, Barna, Janak & Novakova-Marcincinova, 2013) as well as assembly guidance, engineering education and the support of planning tasks (Choi, Jung & Noh, 2015).

AR-based or VR-based solutions are critical drivers to get efficiency from processes particularly in manufacturing and management (Capozzi et al., 2014; Stark et al., 2014). However, according to Capozzi, Lorizzo, Modoni and Sacco, AR tools also create opportunities in different areas of industrial and manufacturing sectors with the aims of (1) increasing value-added facilities; (2) minimizing throwbacks coming from workers and suppliers; (3) decreasing duration of orientation and training for workers; (4) decreasing checkup time for navigating activities or fitting tools, parts and supplies; (5) reducing redundant transposition of people or transportation of assets and (6) enhancing job security and providing benefits from officials or employees. Furthermore, Gavish et al. (2015) evaluated the usage of VR and AR platforms to accomplish complicated operations which demand information requirements, procedures and technical capabilities for every single task, such as maintenance of industrial equipment and training of assembly tasks. The results show that AR platforms ensure substantial opportunities especially when compared to traditional training as it mainly pays particular attention to improve cognitive understanding of the tasks.

Gorecky, Schmitt, Loskyll and Zühlke (2014) presented a common interface between the user and CPS generated through the technologies of VR and AR separately. According to this study, although VR presents a factual impact on manufacturing processes, AR creates additional relevant information on the visual sight of employees and so it brings a new dimension to their understandings. Moreover, Aschenbrenner et al. (2016) presented an advanced localization system via a mobile AR architecture. They compared this system with VR-based systems and video systems. They found that an AR system performs significantly better than VR systems as well as video systems. Accordingly, according to the estimation of Digi-Capital (2016), AR/VR could hit $120 billion revenue by 2020, as $90 billion for AR and $30 billion for VR. AR has the majority of market share and grows quickly as shown in Fig. 1.

Fig. 1: Revenue Forecast of AR and VR. Source: Digi-Capital, 2016.

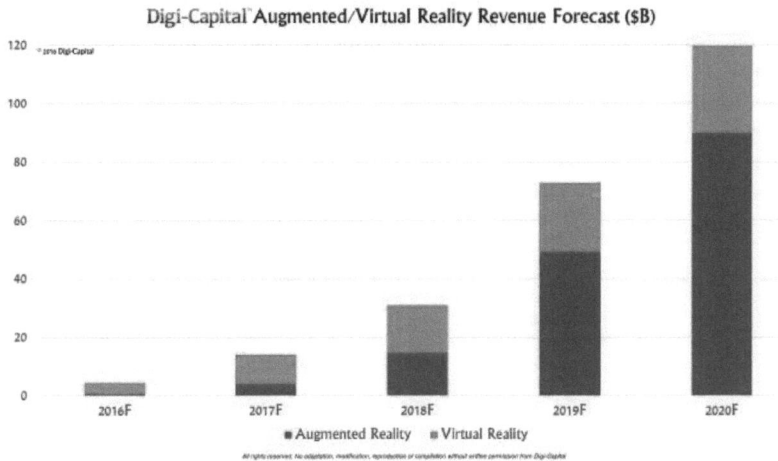

In other respects, Bahrin, Othman, Nor and Azli (2016) highlight the new concept trends in Industry 4.0, as AR is one of the key drivers of Industry 4.0, and also one of the most common technologies transforming industrial production systems, exactly like other Industry 4.0 technologies such as big data analytics, autonomous robots, simulation, cybersecurity, cloud-computing, addictive manufacturing (Rüßmann et al., 2015) and sensor technology (Gilchrist, 2016). Bahrin, Othman, Nor and Azli also define an AR application using a real system that sets apart from data warehouse which transmits instructions to someone in the field using mobile devices or AR glasses. Thus, Industry 4.0 indicates a considerable increase of varied data types, such as scenes, sounds, real-time videos that can be used for triggering maintenance operations (Schmidt et al., 2015).

1.2 The Enterprise AR Ecosystem in a Smart Factory

AR systems help workers in the execution of a task in an industrial environment. In a typical AR ecosystem, creating AR scenes requires a set of cases as follows: (1) The digital content is placed in the user's view by looking at the real world through a camera; (2) through the AR software, it is known how, when and where this image is captured by the camera and a typical AR scene is created; (3) then, AR software covered a number of digital content such as texts, buttons, images, videos and relevant interactions such as instructions, guidance, pop-up windows to the scene and (4) all these processes are provided by a specific content management system that can be made available by the content providers on the user device or can be sent to the user device via web links (Fig. 2).

Fig. 2: Overview of an AR Ecosystem. Source: https://scramboo.com/augmented-reality-explained

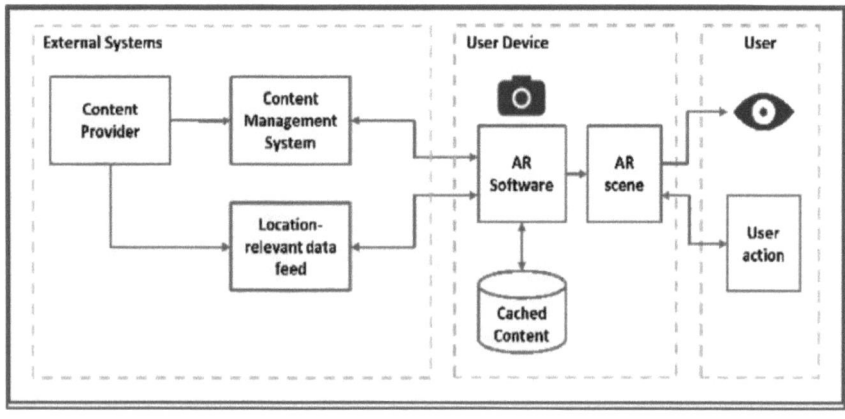

A typical AR view for a smart factory is shown in Fig. 3.

Fig. 3: A Typical AR View for a Smart Factory. Source: https://itunes.apple.com/

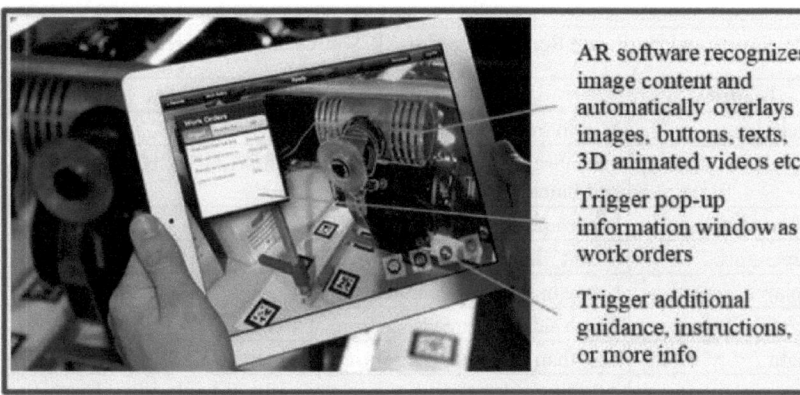

AR is an emerging technology used in many business areas by interacting with customers, employees and on-site workers. There are still several barriers due to the use of AR technologies in smart factories to increase industrial productivity. Mathew (2016) offers four key technologies that will shape the future of this technology as follows: (1) optics—the limited weight, size and power requirements of optical components of the displays; (2) 3D capabilities—the computing power and size requirements of the headsets for 3D; (3) authoring—the weakness in integration with AR software to cover the requirements of various contexts and work environments and (4) interaction—limited human machine interaction including only eye movements, gestures, speech and motion tracking.

The rapid increase in the usage of AR applications across industrial environments resulted in a need for functional requirements with regards to hardware and software. These requirements were initially created through a part of a project between UI (UI stands for University + Industry) LABS and the Digital Manufacturing and Design Innovation Institute. Based on this project, in March 2017, 65 organizations (e.g., Microsoft, Johnson & Johnson, General Electric, Rolls-Royce, Dow Chemical, Intel, the U.S. Air Force, Black & Decker, Newport News Shipbuilding, Stanley, Boeing, Daqri, Upskill, Optech4D, Scope AR, iQagent, Six15, Northrop Grumman, RealWear and others) came together through a workshop to discuss the industrial AR requirements and challenges. This workshop helps to create a shared understanding and language of an AR ecosystem for smart factories (http://thearea.org).

The requirements provided by this project mainly for enterprises and AR providers are classified as hardware and software functional requirements. Functional hardware requirements of a typical AR device are shown in Tab. 1:

Tab. 1: Functional Hardware Requirements of AR. Source: http://thearea.org/

Features	Minimum Hardware Requirements
Battery life	Minimum 12 hours of typical use.
Wireless connectivity	The latest mobile device equipped with Wi-Fi 802.11 standards and Bluetooth.
On-board storage	Minimum 128 GB memory.
Temperature	0–50°C degree of ambient temperature.
Display	Come in sight with ambient light.
Scanning QR code	2 inch sided square visible from at least 5 feet.
Weight	Less than 125 grams (e.g., head-mounted device).
Image target	AR objects do not have to be an image target to be placed in the real-world environment.
Additional features	GPS, accelerometer, three-dimensional design, a wearable Bluetooth device, eye tracking, wireless loudspeaker or microphone, web browser.

Functional requirements of a typical AR software and content generation tool are shown in Tab. 2.

Tab. 2: Functional Software Requirements of AR. Source: http://thearea.org/

Functions	Minimum Software Requirements
Authoring AR content	API links to connect other databases and websites, a user interface, displaying pdf, html, mp4, pptx and animated slideshow files.
Choosing storage medium	From local mobile device, or from web via cloud-based or on-demand server.
Tracking display	It does not depend on a specific style.
Resizing display	Zoom in, zoom out, and 3D rotation.
Collaboration	Allow sharing of images and videos with others.
Taking remote support	Getting directions from an expert beyond, and sharing content with each other.
Allowing footnotes	Communication between users should be open, taking voice call, video call, or file transfer can be made.
Defining job stream	Flow charts can be created to give instructions step by step including 3D videos, images, and texts.

Functions	Minimum Software Requirements
Sending files	QR codes or third party utilities such as documents, images, videos, and 3D videos can be assigned to files.
Logging	AR software can be assigned a note or record to a QR code or any other notification to log into the application for the worker and SME.
Connecting to IoT	Including the third party provided data and also specific locations by creating a localization tags through the use of data on QR code, RFID, and/or Bluetooth. QR codes can be assigned to assets on the production floor.

2. Method

This study is qualitative research involving inductive research processes that guide data analyses and presentation. Inductive research with qualitative design has the potential for generating new concepts and ideas (Gioia, Corley & Hamilton, 2013). During the research process of the study, the relevant literature is examined by categorizing obtained data and reaching inferences. The categories that identify the primary areas of industrial AR applications are explored by authors as manufacturing, logistics, assembly, maintenance, guidance/assistance, training and product creation. Accordingly, the opportunities and challenges when integrating AR into these industrial areas are identified by authors as shown in Tab. 3.

Tab. 3: The Opportunities and Challenges of Industrial AR.

Opportunities	Challenges
Reinforcement	Security and privacy
Cost	Experience
User interfaces	Battery life
Remote troubleshooting	Screen size

3. Findings

3.1 Primary Areas of Industrial AR Applications

AR has the potential to be used in a broad array of industrial activities from many industrial business areas. Some of these areas are identified as follows:

Manufacturing: The use of augmented reality technologies in manufacturing processes named as augmented reality aided manufacturing—such as the programming of manipulation, transferring and warehousing devices, tools, testing, measuring and identifying the parts and assembled product (Novak-Marcincin, Barna, Janak & Novakova-Marcincinova, 2013).

Logistics: The German automaker Volkswagen presents the first use of AR technology and has started using 3D smart glasses for plant logistics by employees at its Wolfsburg factory. This company used an internal camera in the glass as a barcode reader to reduce errors (Sengupta, 2015). Likewise, AR can be used for machine identification systems, for example, via RFID, factory automation systems, wireless sensor networks and other enterprise systems or via on-site data links in the factory of the future linked to each machine or part for their entire lifetime (Berning et al., 2012).

Assembly: A range of studies present background about AR technology employed in support systems within the smart factory environments. The assembly workstation equipped with AR assistance can guide the on-site workers during assembly processes including correct picking as well as skill-sensitive guidance (Mucha, Büttner & Röcker, 2016). Schwerdtfeger and Klinker (2008), as well as Paelke (2014), introduce a head-mounted display to support an assembly process or order picking process. Guo et al. (2014) have tested AR for order picking systems, and they found that 2D graphical representation has more benefits but some more hardware requirements and standby time required to register them for AR. Paelke and Röcker (2015) provide an application in practice of a support system for assembly workers by using augmented reality techniques.

Maintenance: AR tools have the potential for continuous maintenance in Industry 4.0 when integrating context-awareness tools, for example, head-mounted devices, tablets (Zhu, Ong & Nee, 2015). Benbelkacem et al. (2009) present an e-maintenance application on a vision system based on AR concept. This platform, called as ARIMA (stands for Augmented Reality and Image processing in Maintenance Application), provides a solution to technicians during maintenance operation. According to Kans, Galar & Thaduri (2016), AR is the best fit for industrial Internet of Things as Maintenance 4.0. However, this concept has specific benefits not only for technical problems, but also for long-range requirements and socio-economic growth of the organization, smart equipment and processes and even integration of systems as well as system security and interoperability. Fiorentino et al. (2014) present a method to apply AR in industrial maintenance. Accordingly, Webel et al. (2013) worked up on a platform rest upon AR technology that provides training opportunities over maintenance and assembly skills. This platform also examined the efficiency of AR-based training for industrial maintenance activities by making comparison on trainee performance against traditional methods. The findings show that AR is potentially favorable technology for maintenance activities as well as assembly training.

Guidance/assistance: AR devices guide manual processes as well as provide feedback to and from engineers (Pintzos, Rentzos, Papakostas & Chryssolouris, 2014).

Training: AR is a way to utilize human capabilities and allows experts to guide and train maintenance and repair personnel on distance (Kans & Ingwald, 2016). For example, Siemens has developed software that provides a 3D setting with AR glasses to provide training for on-site workers by using operational data and maintenance instructions (Rüßmann et al., 2015). Moreover, AR has the potential to be used in vocational training or workplace-related teaching arrangements. Fehling, Mueller and Aehnelt (2016) introduced an AR application that can be applied in vocational training and on-the-job training situations in operation and maintenance of machines. It can be said that developing training activities and industrial learning tools via AR applications have a very beneficial effect toward the understandings of employees, by reducing mistakes and misunderstandings (Perdikakis, Araya & Kiritsis, 2015).

Product creation: According to Graßler, Taplick and Yang (2016), a product creation process includes various phases to develop a complex product. A learning factory running product creation process has the potential to improve competencies of engineers with problem-solving skills through the practical application of methods, technologies and tools with AR.

3.2 Opportunities and Challenges Faced by Industrial AR Systems

The opportunities when integrating AR into the industrial areas are identified as follows:

Reinforcement: AR has an enormous potential to support industrial processes, as it has widely been used in homogeneous user groups to support workers in industrial scenarios (Grubert et al., 2010).

Cost: AR-based systems are mainly used by manufacturers to improve their maintenance activities and also to reduce costs of having an expert on site (Gilchrist, 2016). However, it should not be forgotten that AR is an expensive technology for the short-term planning (Schwerdtfeger & Klinker, 2008).

User interfaces: User interfaces based on AR paradigm have great potential to interact with CPS, but the development and design process of user interfaces for CPS finds several physical components necessary. Thus, the perceived complexity of information systems by users can be reduced with user interfaces (Paelke & Röcker, 2015). Nowadays, AR contents are usually authoring via desktop user interfaces; however, GUI is required for AR developers to visualize the creation of AR applications (Zhu, Ong & Nee, 2015). The usability and user experience of AR is still a challenge, so it is required to generate a set of validated user interface elements for common applications in industrial area (Paelke, 2014). According to Olshannikova, Ometov, Koucheryavy and Olsson (2015), new approaches, procedures or methods can be developed by software engineers aiming to create the common user interfaces for AR devices.

Remote troubleshooting: AR makes it possible for companies to reduce downtime, minimize the chance of mistakes, and reduce the need for emergency site visits. The capabilities of this technology bring immediate technical expertise on site without calling an engineer (Dutton, 2016).

The challenges when integrating AR into the industrial areas are identified as follows:

Security and privacy: AR applications highlight requirements for security and privacy because data breaches in the virtual world can have consequences in the real world (Baldini et al., 2013). Gaebel, Zhang, Lou and Hou (2016) propose an authentication system for AR combining a wireless signal origin with a user's face location to increase security.

Experience: People who are inexperienced with AR need more time for using the AR-based system efficiently and to become comfortable with AR (Schwerdtfeger & Klinker, 2008). Therefore, users need to be educated and trained on this evolving technology (Olshannikova, Ometov, Koucheryavy & Olsson, 2015).

Battery life: The battery lifetime of AR devices has severely limited their use. Berning et al. (2012) introduce an AR-based system that has critical performance aspects; for example, a limited battery capacity of mobile devices is a crucial aspect for any video-based tracking.

Screen size: Mobile devices or wearable devices such as iPod Touch, tablets, smartphones, AR glasses, or other devices using to develop AR applications usually provide restricted information just because of their screen size. Their screens are frequently pretty uncomfortable to read, especially when it contains many or various complicated information (Pintzos,

Rentzos, Papakostas & Chryssolouris, 2014). Because of their limited screen size, users experience lack of comfort while working with them (Olshannikova, Ometov, Koucheryavy & Olsson, 2015).

4. Discussion and Conclusions

In recent years, Industry 4.0 and the IoT have become one of the most popular concepts in the industrial business areas. These concepts generally create disruptive changes in production/manufacturing, maintenance and assistance systems as well as training activities of industrial environments. However, to enable the full potential of industry 4.0, new technologies, processes and tools need to be developed, allowing for innovative processes. One of these new technologies is AR. Enterprises are increasingly looking for cost-effective maintenance and training activities to support innovative processes during manufacturing, marketing, or after-sales. They mainly need a digital transformation to reshape their businesses or industries. Therefore, enterprises have to adopt these changes in concern with their capacity, motivation, intent and purpose, priority and financial power. However, there is not only one road map that fits all businesses or industries (Gilchrist, 2016) to adopt this digital revolution. Nowadays, AR technology is an important aspect that leads on the flexible benefits for companies in this end: employ less-skilled operations, using up-to-date data, time and cost saving, error rate reduction, registered multimedia content, retaining knowledge in the system not only in people but also adopting a different level of information to the user's skills.

In the present chapter, the enterprise AR ecosystem in a smart factory as well as functional hardware and software requirements to create a typical AR application in these environments are identified. Then, the primary industrial areas (manufacturing, assembly, logistics, training, maintenance, product creation) of AR applications that can be integrated into smart factories are introduced. Taking into account of these areas, opportunities and challenges faced by industrial AR applications are summarized in contexts such as reinforcement, cost, battery, user interfaces, experience, security and privacy and remote troubleshooting. AR technology will continue to advance over the next three years and is supposed to provide a huge benefit to companies adopting Industry 4.0. Nowadays, several companies are already beginning to take advantage of AR. In the following years, AR equipped handheld devices, wearable devices or mobile devices will be broadly used and these devices may be seamlessly integrated into the everyday practices of industrial enterprises. This study provides qualitative research for academics and practitioners and should offer them a roadmap for prioritizing their steps toward Industry 4.0.

References

Aschenbrenner, D., Maltry, N., Kimmel, J., Albert, M., Scharnagl, J., & Schilling, K. (2016). ARTab-using Virtual and Augmented Reality Methods for an Improved Situation Awareness for Telemaintenance. *IFAC-PapersOnLine*, 49(30), 204–209.

Bahrin, M. A. K., Othman, M. F., Nor, N. H., & Azli, M. F. T. (2016). Industry 4.0: A Review on Industrial Automation and Robotic. *Jurnal Teknologi (Sciences & Engineering)*, 78(6–13), 137–143.

Baldini, G., Peirce, T., Handte, M., Rotondi, D., Gusmeroli, S., Piccione, S., … & Serbanati, A. (2013). Internet of Things Privacy, Security and Governance. In *Internet of Things–Converging Technologies for Smart Environments and Integrated ecosystems*, Vermesan, O. & Friess, P. (Eds.), River Publishers, Aalborg, 207–224.

Berning, M., Riedel, T., Karl, D., Schandinat, F., Beigl, M., & Fantana, N. (2012, June). Augmented Service in the Factory of the Future. *Networked Sensing Systems (INSS)*, 2012 Ninth International Conference, IEEE. In 2012 Ninth International Conference on Networked Sensing (INSS) (pp. 1–2), June 11–14, Antwerp, Belgium, IEEE Computer Society: Washington, DC, USA. Doi: 10.1109/INSS.2012.6240571.

Büttner, S., Mucha, H., Funk, M., Kosch, T., Aehnelt, M., Robert, S., & Röcker, C. (2017, June) The Design Space of Augmented and Virtual Reality Applications for Assistive Environments in Manufacturing: A Visual Approach. *In Proceedings of the 10th International Conference on PErvasive Technologies Related to Assistive Environment (pp. 433–440), June 21–23, Island of Rhodes, Greece, ACM Press: New York, USA. Doi: http://dx.doi. org/10.1145/3056540.3076193.*

Benbelkacem, S., Zenati-Henda, N., Belhocine, M., & Malek, S. (2009, march). Augmented Reality System for e-Maintenance application. In *AIP Conference Proceedings*, 1107(1), 185–189 AIP. Doi: https://doi.org/10.1063/1.3106470.

Capozzi, F., Lorizzo, V., Modoni, G., & Sacco, M. (2014, September). Lightweight Augmented Reality Tools for Lean Procedures in Future Factories. In *International Conference on Augmented and Virtual Reality* (pp.232–246).), September 17–20, Lecce, Italy, Springer, Cham. Doi: https://doi.org/10.1007/978-3-319-13969-2_18.

Choi, S., Jung, K., & Noh, S. D. (2015). Virtual Reality Applications in Manufacturing Industries: Past Research, Present Findings, and Future Directions. *Concurrent Engineering*, 23(1), 40–63.

Digi-Capital (2016). Augmented/Virtual Reality revenue forecast revised to hit $120 billion by 2020. Retrieved from https://www.digi-capital.com/news/2016/01/augmentedvirtual-reality-revenue-forecast-revised-to-hit-120-billion-by-2020/#.Wm3icqhl_IU Access on October 5, 2018.

Dutton, G. (2016). Augmented Reality Transforms Biopharma. *Focus*, 1, 15–25.

Fantana, N. K., Riedel, T., Schlick, J., Ferber, S., Hupp, J., Miles, S., ... & Svensson, S. (2013). IoT applications—value creation for industry. Internet of Things: Converging Technologies for Smart Environments and Integrated Ecosystems, 153–206. River Publisher: Aalborg, Denmark.

Fehling, C. D., Mueller, A., & Aehnelt, M. (2016, October). Enhancing Vocational Training with Augmented Reality. In Proceedings of the 16th International Conference on Knowledge Technologies and Data-driven Business, October 18–19, Graz, Austria. ACM Press: New York, USA.

Fiorentino, M., Uva, A. E., Gattullo, M., Debernardis, S., & Monno, G. (2014). Augmented Reality on Large Screen for Interactive Maintenance Instructions. *Computers in Industry*, 65(2), 270–278.

Gaebel, E., Zhang, N., Lou, W., & Hou, Y. T. (2016, October). Looks Good to Me: Authentication for Augmented Reality. In Proceedings of the 6th International Workshop on Trustworthy Embedded Devices (pp. 57–67), October 24–28, Vienna, Austria, ACM Press: New York, NY, USA. Doi:10.1145/2995289.2995295

Gavish, N., Gutiérrez, T., Webel, S., Rodríguez, J., Peveri, M., Bockholt, U., & Tecchia, F. (2015). Evaluating Virtual Reality and Augmented Reality Training for Industrial Maintenance and Assembly tasks. *Interactive Learning Environments*, 23(6), 778–798.

Gilchrist, A. (2016). Industry 4.0. *The Industrial Internet of Things*. Apress, New York.

Gioia, D. A., Corley, K. G., & Hamilton, A. L. (2013). Seeking Qualitative Rigor in Inductive Research: Notes on the Gioia Methodology. *Organizational Research Methods*, 16(1), 15–31.

Gorecky, D., Schmitt, M., Loskyll, M., & Zühlke, D. (2014). Human-Machine-Interaction in the Industry 4.0 Era. In 12th IEEE International Conference on Industrial Informatics (INDIN) (pp. 289–294), July 27–30, Porto Alegre, Brazil. IEEE Computer Society: Washington, DC, USA. Doi: 10.1109/INDIN.2014.6945523

Gräßler, I., Taplick, P., & Yang, X. (2016). Educational learning factory of a holistic product creation process. Procedia CIRP, 54, 141–146.

Grubert, J., Hamacher, D., Mecke, R., Böckelmann, I., Schega, L., Huckauf, A., ... & Tümler, J. (2010). Extended Investigations of User-Related Issues in Mobile Industrial AR. . In 9th IEEE International Symposium on Mixed and Augmented Reality (ISMAR) (pp. 229–230), October 13–16, Seoul, South Korea. IEEE Computer Society: Washington, DC, USA. DOI: 10.1109/ISMAR.2010.5643581

Guo, A., Raghu, S., Xie, X., Ismail, S., Luo, X., Simoneau, J., ... & Starner, T. (2014). A Comparison of Order Picking Assisted by Head-Up Display (HUD), Cart-Mounted Display (CMD), Light, and Paper Pick List. In Proceedings of the 2014 ACM International Symposium on Wearable Computers (pp. 71–78), September 13 – 17, Seattle, Washington, ACM Press: NY, USA.

Kans, M., & Ingwald, A. (2016). Business Model Development Towards Service Management 4.0. *Procedia CIRP*, 47, 489–494.

Kans, M., Galar, D., & Thaduri, A. (2016). Maintenance 4.0 in Railway Transportation Industry. In: Koskinen K. et al. (eds) Proceedings of the 10th World Congress on Engineering Asset Management (WCEAM 2015) (pp. 317– 331). Lecture Notes in Mechanical Engineering. Springer, Cham. Doi:https://doi.org/10.1007/978-3-319-27064-7_30

Mucha, H., Büttner, S., & Röcker, C. (2016). Application Areas for Human-Centered Assistive Systems. In 2nd Workshop on Human-Computer Interaction Perspectives on Industry 4.0, co-located with i-KNOW 2016, October 11–12, Graz, Austria.

Novak-Marcincin, J., Barna, J., Janak, M., & Novakova-Marcincinova, L. (2013). Augmented Reality Aided Manufacturing. *Procedia Computer Science*, 25, 23–31.

Olshannikova, E., Ometov, A., Koucheryavy, Y., & Olsson, T. (2015). Visualizing Big Data with Augmented and Virtual Reality: Challenges and Research Agenda. *Journal of Big Data*, 2(1), 22.

Ong, S. K., Pang, Y., & Nee, A. Y. C. (2007). Augmented Reality Aided Assembly Design and Planning. *CIRP Annals-Manufacturing Technology*, 56(1), 49–52.

Paelke, V. (2014). Augmented Reality in the Smart Factory: Supporting Workers in an Industry 4.0 Environment. In Emerging Technology and Factory Automation (ETFA) (pp. 1–4), September 16–19, Barcelona, Spain. IEEE Computer Society: Washington, DC, USA. DOI: 10.1109/ETFA.2014.7005252

Paelke, V., & Röcker, C. (2015). User Interfaces for Cyber-Physical Systems: Challenges and Possible Approaches. In Marcus A. (eds) Design, User Experience, and Usability: Design Discourse. DUXU 2015: Design, User Experience, and Usability: Design Discourse (pp. 75–85). Lecture Notes in Computer Science, vol 9186. Springer, Cham. DOI: https://doi.org/10.1007/978-3-319-20886-2_8

Perdikakis, A., Araya, A., Kiritsis, D. (2015). Introducing Augmented Reality in Next-Generation Industrial Learning Tools: A Case Study on Electric and Hybrid Vehicles. *Procedia Engineering*, 132, 251–258.

Pintzos, G., Rentzos, L., Papakostas, N., Chryssolouris, G. (2014). A Novel Approach for the Combined Use of AR Goggles and Mobile Devices as Communication Tools on the Shopfloor. *Procedia CIRP*, 25, 132–137.

Mathew, S. (2016). 8 drivers that will shape the future of virtual/augmented reality. Retrieved October 5, 2018 from https://yourstory.com/2016/07/future-virtual-augmented-reality/

Rodriguez, L., Quint, F., Gorecky, D., Romero, D., Siller, H. R. (2015). Developing a Mixed Reality Assistance System Based on Projection Mapping Technology for Manual Operations at Assembly Workstations. *Procedia Computer Science*, 75, 327–333.

Rüßmann, M., Lorenz, M., Gerbert, P., Waldner, M., Justus, J., Engel, P., Harnisch, M. (2015). Industry 4.0: The Future of Productivity and Growth in Manufacturing Industries. *Boston Consulting Group*, : Boston, USA. Retrieved October 5, 2018 from http://www.inovasyon.org/pdf/bcg.perspectives_Industry.4.0_2015.pdf

Schmidt, R., Möhring, M., Härting, R. C., Reichstein, C., Neumaier, P., Jozinović, P. (2015). Industry 4.0-Potentials for Creating Smart Products: Empirical Research Results. In: Abramowicz W. (eds) Business Information Systems. BIS 2015: Business Information Systems (pp. 16–27). Lecture Notes in Business Information Processing, vol 208. Springer, Cham. DOI: https://doi.org/10.1007/978-3-319-19027-3_2

Schwerdtfeger, B., & Klinker, G. (2008). Supporting Order Picking with Augmented Reality. *In Proceedings of the 7th IEEE/ACM international Symposium on Mixed and Augmented Reality (pp. 91–94), September 15–18, IEEE Computer Society: Washington, DC, USA.*

Sengupta, R. (2015). International Business Times, November 27, 2015,). Volkswagen factory workers using augmented reality tech and 3D smart glasses, International Business Times on November 27, 2015. Retrieved October 5, 2018 from http://www.ibtimes.com.au/volkswagen-factory-workers-using-augmented-reality-tech-3d-smart-glasses-1487284

Stark, R., Grosser, H., Beckmann-Dobrev, B., Kind, S., INPIKO Collaboration. (2014). Advanced Technologies in Life Cycle Engineering. *Procedia CIRP*, 22, 3–14.

Quint, F., Sebastian, K., & Gorecky, D. (2015). A mixed-reality learning environment. Procedia Computer Science, 75, 43–48.

Valdeza, A. C., Braunera, P., Schaara, A. K., Holzingerb, A., Zieflea, M. (2015). Reducing Complexity with Simplicity-Usability Methods for Industry 4.0. In Proceedings of 19th Triennial Congress of the IEA, August 9–14, Melbourne, Australia. Retrieved October 5, 2018 from http://ergonomics.uq.edu.au/iea/proceedings/Index_files/papers/1288.pdf

Zhu, J., Ong, S. K., Nee, A. Y. (2015). A Context-Aware Augmented Reality Assisted Maintenance System. *International Journal of Computer Integrated Manufacturing*, 28(2), 213–225.

Webel, S., Bockholt, U., Engelke, T., Gavish, N., Olbrich, M., Preusche, C. (2013). An Augmented Reality Training Platform for Assembly and Maintenance Skills. *Robotics and Autonomous Systems*, 61(4), 398–403.

Umut Şener*, Ebru Gökalp and P. Erhan Eren

Toward A Maturity Model for Industry 4.0: A Systematic Literature Review

1. Introduction

The term Industry 4.0 was first used at the 2011 Hannover Fair to refer to the German Federal Government's 2020 development plan for advanced technology, and was further described as a new revolution of today's industry aiming a fully digital transformation in production (Kagermann, Wahlster & Held, 2012).

The first industrial revolution emerged through the employment of hydro energy and steam power in production. The second industrial revolution is related to production automation and mass production with electrical energy. As for the third Industrial Revolution, it includes the transformation from analog to digital technology by using Information and Communication Technologies (ICT) in production. The fourth industry revolution, called Industry 4.0, is predicted to be a period of industrialization that will bring about radical changes in industrial production, with the use of Cyber-Physical Systems (CPS), which combine the Internet of Things (IoT) with the production ecosystem (Shrouf, Ordieres & Miragliotta, 2014). Within the scope of this revolution, it is aimed that physical processes are monitored by CPS, which create a copy of the physical world in a virtual environment, where processes and systems are interconnected, and they have intelligent and self-controlled structure (Reiner, 2014). It is anticipated that this revolution provides manufacturing companies the capability to achieve increased efficiency, transparency and flexibility, to reduce costs and to create opportunities for the development of innovative and value added services and business models (Burmeister, Luettgens & Piller, 2015; Gökalp et al., 2016; Kagermann, Wahlster & Held, 2012; Lemke & Brenner, 2014; Porter & Heppelmann, 2014; Posada et al., 2015).

With the Industry 4.0 revolution, enterprises have begun to search for possible ways to restructure and improve their business processes in order to adapt to competitive markets and stay in the market. For this purpose, enterprises have

---

*   Corresponding author: Umut Şener, METU Informatics Institute, Ankara/Turkey, 0312 210 6872.

started to develop corporate strategies within the scope of Industry 4.0 by evaluating basic business structures. They try to re-shape their processes based on new technologies of Industry 4.0 in order to have competitive advantages. Because of the fact that Industry 4.0 is still in the initial stages, it is essential to describe implementation guidelines for Industry 4.0 to guide enterprises which are transitioning to Industry 4.0 applications for improving their capabilities.

The Maturity Model (MM) is a framework based on the successful implementation of organizational capabilities, and includes a steady stream of objectives and sequential levels or stages. The basic assumption of the use of MM is the higher level of maturity, the greater business perfection, and the better operations that the organization performs. The MM approach is commonly employed to determine the current maturity level and generate a roadmap that should be followed to move to the next level (Röglinger, Pöppelbuß & Becker, 2012). Therefore, MM is selected as the baseline framework for this study.

In the scope of the study, a systematic literature review is conducted in order to identify similar existing studies and also, existing maturity models are determined and comprehensively analyzed. According to the systematic literature review, it is concluded that the maturity of Industry 4.0 applications are not fully explored and there is a need for a MM in the context of the fourth industry revolution. Therefore, Industry 4.0-MM is presented with the aim of satisfying this need, and it has been developed to provide guidance on what should be done for the improvement of the current level by assessing the capabilities of enterprises within the context of Industry 4.0 in a standard, consistent, repeatable and appropriate manner.

In this chapter, first, the research method that forms the basis of the study is explained. Secondly, the findings of the systematic literature review are summarized. Accordingly, the main structure of the proposed maturity model is explained. Finally, the concluding remarks are presented.

2. Method

A systematic literature review, as proposed by Kitchenham (2004), is followed to identify similar studies. The followed steps in the literature review process and their detailed description are stated in Tab. 1.

Tab. 1: The Systematic Review Steps.

Steps	Explanation
Starting point	The research topic was selected.
Language	The search language was selected as English.
Search terms	The keywords of the search were the terms "Industry 4.0", "Industry Internet of Things", "Industry Internet", "Industrial Internet", "Cloud-based Manufacturing", "Digitization", "Smart Manufacturing", "Cyber-physical systems", "Smart Factory", "Ubiquitous Manufacturing" AND "Maturity Model", etc.
Databases	The systematic review was conducted on Scopus (www.scopus.com/search/form.url), Aisel (www.aisel.aisnet.org) and Web of Science (http://apps.webofknowledge.com/).
Checking the reference list	88 studies were obtained from two databases (34 in Scopus, 36 in Aisel and 18 in Web of Science) after applying the search query. Furthermore, references of these articles were reviewed as well.
Citation search	Journals indexed with SSCI, SCI and AIS were identified through the search. Series, meetings and reviews were excluded(Conference proceedings were included because there were only a limited number of journals that investigates maturity models in the context of the fourth industry revolution.).
Management of results	The search results and findings were stored in a database.
Selection of primary studies	Initial exclusion: Keywords, titles and abstracts of the search result were examined for identifying the relevance of the articles before fully reading them.
	Year: Papers with publication years after 2000 were included. After applying these steps, 18 studies remained.
Study quality assessment	18 studies were reviewed and 7 of them investigate the MM for Industry 4.0. 11 of the studies are mainly related to Industry 4.0 or MMs applied in the IT sector, but not specifically in the context of Industry 4.0, and they only contribute to the phase of defining the dimensions of the proposed MM. Thus, the remaining 18 studies (primary studies) are considered as the baseline for this study.
Data synthesis	In the light of the primary studies, a MM for Industry 4.0 is constructed. The MM was reviewed by 3 experts who are knowledgeable about information systems. The MM was discussed through a series of meetings conducted for eliminating conflicts. Finally, they reached an agreement on the dimensions of the proposed model and its structure.

As a result of the systematic literature review, seven existing MMs are identified. In order to objectively analyze these MMs, the evaluation criteria are defined, as described in Tab. 2. Then, the strengths and weaknesses of each MM are systematically assessed. Thus, the sufficiency of existing models, the applicability in the context of the Industry 4.0, and the need for a new MM are questioned.

Tab. 2: Assessment Criteria for Gap Analysis.

Criteria #	Criteria Name	Definitions
C1	Research context	The research context of the study.
C2	Fitness for purpose	The level of suitability of the study in terms of the maturity assessment of Industry 4.0.
C3	Completeness of aspects	The level of completeness of aspects in terms of addressing all or a subset of major aspects in the context of Industry 4.0
C4	Number of maturity levels	The number of maturity levels that the MM has.
C5	Definitions of maturity levels	The level of completeness of explanations regarding the maturity levels.
C6	Number of dimensions	The number of maturity dimensions that the MM has.
C7	Definitions of dimensions	It questions if the corresponding MM provides the description of the maturity dimensions.
C8	Granularity of dimensions	The degree of granularity of the maturity dimensions.
C9	Definition of measurement attributes	The level of completeness of explanations regarding the measurement attributes.
C10	Description of assessment method	The level of completeness of explanations regarding the evaluation approach of the corresponding MM.
C11	Objectivity of the assessment method	The degree of objectivity of the evaluation approach of the MM. The descriptions of the attributes, practices and each maturity level should be stated unambiguously. And the overall maturity level should correctly reflect the number of positively answered questions.
C12	The method of the study	It questions whether the study is empirical or conceptual.
C13	Origin of the study	It questions whether the study is Academia-based (Acd.) or Practioner-based (Prc.)
C14	Access	It questions if the study is accessible for free or paid.

Although there are no standard or published evaluation criteria for the scope of Industry 4.0, there are some standards related to constructing a MM such as Software Process Improvement and Capability Determination (SPICE) (ISO/IEC 33004:2015, 2016) and Capability Maturity Model Integration (CMMI) (Team, 2006). Rout, Tuffley & Cahill (2001) criticize "the purpose, the scope, the elements and the indicators" of CMMI and mapping capability of CMMI with SPICE and maturity results' verifiability based on *completeness-clearness-unambiguity* criteria. SPICE (ISO/IEC, 2003) defines the purpose of the part of The Assessment of Organizational Maturity as "ensuring that the results are *objective, impartial, consistent, repeatable, comparable, and representative* of the assessed organizational units". Consequently, the assessment criteria, described in Tab. 2, are identified as compatible with these descriptions. Then, each MM is assessed based on those criteria.

Subsequently, the strengths and weaknesses of existing studies are identified in a systematic way. The analysis of the primary studies was conducted by experts with knowledge in the field of Industry 4.0 and MMs. First, each expert independently assessed the current MMs according to the criteria listed in Tab. 2. Subsequently, a meeting was held among the researchers and a consensus was obtained on the conflicting views. At the end of the meeting, the results were briefly summarized, as shown in Tab. 3.

The qualification of the criteria, such as Definitions of dimensions and Granularity of Dimensions are stated by using four levels corresponding to the score that symbolizes the degree of achievement of the criteria. "NA" that stands for "Not Achieved" refers to the degree of achievement from 1 % to 15 %; while "PA" is the abbreviation of "Partially Achieved" and means that the extent of achievement is from 16 % to 50 %; similarly "LA" represents "Largely Achieved," which symbolizes the achievement degree from 51 % to 85 %. If the criterion has the degree of achievement between 86 % and 100 %, it is stated as "FA," meaning "Fully Achieved".

3. Findings

As a result of the literature review and analysis of existing studies, the findings are as follows:

- **MM1:** The Connected Enterprise Maturity Model (Rockwell automation, 2014) examines the technological readiness of organizations. The evaluation is based on 4 dimensions; but there is no detailed explanation of dimensions and related items.

- **MM2:** IMPULS – Industrie 4.0 Readiness (Lichtblau et al., 2015) presents an evaluation method with six dimensions for assessing the readiness for Industry 4.0. IMPULS presents an action plan to increase readiness for Industry 4.0. However, according to the evaluation method of this model, if an organization in the same sector performs the IMPULS evaluation, the maturity level of the rival is defined; otherwise the rivals are discarded. Therefore, the evaluation method should be improved for evaluating the performance of rivals.
- **MM3:** Empowered and Implementation Strategy for Industry 4.0 (Lanza et al. 2016) presents a set of strategies for companies. The study covers a MM, but there is no information about the MM presented.
- **MM4:** PricewaterhouseCoopers Company (PricewaterhouseCoopers, 2016) offers an application tool called Industry 4.0/Digital Operations Self-Assessment. It provides a web-based evaluation tool for identifying maturity levels of Industry 4.0 applications of the interested companies, but this model includes six dimensions and investigates Digital Readiness for the fourth industrial revolution. The creation process of dimensions is not shared with the users.
- **MM5:** A maturity model Industry 4.0 Readiness (Schumacher, Erol & Sihn, 2016) suggests eight dimensions for evaluation. The evaluation method is based on the scoring of each dimension using the Likert scale. While the method is simple, this model does not provide an action plan that improves the maturity level of the company.
- **MM6:** Toward a Maturity Model for Industrial Internet (Menon, Kärkkäinen & Lasrado, 2016) is developed based on Mettler's template (Mettler, 2009). They present a design framework for maturity model development. Since the study has not been completed, the recommended MM has not been published.
- **MM7:** SIMMI 4.0 (Leyh, Schäffer, Bley & Forstenhäusler, 2016; 2017) provides a framework that evaluates the organizational capabilities in terms of the software and technology. However, organizational dimensions (i.e., personnel, corporate strategy, etc.) and environmental dimensions (competitors, market structure, etc.) are ignored in evaluating the maturity level.

Tab. 3: Analysis of Existing MMs in the Context of Industry 4.0.

MM#	C1	C2	C3	C4	C5	C6	C7	C8	C9	C10	C11	C12	C13	C14
MM1	IT readiness	NA	PA	-	NA	4	NA	NA	NA	NA	NA	Both	Prc.	Free
MM2	Industry 4.0 readiness	PA	PA	5	LA	6	FA	PA	LA	FA	LA	Both	Acd.	Paid
MM3	Implementation strategies of Industry 4.0	NA	NA	-	NA	-	-	NA	NA	NA	NA	Conc.	Acd.	Free
MM4	Digital readiness for Industry 4.0	PA	PA	5	FA	6	PA	PA	PA	NA	PA	Both	Prc.	Paid
MM5	Industry 4.0 maturity	PA	PA	5	PA	8	LA	PA	PA	PA	PA	Both	Acd.	Free
MM6	Industrial internet maturity	PA	NA	5	-	-	-	-	-	NA	NA	Conc.	Acd.	Free
MM7	Industry 4.0 maturity	PA	PA	5	FA	3	LA	PA	PA	LA	PA	Conc.	Acd.	Free

These existing MMs are analyzed according to the assessment factors given in Tab. 2 to assess their strengths and weaknesses systematically. The assessment result is given in Tab. 3. As an example, for the assessment of MMs, MM7 is evaluated based on the assessment criteria as follows. The model investigates the maturity for the success of Industry 4.0, but only focuses on the software and technological aspects of the organizations that apply the digital transformation. Thus, C2 is stated as PA that means the suitability for the research objective is partially achieved (PA), similarly C3 is rated as PA that means completeness of aspect is below 86 % achievement, since the model ignores the organizational aspect for the assessment. Furthermore, the model has 5 maturity levels, and the detailed description of the maturity levels is provided. Therefore, C4 is stated as 5, and C5 is represented as fully achieved (FA) which means descriptions of maturity levels is provided to the extent of the achievement between 86 % and 100 %. There is no satisfying achievement from C7 to C11; thus, these criteria are rated as NA for MM1, which is a Practitioner-based, both Empirical and Conceptual, and free model.

As a result of the systematic literature review, it has been determined that there is a limited number of studies regarding the development of MM for Industry 4.0, so there is a lack of research in this domain, even though it seems that there is a research stream for investigating the subject in recent years. Moreover, none of the existing MMs for Industry 4.0 in the literature fully satisfy (Fully Achieved [FA]) the evaluation criteria. For this reason, our aim is to develop an MM for Industry 4.0 that fills this research gap.

3.1 The Proposed Industry 4.0-Maturity Model: Industry 4.0-MM

Although there are many studies on MMs for Information Technologies, there are only a few MMs for the manufacturing industry. Product-Service Systems (Neff et al., 2014), Product Life Cycle Management (PLM) (Vezzetti, Violante & Marcolin, 2014) are analyzed in detail. These models generally examine success factors in production processes, and it has been determined that they do not specifically target the manufacturing industry. There are also some studies that provide a template for the development of a MM such as Mettler's template (Mettler, 2009), ADR (Sein et al., 2006) and TOGAF (TOGAF, 2011). Today's best-known MMs are software-specific and process-oriented standards such as ISO/IEC 15504-SPICE and CMMI-DEV (Team, 2006).

The benefits provided by these models are as follows: increased predictable quality and productivity; increased performance; decreasing error; increased employee satisfaction, more staff involvement; increased return on investment and increased customer satisfaction (Goldenson & Gibson, 2003). Since they offer many benefits, the software industry-specific maturity models have been adapted to other sectors, such as Automotive-SPICE (Automotive SIG., 2010), MediSPICE (M. SPICE, 2011), Enterprise SPICE (E. SPICE, 2010), Gov-PCDM (Gökalp & Demirors, 2017).

Along with the increasing development of models based on standards such as SPICE and CMMI, the ISO/IEC 15504 model has been updated to ISO 3300x series that has a structure covering 16 different standards including guidance on how to develop a MM. Thus, the MM development process is intended to have a standard structure. Therefore, the proposed MM, entitled as Industry 4.0-MM, is developed based on the SPICE standard, which provides a general framework for evaluating and improving process characteristics. The model structure is constructed according to the ISO/IEC 33000-SPICE (ISO/IEC 33004:2015, 2016) standard, since it provides widely accepted guidelines on how to design a MM structure.

Industry 4.0-MM focuses on the evaluation of enterprise capabilities in the domain of Industry 4.0. As shown in Fig. 1, Industry 4.0-MM consists of dimensions and maturity levels of each dimension. The dimensions are defined as Asset Management, Data Governance, Application Management, Process Transformation and Organizational Alignment, and each dimension is expressed by 6 levels which are adapted from the ISO / IEC 3300x series. The detailed explanations regarding the aspect and capability dimensions of the Industry 4.0-MM are given in (Gökalp, Şener & Eren, 2017).

Fig. 1: The Proposed Industry 4.0 Maturity Model.

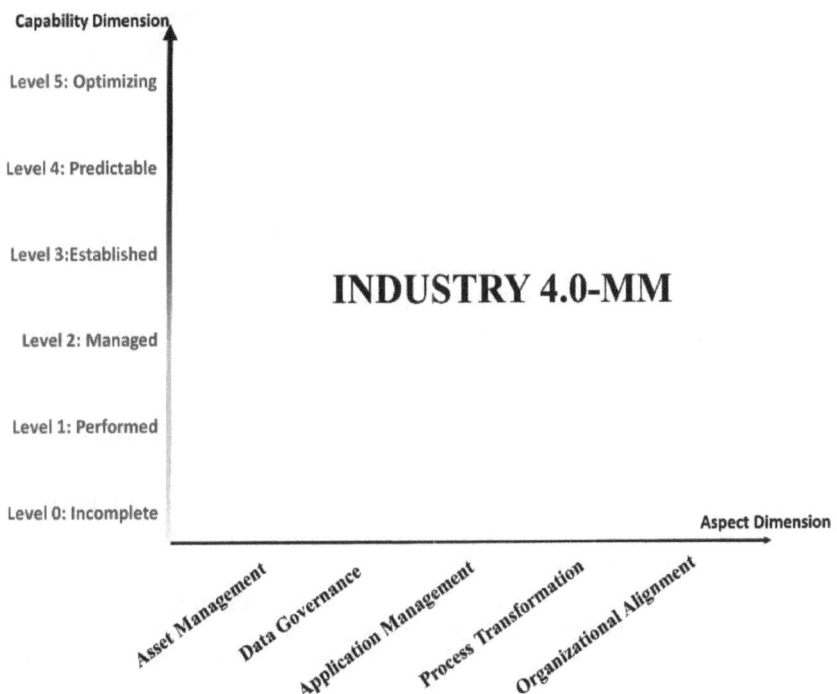

4. Conclusion

In this study, a systematic literature review is carried out on MMs for Industry 4.0. The findings of this review are analyzed and existing studies are examined according to a set of criteria, such as scope, purpose, completeness, clearness and objectivity. Accordingly, the suitability of existing MMs and their applicability within Industry 4.0 have been questioned. The results indicate that there is a need for a MM for Industry 4.0 applications. In order to satisfy this need, a new MM named as Industry 4.0-MM is presented.

The aim of Industry 4.0-MM is to help organizations in their transition to the Industry 4.0 environment by providing a guideline regarding how to pursue the digital transformation in a standardized, objective and repeatable way. The model measures the level of maturity of existing capabilities within the organizations, and accordingly provides a road-map to improve each capability in the existing situation.

As future work, the attributes of maturity dimensions and the corresponding action list for each maturity level of the Industry 4.0-MM will be defined. In addition, an exploratory case study is planned for demonstrating the applicability of the model.

References

Automotive SIG. (2010). The SPICE User Group Automotive Special Interest Group, Automotive SPICE Process Reference Model. Retrieved 20.02.2018 from http://www.automotivespice.com

Burmeister, C., Luettgens, D., & Piller, F. T. (2015). Business Model Innovation for Industrie 4.0: Why the "Industrial Internet" Mandates a New Perspective on Innovation. Die Unternehmung, 70(2), 124–152.

Enterprise SPICE. (2010). An Integrated Model for Enterprise-wide Assessment and Improvement (No. 1, p. 183). Technical Report. Retrieved 20.02.2018 from http://enterprisespice.com

Gökalp, E., & Demirörs, O. (2017). Model Based Process Assessment for Public Financial and Physical Resource Management Processes. *Computer Standards & Interfaces*, 54, 186–193.

Goldenson, D., & Gibson, D. L. (2003). Demonstrating the Impact and Benefits of CMMI: An Update and Preliminary Results. (No. CMU/SEI-2003-SR-009). CARNEGIE-MELLON UNIV PITTSBURGH PA SOFTWARE ENGINEERING INST.

Gökalp, M. O., Kayabay, K., Akyol, M. A., Eren, P. E., & Koçyiğit, A. (2016, December). Big Data for Industry 4.0: A Conceptual Framework. In *Computational Science and Computational Intelligence (CSCI), 2016 International Conference on* (pp. 431–434). IEEE. Las vegas, NV, USA, 15–17 December 2016.

Gökalp, E., Şener, U., & Eren, P. E. (2017, October). Development of an Assessment Model for Industry 4.0: Industry 4.0-MM. In *International Conference on Software Process Improvement and Capability Determination* (pp. 128–142). SPICE 2017, Palma de Mallorca, Spain, October 4–5, 2017.

Haren, V. (2011). TOGAF Version 9.1 10th. Van Haren Publishing

ISO/IEC: ISO/IEC 33004:2015 – Information technology – Process assessment – Requirements for process reference, process assessment and maturity models, 9 p. (2015). Retrieved from www.iso.org. (Accessed on 20.02.2018)

ISO/IEC TR 15504–7:2008, Information technology – Process assessment – Part 7: Assessment of Organizational maturity, (2008). Retrieved from www.iso.org. (Accessed on 20.02.2018)

Lanza, G., Nyhuis, P., Ansari, S. M., Kuprat, T., & Liebrecht, C. (2016). Befähigungs- und Einführungsstrategien für Industrie 4.0. *ZWF Zeitschrift für wirtschaftlichen Fabrikbetrieb*, 111(1–2), 76–79.

Lichtblau, K., Prof. Stich, V., Dr. Bertenrath, R., Blum, M., Bleider, M., Millack, A., ... & Schröter, M. (2015). IMPULS—Industrie 4.0 Readiness, Aachen-Köln. Retrieved 20.02.2018 from https://www.industrie40-readiness.de/?lang=en

Kagermann, H., Wahlster, W., Helbig, J., Hellinger, A., & Karger, R. (2012). Im Fokus: Das Zukunftsprojekt Industrie 4.0: Handlungsempfehlungen zur Umsetzung. Bericht der Promotorengruppe Kommunikation. Forschungsunion.

Kitchenham, B. (2004). Procedures for Performing Systematic Reviews. Keele, UK, Keele University, 33(TR/SE-0401), 28. http://doi.org/10.1.1.122.3308

Lemke, C., & Brenner, W. (2014). Einführung in die Wirtschaftsinformatik: Band 1: Verstehen des digitalen Zeitalters. Springer, Heidelberg.

Leyh, C., Schäffer, T., Bley, K., & Forstenhäusler, S. (2016). SIMMI 4.0—A Maturity Model for Classifying the Enterprise-wide IT and Software Landscape Focusing on Industry 4.0, 8, 1297–1302. http://doi.org/10.15439/2016F478

Leyh, C., Schäffer, T., Bley, K., & Forstenhäusler, S. (2017). "Assessing the IT and Software Landscapes of Industry 4.0-Enterprises: The Maturity Model SIMMI 4.0," in Information Technology for Management: New Ideas and Real Solutions (Lecture Notes in Business Information Processing, LNBIP, Vol. 277), E. Ziemba (ed.), Heidelberg, Germany: Springer, pp. 103–119 (doi: 10.1007/978-3-319-53076-5_6).

Medi SPICE (2011). Portal of the Medi SPICE Project. Retrieved 20.02.2018 from http://medispice. ning. com

Menon, K., Kärkkäinen, H., & Lasrado, L. A. (2016). Towards a Maturity Modeling Approach for the Implementation of Industrial Internet. In PACIS (p. 38). 20th Pacific Asia Conference on Information Systems (PACIS 2016). 27 June- 1 July 2016, Chiayi, Taiwan.

Mettler, T. (2009). A Design Science Research Perspective on Maturity Models in Information Systems. Universiteit St. Gallen, St. Gallen, Switzerland, Technical Report BE IWI/HNE/03, 41(0). Retrieved 20.02.2018 from http://ehealth.iwi. unisg.ch/fileadmin/hne/downloads/20090512_Maturity_Model_Design.pdf

Neff, A. A., Hamel, F., Herz, T. P., Uebernickel, F., Brenner, W., & vom Brocke, J. (2014). Developing a maturity model for service systems in heavy equipment manufacturing enterprises. Information & management, 51(7), 895–911.

Porter, M. E., & Heppelmann, J. E. (2014). How Smart, Connected Products Are Transforming Competition. *Harvard Business Review*, 92(11), 64–88.

Posada, J., Toro, C., Barandiaran, I., Oyarzun, D., Stricker, D., deAmicis, R., ... & Vallarino, I. (2015). Visual Computing as a Key Enabling Technology for Industrie 4.0 and Industrial Internet. *IEEE Computer Graphics and Applications*, 35(2), 26–40.

PricewaterhouseCoopers. (2016). The Industry 4.0/Digital Operations Self Assessment. Retrieved 20.02.2018 from https://i4-0-self-assessment.pwc.nl/i40/landing/

Reiner, A. (2014). Industrie 4.0—Advanced Engineering of Smart Products and Smart Production. *International Seminar on High Technology*, (January), 1–14. http://doi.org/10.13140/2.1.1039.4406

Rockwell Automation. (2014). The Connected Enterprise Maturity Model, 12. Retrieved 20.02.2018 from http://www.rockwellautomation.com/rockwellautomation/innovation/connected-enterprise/maturity-model.page?

Röglinger, M., Pöppelbuß, J., & Becker, J. (2012). Maturity Models in Business Process Management. *Business Process Management Journal*, 18(2), 328–346.

Rout, T., Tuffley, A., & Cahill, B. (2001). CMMI Evaluation: Capability Maturity Model Integration Mapping to ISO/IEC 15504 2: 1998. Software Quality Institute, Griffith University Brisbane, 2001.

Schumacher, A., Erol, S., & Sihn, W. (2016). A Maturity Model for Assessing and Maturity of Manufacturing Enterprises. Procedia CIRP, 52, 161–166. http://doi.org/10.1016/j.procir.2016.07.040

Sein, M. K., Henfridsson, O., Purao, S., Rossi, M., & Lindgren, R. (2006). Action Design Research. *MIS Quarterly*, 30(3), 611–642. http://doi.org/10.1080/0268396022000017725

Shrouf, F., Ordieres, J., & Miragliotta, G. (2014). Smart Factories in Industry 4.0: A Review of the Concept and of Energy Management Approached in Production Based on the Internet of Things Paradigm. *IEEE International Conference on Industrial Engineering and Engineering Management*, 2015, January, 697–701. IEEE. 9–12 December 2014, Malaysia. Retrieved 20.02.2018 from http://doi.org/10.1109/IEEM.2014.7058728

Team, C. P. (2006). CMMI' for Development, Version 1. 2. Framework, (August), 573. Retrieved 20.02.2018 from http://doi.org/CMU/SEI-2006-TR-008 ESC-TR-2006-008

Vezzetti, E., Violante, M. G., & Marcolin, F. (2014). A Benchmarking Framework for Product Lifecycle Management (PLM) Maturity Models. *International Journal of Advanced Manufacturing Technology*, 71(5–8), 899–918. http://doi.org/10.1007/s00170-013-5529-1

Gülay Ekren[*], Alptekin Erkollar and Birgit Oberer

Toward Industry 4.0: Challenges of ERP Systems for SMEs

1. Introduction

Recent developments in information and communication technologies and extensive implementation of sensors, computer networks and cloud computing have prepared the infrastructure for Industry 4.0. Industry 4.0 is a research priority for the European Research Council, mainly under the Horizon 2020 program (Bagheri, Yang, Kao & Lee, 2015). Today, many enterprises are hostile as well as skeptical toward technologies and tools coming together with this advanced approach. In this context, ERP (Enterprise Resource Planning) in Industry 4.0 is a new system providing many opportunities, but also a wide variety of challenges.

An ERP system is an integrated software solution which consists of a range of business processes in a single database, a single application with a unified interface. Using ERP systems can be seen as significant for a future competitive strategy of a company (Ehie & Madsen, 2005). To keep up with Industry 4.0 in a global perspective, large-scale enterprises, as well as small- and medium-sized enterprises (SMEs), are embarking upon to transfer their traditional systems to ERP systems. Because, ERP systems offer a wide variety of opportunities for increasing productivity, quality, the amount of work produced, overall business performance, as well as several challenges. Nevertheless, ERP systems are one of the solutions for SMEs to face global challenges (Deshmukh, Thampi & Kalamkar, 2015). However, because of their limited resources and characteristics, SMEs must be considered differently in comparison to large-scale enterprises. ERP systems may impose an inflexible structure on a company that threatens the current business mood of many SMEs (Olsen & Saetre, 2007). Olsen and Saetre proposed to develop proprietary software that matches their needs, as a solution.

It is clear from previous studies that SMEs face many challenges in the adaptation of ERP systems. For example, Raymond and Uwizeyemungu (2007)

---

[*] Corresponding author: Gülay Ekren, gekren@sinop.edu.tr; Sinop University, Ayancık Meslek Yüksekokulu, 57400 Ayancık/Sinop, Turkey; +90 368 613 3436 (ext. 6915)

presented a typological form of SMEs in manufacturing area regarding their adaptation of an ERP system by their technological, environmental and organizational contexts. Moreover, Venkatraman and Fahd (2016) explored barriers faced with ERP systems by Australian SMEs, and found out that the obstacles to the successful adoption of ERP systems are such as customization concerned with challenges, redesigning business processes, training, upgrade/update costs, limited flexibility in adapting business processes, maintenance and integration costs. Furthermore, Haddara and Zach (2012) conducted a study in Egypt to find out the ERP's readiness for the smart factories of Industry 4.0. The findings show that the existing ERP systems are technically ready for automation but not for full automation. They emphasize costly client organizations as well as the need for the communication protocols and standards, especially the necessity for unified sensor-communication standards. Also, Deshmukh, Thampi and Kalamkar (2015) collected data from 95 SMEs in India and highlighted quality benefits of ERP implementation in SMEs. This study presents an action plan to transform an organizational system from legacy system to more economical ERP systems such as cloud-based, open source systems for SMEs. However, there is no research that distinguishes the critical influencing factors as well as challenges in the view of different phases (design, implementation, adaption, maintenance) of ERP systems, especially for SMEs.

In this study, challenges and possible solutions for ERP systems of SMEs under the vision of Industry 4.0 are presented by reviewing literature. Then the identified challenges and solutions are organized regarding four different phases of an ERP system, such as system preparation and design, system installation and implementation, system adaption, system operations and maintenance. This study makes an effort to make some contribution to the use of ERP systems by SMEs. The results of this study are expected to support future works with regards to ERP adaption for SMEs, a variety of solutions to a range of challenges that are faced by ERP systems for SMEs are presented.

1.1 Industry 4.0

Industry 4.0 is one of the strategic initiatives of the German government, which was adapted to the 2011 action plan of High-Tech Strategy 2020 (Weyer, Schmitt, Ohmer & Gorecky, 2015). The first industrial revolution was the mechanization of manufacturing equipment powered by water and steam; the second industrial revolution introduced mass production with the help of electrical power, and third industrial revolution uses electronics and computers to automate manufacturing further. Nowadays, the next industrial revolution, as a young term named

Industry 4.0, uses communications technology such as low power Bluetooth, low power Wi-Fi, Thread, Digimesh, WirelessHart, 802.11, LoHoWAN, HaLow, DASH7, and also ZigBee concerned with machine-to-machine data communications over short distances (Gilchrist, 2016). Industry 4.0 describes the enhanced integration of information and communication technologies into production environment as it referred to a wide variety of concepts such as Cyber-Physical Systems (CPS), smart machines, smart factories, smart data storage systems, Internet of Things (IoT), and human–environment interaction (Kolberg & Zühlke, 2015; Posada et al., 2015). IoT refers to the interconnections of networked physical objects (so-called smart objects or smart products) via the Internet. By this means, the gap between real and virtual world is becoming smaller by transforming factories into intelligent environments (Weyer et al., 2016); these factories usually are called smart factory or factory of the future.

Industry 4.0 creates an entirely digital value chain based on highly intelligent and connected systems (Haddara, & Elragal, 2015), that is, CPS. CPS helps to reach this advanced environment by making a connection between real objects and virtual processes via microcontrollers, actuators, sensors and human–machine interfaces. In other words, CPS is an integrated with integrated computation, communication and control systems (cyberspace) with natural and artificial systems (real space). The integration of CPS in various sectors, such as automotive, railways, medical and manufacturing, would provide factories self-configure for resilience, self-adjust for variation, and self-optimize for disturbance to transform today's plants into an Industry 4.0 factory or so-called smart factory (Bagheri, Yang, Kao & Lee, 2015). The CPS also allows the integration of manufacturing processes and business processes which is guiding the way to Industry 4.0. It aims to produce more customized, diversified and mass-produced products, as well as seamlessly integrated business processes. For a manufacturing system, enterprise-wide knowledge integration requires integration with manufacturing related information systems like ERP systems (Majstorovic, Macuzic, Sibalija & Zivkovic, 2015).

1.2 Challenges Faced by ERP Systems for SMEs

ERP is an enterprise-wide software package integrating business functions such as manufacturing, inventory, shipping, logistics, distribution, invoicing, accounting, sales, marketing, billing, production, human resources and quality control into a unified system through a shared database. These functions can be customized up to a certain limit according to the specific needs of each organization. |In addition, the integration of all data related to products, services, customers and suppliers into a single and centralized database provides ubiquitous access to the companies' all resources not only by staff but also stakeholders (Sadrzadehrafiei, Chofreh, Hosseini & Sulaiman , 2013). Consequently, common characteristics of a fully functioning ERP system can be summarized as using a centralized database, variety of integrated modules, scalability in line with the size of the organization, and providing user customization. Like ERPs, SMEs also have common characteristics such as such as fewer employees, less order or customers, low revenues. Venkatraman and Fahd (2016) emphasize the specific characteristics of SMEs and provide primary criteria defined by different countries regarding employee size and revenue as follows (Tab. 1):

Tab. 1: The Employee Size and Revenue of SMEs by Different Countries (Venkatraman & Fahd, 2016).

Country	Employee Size	Revenue
China	<2000	CN¥300 million
India	<1000	Rs. 25 Lakh– Rs. 10 Crore
Canada	<500	$25 million
Japan	<100 or <300	¥50 million
European Commission	<250	€50 million
Singapore	<200	$100 million
Australia	<200	$20 million
United States	No standard definition exists	

The adoption and implementation of ERP systems by SMEs are considered as technological as well as organizational innovation (Antoniadis, Tsiakiris & Tsopogloy, 2015). However, the implementation and maintenance phases of ERP systems are usually vendor-driven. ERP vendors usually continue to hand out software patches for bug fixes, or new enhanced releases. Moreover, the efforts on post-implementation activities (maintenance and support) impact the overall success of the ERP life-cycle (Law, Chen & Wu, 2010). Antoniadis, Tsiakiris and Tsopogloy. examined critical factors affecting the successful adoption of an ERP system by SMEs. These factors are (1) a collaboration between the departments of the enterprise, (2) required skills for implementing any ERP system, (3) tracking and evaluation process of implementing ERP, (4) software training due to ERP system implementation, (5) requirement analysis of enterprise or system specifications, (6) clarity of objectives, (7) convenience in technological infrastructure of the enterprise, (8) the technical support of software provider, (9) capability of management and staff, (10) existing financial and human resources of enterprise, (11) confidence level of staff and (12) quality of service provided. Therefore, it is difficult for SMEs to adopt advanced applications and technologies of Industry 4.0, because they usually have not got enough workforce to enter new areas, and also do not have enough financial capacity to invest in emerging technologies as well as a skeptical attitude toward technology strategies. Furthermore, they require knowledge/training and support for demanding applications and technologies to be able to remain alive in an increasingly globalized world (Faller & Feldmüller, 2015). On the other hand, Industry 4.0 is required to develop technological and organizational solutions that are adapted to the needs of SMEs (Kagermann et al., 2013).

A range of challenges, as well as solutions faced by ERP systems for SMEs in the vision of Industry 4.0, are specified as follows: lack of unified standards and protocols, lack of using business intelligence capabilities of ERP systems, ERP as a considerable investment for companies, complexity of ERP systems, changes and uncertainty during ERP system implementation.

1.3 Lack of Unified Standards and Protocols

Nowadays, it is crucial for companies to create an end-to-end integration between business processes (i.e., engineering workflows, services) and CPS by using ERP systems since these systems improve the competitiveness of companies and support the dynamics of their development. Therefore, ERP systems are emphasized to be one of the backbones of Industry 4.0. Haddara and Elragal (2015) conducted interviews with manufacturers, ERP vendors and partners to analyze the presence of ERP systems in Industry 4.0. The findings state that ERP systems are considered as technologically and functionally on hand for this new industrial revolution. However, there are some challenges about the lack of unified standards and protocols in machine-to-machine and machine to ERP systems. For example, proprietary communication protocols prevent machine-to-machine collaboration by locking vendors to access basic concepts of a smart factory. This is why unified global communication standards are required to set the stage for Industry 4.0; the success of Industry 4.0 depends on standardization actions between technology providers, integrators and end-users (Weyer, Schmitt, Ohmer, & Gorecky, 2015).

All ERP systems have not got a standard interface, as this makes them difficult to use considering legacy systems and increase maintenance costs (de Castro Silva & de Oliveira, 2015). Since the 1960s, legacy systems were designed, developed and implemented as centralized computing systems based on programming languages (i.e., COBOL, FORTRAN) (Rashid, Hossain & Patrick, 2002). In these systems, people regularly have to refresh their skills, and then their unwillingness makes them challenging to use ERP systems (Amoako-Gyampah, 2004).

Kolberg and Zühlke (2015) proposed to transfer existing Industry 4.0 solutions into Lean Production by creating CPS-equipped working stations since there are required communications protocols to standardize building blocks as well as hardware solutions with CPS as an interface for operating stations. Also, the usage of web standards together with automation systems or CPS must be considered to understand as well as taking the advantages of CPS, such as web services of engineering tools (i.e., PLC, HMI) (Faller & Feldmüller, 2015).

1.4 Lack of Using Business Intelligence Capabilities of ERP Systems

Different from ERP I, an ERP II system is a kind of advanced information system which gathers customers, suppliers, supply chain management, customer relationship management and ERP under a single roof, and can be labeled as business intelligence (Rajnoha, Kádárová, Sujová & Kádár, 2014). SMEs are still struggling to strengthen the advantages of ERP systems. According to research results conducted by Antoniadis, Tsiakiris and Tsopogloy (2015), business intelligence capabilities (i.e., data integration, controlling activities, flexible decision making) offered by ERP systems are underused, as SMEs do not correctly draw from the advantages generated by ERP systems such as improved productivity, reducing costs, reliable performance, reductions in paperwork or saving time, because of their shared characteristics (less manpower to enter new areas, less money to invest technology, or skeptical attitude towards technology). Nowadays, management of companies needs to plan and simulate business processes and also establish a list of requirements for the application of ERP systems based on business intelligence capabilities.

1.5 ERP as a Considerable Investment for Companies

The implementation of proprietary ERP systems often requires high investments, and corporate time and resources (i.e., infrastructure, qualified staff). Initial costs for setup and support, the cost and time needed for training staff to handle the new system are seen as limitations for adopting proprietary ERP systems (Antoniadis, Tsiakiris & Tsopogloy, 2015). Proprietary ERP systems have significant players such as SAP, Oracle ERP Cloud, Microsoft Dynamics, the SAGE grouper Solutions, and Infor LN Global ERP. They are usually being used by large organizations but also SMEs. However, because of their expensive licenses (for both server and web) as well as costly installation steps, open-source ERP systems offer an attractive alternative for organizations without customization of every implementation (Fougatsaro, 2009). Although they are required high costs, return on investment (ROI) of ERP systems are considered to be invaluable for both medium size and large companies (Rajnoha, Kádárová, Sujová & Kádár, 2014). However, it is apparently evident that ROI comes in the medium or long term. Also, ROI of an ERP system is usually complicated to calculate (Berchet & Habchi, 2005).

1.6 Complexity of ERP Systems

ERP systems are highly complex business information systems. SMEs, as well as cross-national companies, are often challenged by the growing complexity of ERP systems, but usability may be one of the key factors to overcome this problem. End-users usually need more time to complete tasks with more complex ERP interfaces. Mittelstädt, Brauner, Blum and Ziefle, (2015) examined the effects of information complexity (i.e., data amount and task complexity) and visual presentation as a critical aspect of usability and human factors on the decision efficiency and effectivity in ERP systems. According to Mittelstädt, Brauner, Blum and Ziefle, software developers must pay attention to the factors such as age and cognitive abilities rising from user diversity, as these factors substantially help to create more productive and competitive companies by increasing effectivity, efficiency and satisfaction.

Therefore, when reviewing the technical requirements of ERP systems, the sophistication of the integration and the cost of workforces and advantages must be considered based on architecture of an ERP system, the design of ERP prepared for production line and the ERP customizing issues for production line etc. (Faller & Feldmüller, 2015). Schuh, Gartzen, Rodenhauser and Marks (2015) indicated the requirements due to the increasing complexity and dynamics of products and processes in Industry 4.0, and emphasize significant challenges that occurred on the qualification of employees. Employees usually need to facilitate advanced cognitive acquisition and utilization due to a significant amount of data and its complex relationships. Therefore, supportive learning materials such as texts, pictures, videos, animations, etc. that provided for employees have to be displayed at the right level of detail, inappropriate language, and the best way of presenting with the adequate medium. Employees also must be unburdened from routine activities and focus on real works and learning tasks. Furthermore, according to Deshmukh, Thampi and Kalamkar (2015), SMEs often have complicated business processes, and Information Technology (IT) needs. Therefore, before starting to implement an ERP system, some factors must be considered, such as resource planning of infrastructure, training facilities, human resource planning, the support of top management and choosing right people for the overhauling.

1.7 Changes and Uncertainty during ERP System Implementation

The implementation of an ERP system usually requires a full variety amendment in processes and bring along various types of uncertainty. Changes and uncertainty during ERP system implementations considerably affect ERP system's success, so creating a risk management schedule can be a solution (Rajnoha, Kádárová, Sujová & Kádár., 2014). Besides, effective communication can be a solution to explain changes (Amoako-Gyampah, 2004).

Uncertainty refers to an unexpected event during the ERP implementation. For example, if uncertainty during ERP system implementation has occurred, such as late delivery of required hardware, then it will affect operations such as transferring and the integration of modules (Loh & Koh, 2004). Loh and Koh defined several uncertainties met by SMEs during ERP implementation; for example, knowing nothing on whether the required estimates are acceptable, inexact preferred software, immature project leaders, cloudy roles and responsibilities for vendors and internal staff, poor teamwork, poor project planning, unstable project partners, incorrect software configuration and system integration and drawbacks in defining the business processes within the SME that required change over etc. Additionally, Shanks et al. (2000) emphasized that cultural issues must be considered before ERP implementation, since a typical implementation process can adopt one culture but can differ in another culture (Haddara & Zach, 2012). Therefore, consulting organizations can assist SMEs when applying ERP systems.

2. Method

This study is qualitative research designed by a descriptive analysis through an extensive literature review. The qualitative process is being pursued to reveal perceptions and events naturally and realistically (Yıldırım & Şimşek, 2016). In this study, challenges and possible solutions for ERP systems in the face of SMEs are evaluated by reviewing the literature and are presented by descriptive analysis. Then, they are organized regarding four specific phases of an ERP system. The particular phases of an ERP system identified by the authors are system preparation and design, system installation and implementation, system adaption, system operations and maintenance (Fig. 1).

Fig. 1: Specific Phases of an ERP System.

Within the scope of this study, the specific phases of an ERP system are defined as follows:

2.1 First Phase: System Preparation & Design

In this phase, requirements, specifications and objectives are defined, ERP vendors and software packages (an ERP package that best fit with existing business processes and organizational needs) are selected, contacts with consulting firms are established, required budget is elaborated and relevant standards and protocols (e.g., ISO 9001, communication standards, web standards) are defined and analyzed according to different flows and production processes. In the design stage, various operations of the company and ERP functions are defined, specific development and prototyping are used (Berchet & Habchi, 2005).

2.2 Second Phase: System Installation & Implementation

This phase concerns the installation of ERP modules including unit testing, determining the authorizations for end-users, identification of the beginning or the initial procedures, preparing guides for users, planning the training of users, building test scenarios for integration. Implementation stage concerns testing, evaluation and ultimate validation of the prototype. Appropriate screening for modules and integration tests for inter-modules are designed. Operational starting will be initiated in this phase (Berchet & Habchi, 2005). Additionally, implementation phase consists of the physical installation of ERP system, Business Process Reengineering (BPR) and customization of processes, and other ordinary operations that align with the organization requirements of the system (Haddara & Zach, 2012).

2.3 Third Phase: System Adaption

Users' understanding and assimilation of the new system is crucial in this phase. Key users are defined to control the applications in their department regarding benefits and limitations of ERP system. They are one of the performance criteria for building an objective assessment. After all, it is aimed to use ERP system autonomously by end-users. Customers' and suppliers' demands are taken into account in this phase (Berchet & Habchi, 2005). On the other hand, BPR pushes radical change in business processes and contributes significantly to the successful adaption of ERP (Hong et al., 2016).

2.4 Fourth Phase: System Operations & Maintenance

Processes and physical, informational, decisional and financial flows are actively improved on departments such as sales, purchasing, accounting, planning, controlling and supplying this phase. In this phase, the company's operational activities (information and financial flows) are also monitored throughout the different modules during the ERP deployment process (Berch and Habchi, 2005), and the company has an idea about the acceptance, availability and sustainability of the system by keeping up with the satisfaction of the users, and their realization to the efficacy of the system.

3. Findings

The challenges and solutions classified according to the first phase of an ERP system called "System preparation and design" are summarized in Tab. 2.

Tab. 2: Possible Challenges and Solutions for the First Phase

First Phase: System Preparation & Design
Possible challenges
• ERP systems have not got a standard interface. Also, it is difficult to use legacy systems. • The lack of unified standards and protocols between machine and machine and even machine and ERP systems. • Unified global communication standards are required to set the stage for Industry 4.0. • ERP systems are a massive investment for companies, that is, high initial costs for setup and support.
Possible solutions
• The use of web standards together with automation systems or CPS (Faller & Feldmüller, 2015). • Transferring existing Industry 4.0 solutions into Lean Production by creating CPS-equipped working stations (Kolberg & Zühlke, 2015). • Open source ERP systems as an alternative without customization for every implementation (Fougatsaro, 2009).

The challenges and solutions classified according to the second phase of an ERP system called "System installation & implementation" are summarized in Tab. 3.

Tab. 3: Possible Challenges and Solutions for the Second Phase

Second Phase: System installation & implementation
Possible challenges
• ERP systems are considerably complicated information systems. • Need more time to complete tasks with more complex ERP interfaces. • Considering the cost of efforts and benefits for integration. • Considering changes and uncertainty during system implementations. • Customizing is extremely difficult and requires more time. • Software bugs and technical difficulties in configuration. • Poor top management commitment and unclear assignment of liabilities.
Possible solutions
• Developing and providing appropriate teaching and research platforms (Weyer et al., 2015). • Staff in the company required to become experienced and qualified in using an ERP system (Ziemba & Obłąk, 2013). • Considering user diversity such as age, cognitive abilities (Mittelstädt et al., 2015), language, skills, learning types. • Building a risk management plan (Rajnoha et al., 2014). • Effective communication with staff to explain the benefits of the system (Amoako-Gyampah, 2004). • Customization should be minimized (Hong et al., 2016). • Using BPR to fit the packages rather than doing modifications to the software to provide the organizations' workflow (Hong et al., 2016).

The challenges and solutions classified according to the third phase of an ERP system called "System adaption" are summarized in Tab. 4.

Tab. 4: Possible Challenges and Solutions for the Third Phase

Third Phase: System adaption
Possible challenges
• High costs and more time are needed to get training for staff and to make a success of the new system. • ERP systems are unique, sophisticated information systems. • The extent of efforts and time spent in process change to align with ERP.
Possible solutions
• To reduce complexity for employees, learning materials such as texts, pictures, videos, animations, etc. can be provided (Schuh et al., 2015). • Evaluating organizational fit of ERP system based on relevant ERP knowledge (Hong & Kim, 2002; Raymond & Uwizeyemungu, 2007). • BPR implementation has a positive effect on ERP performance (Hong et al., 2016).

The challenges and solutions classified according to the fourth phase of an ERP system called "System operations & maintenance" are summarized in Tab. 5.

Tab. 5: Possible Challenges and Solutions for the Fourth Phase

Fourth Phase: System operations & maintenance
Possible challenges
• The benefits from the business intelligence capabilities being offered by ERP systems usually remain unused. • ERP systems are usually required high maintenance costs. • Specialized resources required for ERP maintenance such as infrastructure for hardware, software and other resources.
Possible solutions
• A requirement to plan and simulate business processes (Antoniadis et al., 2015). • Establishing a list of requirements for the implemented ERP system based on business intelligence capabilities (Antoniadis et al., 2015). • Using on-site learning between workstations and production line to display and register a distant relationship between field level and ERP level (Faller & Feldmüller, 2015). • Collaborating with a trustworthy consulting firm (Serrano & Sarriei, 2006). • Hosting at vendors' infrastructure prevents data loss, permits data access at 7/24, and reduces specialized resources required for ERP maintenance (Venkatraman & Fahd, 2016).

4. Discussion and Conclusions

The transformation of today's industrial factories into smart factories takes the manufacturing industry primarily into the next level of industrial evolution, which is called Industry 4.0. However, Industry 4.0 will not only strengthen companies' competitive position but also drive solutions to both global and national challenges (Wan & Zhou, 2015; Deshmukh, Thampi & Kalamkar, 2015). To address this, a set of challenges and available solutions of ERP systems in the face of SMEs are presented in this study. To sum up, the challenges faced by ERP systems for SMEs are the lack of unified standards and protocols, the lack of using business intelligence capabilities of ERP systems, ERP as a considerable investment for companies, the complexity of ERP systems, and also changes and uncertainty during ERP system implementation. Possible solutions presented in this study can be developed further, as alternative solutions. Also, with regards to the potential solutions shown here, an ERP system might be tailored to the requirements of SMEs to meet their business processes. A set of recommendations are proposed for ERP vendors to cope with the challenges faced by ERP systems in the case of SMEs are as follows:

- ERP systems should be kept up with the diversified needs of SMEs.
- Monetary limitations play leading role in the use of ERP systems for SMEs. Thus, there is an urgent need for flexible pricing policies.
- Open source ERP systems may be an alternative solution for customization as well as reduction of high initial costs for setup and support.
- End-users of ERP systems in SMEs should be trained to gain more experience and competence as well as to handle new system and reduce complexity.
- The functionality of the ERP systems must be in smaller size, less complex as well as having unique features to fit common characteristics of SMEs.
- ERP software usually forces a rigid structure, but it is required to be more flexible for SMEs.

ERP systems are seen technically and functionally stand-by for Industry 4.0 (Haddara & Elragal, 2015) as well as in hand for SMEs (Venkatraman & Fahd, 2016). These systems are also one of the solutions for SMEs to face the global challenges. However, their limitations and difficulties perceived affecting the successful adoption, implementation or usage for SMEs must be considered. The findings summarized in this study are supposed to support the managers and practitioners in SMEs for implementing and adapting ERP systems successfully. The results of this study are expected to support future works with regards to ERP adaption for SMEs, a variety of solutions to a range of challenges that is faced by ERP systems for SMEs are presented; also, the suggestions given here show that more future studies are required on the adaption of ERP systems for SMEs.

References

Amoako-Gyampah, K. (2004). ERP Implementation Factors: A Comparison of Managerial and End-User Perspectives. *Business Process Management Journal*, 10(2), 171–183.

Antoniadis, I., Tsiakiris, T., & Tsopogloy, S. (2015). Business Intelligence during Times of Crisis: Adoption and Usage of ERP Systems by SMEs. *Procedia-Social and Behavioral Sciences*, 175, 299–307.

Bagheri, B., Yang, S., Kao, H. A., & Lee, J. (2015). Cyber-Physical Systems Architecture for Self-Aware Machines in Industry 4.0 Environment. *IFAC-PapersOnLine*, 48(3), 1622–1627.

Berchet, C., & Habchi, G. (2005). The Implementation and Deployment of an ERP System: An Industrial Case Study. *Computers in Industry*, 56(6), 588–605.

de Castro Silva, S. L. F., & de Oliveira, S. B. (2015). Planning and Scope Definition to Implement ERP: The Case Study of Federal Rural University of Rio de Janeiro (UFRRJ). *Procedia Computer Science*, 64, 196–203.

Deshmukh, P. D., Thampi, G. T., & Kalamkar, V. R. (2015). Investigation of Quality Benefits of ERP Implementation in Indian SMEs. *Procedia Computer Science*, 49, 220–228.

Ehie, I. C., & Madsen, M. (2005). Identifying critical issues in enterprise resource planning (ERP) implementation. Computers in industry, 56(6), 545–557.

Faller, C., & Feldmüller, D. (2015). Industry 4.0 Learning Factory for Regional SMEs. *Procedia CIRP*, 32, 88–91.

Fougatsaro, V. G. (2009). *A Study of Open Source ERP Systems*. Thesis for the Master's Degree in Business Administration, School of Management, Blekinge Institute of Technology, Sweden.

Gilchrist, A. (2016). Introducing Industry 4.0. In *Industry 4.0* (pp. 195–215). Apress. Berkeley, CA.

Haddara, M., & Elragal, A. (2015). The Readiness of ERP Systems for the Factory of the Future. *Procedia Computer Science*, 64, 721–728.

Haddara, M., & Zach, O. (2012). ERP Systems in SMEs: An Extended Literature Review. *International Journal of Information Science*, 2(6), 106–116.

Hong, K. K., & Kim, Y. G. (2002). The Critical Success factors for ERP Implementation: An Organizational Fit Perspective. *Information & Management*, 40(1), 25–40.

Hong, S. G., Hong, S. G., Siau, K., Siau, K., & Kim, J. W. (2016). The Impact of ISP, BPR, and Customization on ERP Performance in Manufacturing SMEs of Korea. *Asia Pacific Journal of Innovation and Entrepreneurship*, 10(1), 39–54.

Kagermann, H., Wahlster, W., & Helbig, J. (2013). Recommendations for implementing the strategic initiative INDUSTRIE 4.0, Final report of the Industrie 4.0, Final report of the Industrie 4.0 Working Group, Acatech. URL: https://www.din.de/blob/76902/e8cac883f42bf28536e7e8165993f1fd/recommendations-for-implementing-industry-4-0-data.pdf.

Kolberg, D., & Zühlke, D. (2015). Lean Automation Enabled by Industry 4.0 Technologies. *IFAC-PapersOnLine*, 48(3), 1870–1875.

Law, C. C., Chen, C. C., & Wu, B. J. (2010). Managing the Full ERP Life-Cycle: Considerations of Maintenance and Support Requirements and IT Governance Practice as Integral Elements of the Formula for Successful ERP Adoption. *Computers in Industry*, 61(3), 297–308.

Loh, T. C., & Koh, S. C. L. (2004). Critical Elements for a Successful Enterprise Resource Planning Implementation in Small- and Medium-Sized Enterprises. *International Journal of Production Research*, 42(17), 3433–3455.

Majstorovic, V., Macuzic, J., Sibalija, T., & Zivkovic, S. (2015). Cyber-Physical Manufacturing Systems-Manufacturing Metrology Aspects. *Proceedings in Manufacturing Systems*, 10(1), 9.

Mittelstädt, V., Brauner, P., Blum, M., & Ziefle, M. (2015). On the Visual Design of ERP Systems—The Role of Information Complexity, Presentation, and Human Factors. *Procedia Manufacturing*, 3, 448–455.

Olsen, K. A., & Saetre, P. (2007). ERP for SMEs—Is Proprietary Software an Alternative? *Business Process Management Journal*, 13(3), 379–389.

Posada, J., Toro, C., Barandiaran, I., Oyarzun, D., Stricker, D., de Amicis, R., ... & Vallarino, I. (2015). Visual Computing as a Key Enabling Technology for Industry 4.0 and Industrial Internet. *IEEE Computer Graphics and Applications*, 35(2), 26–40.

Rajnoha, R., Kádárová, J., Sujová, A., & Kádár, G. (2014). Business Information Systems: Research Study and Methodical Proposals for ERP Implementation Process Improvement. *Procedia-Social and Behavioral Sciences*, 109, 165–170.

Raymond, L., & Uwizeyemungu, S. (2007). A Profile of ERP Adoption in Manufacturing SMEs. *Journal of Enterprise Information Management*, 20(4), 487–502.

Sadrzadehrafiei, S., Chofreh, A. G., Hosseini, N. K., & Sulaiman, R. (2013). The Benefits of Enterprise Resource Planning (ERP) System Implementation in Dry Food Packaging Industry. *Procedia Technology*, 11, 220–226.

Schuh, G., Gartzen, T., Rodenhauser, T., & Marks, A. (2015). Promoting Work-Based Learning through Industry 4.0. *Procedia CIRP*, 32, 82–87.

Serrano, N., & Sarriei, J. M. (2006). Open Source Software ERPs: A New Alternative for an Old Need. *IEEE Software*, 23(3), 94–97.

Shanks, G., Parr, A., Hu, B., Corbitt, B., Thanasankit, T., Seddon, P. (2000). Differences in Critical Success Factors in ERP Systems Implementation in Australia and China: A Cultural Analysis. *ECIS 2000 Proceedings*, 537–544.

Venkatraman, S., & Fahd, K. (2016). Challenges and Success Factors of ERP Systems in Australian SMEs. *Systems*, 4(2), 20.

Yıldırım, A., & Şimşek, H. (2016). Sosyal bilimlerde nitel araştırma yöntemleri. Ankara: Seçkin Yayınları.

Wan, J., Cai, H., & Zhou, K. (2015). Industrie 4.0: Enabling Technologies. In Proceedings of 2015 International Conference on Intelligent Computing and Internet of Things.

Weyer, S., Meyer, T., Ohmer, M., Gorecky, D., & Zühlke, D. (2016). Future modeling and simulation of CPS-based factories: an example from the automotive industry. IFAC-PapersOnLine, 49(31), 97–102.

Weyer, S., Schmitt, M., Ohmer, M., & Gorecky, D. (2015). Towards Industry 4.0-Standardization as the Crucial Challenge for Highly Modular, Multi-Vendor Production Systems. *IFAC-PapersOnLine*, 48(3), 579–584.

Ziemba, E., & Obłąk, I. (2013). Critical Success Factors for ERP Systems Implementation in Public Administration. *Interdisciplinary Journal of Information, Knowledge, and Management*, 8(1), 1–19.

Abdulkadir Hızıroğlu, Dilan Özcan Kalfa[*], Ourania
Areta and Musab Talha Akpınar

Big Data on Cloud for Telecommunications Industry

1. Introduction

Business analytics refers to applying various analytics techniques to data that may be generated through the internal business processes, the operational data stores or could be acquired through external and open data sources (Gandomi & Haiza, 2015). These data sources can be in structured or semi/unstructured form. Considering the fact that huge amount of data has been generated by businesses, especially by service industries, the problem that needs to be tackled is to handle such a big corporate asset (aka big data) and to utilize them in line with the strategic focus of the organizations in order to extract valuable knowledge.

Big data could be defined as the massive data sets. In these data sets, the term "big" depends also on the complex context inside. The term "big" is also referring to the very large size. For this reason, most of the big data sets are hard to fit into the computer's memory in one execution time. The big size depends on the content source(s) of the data. These content sources are created from three main sources. One of the main sources is traditional enterprise data in which it is basically created and managed by the standard SQL structure transactions. CRM systems, transactional ERP data and similar enterprise and transaction-oriented system could be given as some examples. The other one is machine-generated/ sensor data that is depending on the systems; automated data sets could be given as example. Call detail records and GPS-based tracking system could be given as examples. The last one is social data, which is data created in the social networks such as Twitter, Facebook and LinkedIn. Another important concept is how these sources' context became meaningful. This process basically depends on the characteristics of the big data. Big data characteristics are defined with the 4Vs (Volume, Velocity, Variety and Value) dimensions. Volume shows the size of the data. Machine generated data could be given as an example. Velocity is the defined

---

[*]   Corresponding author: Dilan Özcan Kalfa, Ankara Yıldırım Beyazıt University Department of Management Information Systems Research Assistant, Ankara Turkey.

by the frequency of data. Tweets could be given as example, under a single topic; lots of data could be generated in a single moment of time. Variety shows the different types of data from simple database record such as employee record to YouTube video of the related employee. Value is the economic importance of the data, such as the economic value of the users; Facebook's like button has a higher value than single user nickname.

Big data could be private. However, there is much higher content in the publicly available part. Public data are collected by the governmental organizations and their public bodies. Public data can be used by private businesses to analyze trends in consumer behavior. It can be served as metrics for financial and operational decisions. Open data is defined as the data released to the public share. Open data can help businesses to measure consumer demand as well as the level and type of competitive supply in geographic markets. This could be officially collected data, such as collected by the national statistical institute or social network data open for the public access. This term also referred to the scientific data. Under the UK government's open data initiative, open data is given as the establisher of accountability, community and driver of economic growth. Demographics of consumers across the country, consumer preferences and budget allocation, business sector indices and public procurement statistics are some examples of open data that can be accessed through European Union Open Data Portal and National Open Data Portals. Big data could include open data or it could contain whole open data. However, to turn any kind of data into information and knowledge, more complicated solutions would be required. Relevant public data and big data can be fed into a platform by business analytics function in order to assist participating organizations with analytics and reporting in the sales and market analytics, customer analytics and campaign analytics and optimization areas.

A typical business analytics solution consists of several layers, including Data Infrastructures layer, Data Integration layer, Online Analytical Processing and Analytical Environment (Acker, Blockus & Pötscher, 2013; Bose, 2009; Brown, Chui & Manyika, 2011).

The processing and analysis of data generated through business processes have been of great importance for acquiring competitive. This is specifically important for certain European services industries, which produce a substantial amount of value-added services to the pertaining economies. Due to the crucial place in the Europe economy, we focus on the telecommunication sector, one of Europe's most important economic sectors. According to the ETNO's 2013 annual economic report, although the total revenue in EU's telecommunications sector amounts to € 274.7 billion, there is a negative growth trend of the region's telecom revenues since 2009. These figures show the importance of telecommunication companies' strategic planning to increase Europe's position on global telecoms market. However, some telecommunications companies that have a potential to be multi-national organizations are in greater need of assistance to be more competitive and more agile in reaching their customers and expanding their operations and services. As we mentioned above, the velocity and veracity of the data generated by these organizations have come to a state that makes it difficult to utilize the data as an organizational source of a company. Big data analytics and knowledge extraction on contemporary business intelligence or analytics systems can be rather expensive, and are not tailored to the specific requirements of the telecommunications industry (Gandomi & Haiza, 2015). Also, the numbers of data and variances in structure, format, language, etc. have vastly increased, and thus on one hand, many organizations do not have required capability for implementing such programs, while on the other hand, they have generally been dependent on third-party suppliers to transform into an analytics-oriented business cycle that could enable the organizations to take quick and effective actions using the knowledge extracted from big data sources. The proposed model will solve these issues by providing a cloud-based open platform to the telecommunications industry.

The rest of the paper is organized as followings. Firstly, we presented some frameworks on literature that provide big data solution to telecommunications industry. After that, we proposed our conceptual model and its components. Lastly, we concluded our research with discussing potential advantages of our conceptual framework.

2. Big Data Analytics for Telecommunications Industry

The world of communications has seen unprecedented data growth in the last few years. The advent of smartphones, mobile broadband, peer-to-peer traffic, growth in mobile data volumes, social media chatter and the increase in video-based services have all contributed to data volume while increasing consumption of media and content-based services and bandwidth consumption. With the help of business analytics solution, telecommunications sector can collect strategic data about their customer (Vesset et al., 2012). Moreover, they can deal with customer retention, get information about their customers' mood and preferences. Although proactive data brings these advantages, it needs complex analysis of unstructured and semi-structured data. The difficulties that telecommunications industry deals with data integration is shown in Fig. 1.

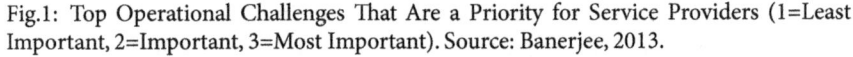

Fig.1: Top Operational Challenges That Are a Priority for Service Providers (1=Least Important, 2=Important, 3=Most Important). Source: Banerjee, 2013.

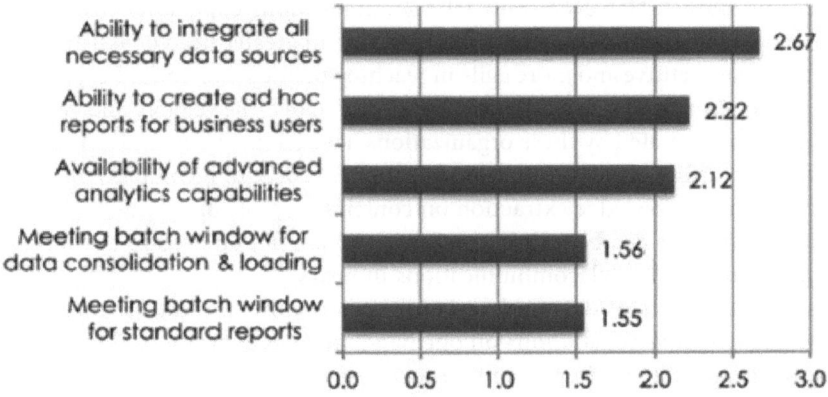

However, finding and analyzing information quickly and effectively is becoming a difficult task because companies' need for using of various (semi/unstructured) data from different sources has grown dramatically. In addition, managing data, for example, archiving, retrieving and analyzing, creates a substantial burden on Information Technology (IT) technologies and technical expertise, which makes deploying effective ways to collect and analyze important data a very demanding task.

According to Merrill Lynch, more than 85 % of business information appeared in semi-structured and unstructured forms (Blumberg & Atre, 2003b). These data contain a large quantity of information needed for analyzing and decision making. Existing business analytics applications have the difficulty of searching, finding and assessing semi-structured and unstructured data and they do not have the ability to use this enormous amount of information. This inadequately informed environment influences strategic management process negatively.

Once a big data analytic solution is in place, communications service providers (CSPs) can more easily introduce new capabilities that will increase revenue and customer satisfaction, and reduce costs. Communications service providers can accurately determine a customer's location in real-time. There are three kinds of location data accessible to CSPs: passive mobile positioning data, call detail records and global positioning system (GPS) data (Zhang, Cheng & Boutaba, 2010). Mobile positioning data is gathered periodically when a phone is powered on, regardless of whether calls are being placed or received. Call detail records are generated each time a call is placed and contain information such as who made the call, who received the call and the start and end time of the call. GPS provides a higher degree of accuracy data by delivering coordinates that are within a few feet of a person's actual location.

Academic researchers have shown the importance of big data for the telecommunication industry. In 2013, Dam (2013) defined major telecommunication industry big data sources. Transactions, log data, audio, events, social media, geospatial, emails, external feeds, free-form text and sensors are given as main sources. However, implementation of big data solutions requires big computational powers. For this reason, it is not possible to implement single-machine based solutions in a single environment. Therefore, choosing cheaper and simpler solutions could be a remedy for that. Another technological solution could be the solution packages currently provided by Information and Communication Technology (ICT) vendors. Although some of the already available products could be considered as alternative solutions, they are not affordable in terms of pricing; much cheaper versions of these technologies should be made available to the pertaining organizations. Big data analytics solutions will enable telecommunication industry to better understand their customers and develop subscriber profiles that can be used to create more intelligent marketing campaigns.

3. Assessment of Current Analytics Solution from Telecommunications Industry Perspective

A big data analytics infrastructure is not a replacement for the traditional analytics infrastructure; rather, it is designed to correct traditional analytics mistakes and provide an enhanced platform. As such, information is requested and consumed either to make better decisions or to create new products or applications, and the overall infrastructure evolves to better serve the demand. This evaluation creates an organic relationship between service provider's business, network and IT territories, gradually effacing the infamous silos. Key benefits that service providers will obtain from savings and operational efficiency could be summarized like this (Turban, Sharda, Aronson & King, 2008):

- Decreasing data cost
- High data load speed
- Reduction in administration cost
- Decreasing data query time
- Needing less memory because of advanced compression techniques
- Leverage commodity hardware

Due to the reasons stated above, telecommunication organizations certainly need external help when it comes to analytics. Important information and knowledge companies such as Oracle, Google, etc., open potential opportunities in the ocean of big data. With its SQL and NoSQL based system, Oracle built its structure over Hadoop technology. Oracle's analytical solution is structured over enterprise architecture via data mining, text mining and graph analysis capabilities.

Cloud-based analytics offer an infrastructure for big data solutions by eliminating hardware requirements and handling computing power limitations. It provides massive cloud computing and storage facilities to the companies. Companies can deal with big data without setting up new IT systems and technical staff to operate them with the help of cloud-based big data solutions. For example, Google Analytics brought an important perspective into the market. The data from the company and customer sides are collected by Google Services and distributed to the clients to satisfy analytics requirements through a huge cloud-based infrastructure. In addition, the cloud-based analytics systems employ various tools and methods to quickly collect and systematically analyze structured and unstructured data in various forms from different sources without any need for in-house analytics expertise. Cloud-based technologies have been successfully and effectively used in various applications, for example, climate analysis, biomedical image processing and satellite data analysis, requiring high computational resources (Raisinghani, 2003). The very recent technologies like SuperDataHub and Space-Time Research clearly indicate that cloud-based systems are emerging technologies, which will play an important role in "big data analytics" in near future (Vercellis, 2012). Publicly available data is the most important material for the implementation of the model. The analytics tools will provide the potential meaning that will be extracted from the publicly available data. However, considering the volume of the data, the implementation is not possible on a single machine. Cloud computing will become handy for inferences from the big data. In national bases, there are some attempts for implementing solo parts of the systems such as sentiment analysis and cloud-based system builds. However, they cannot be considered as under the concept of a patent. For an organization, successful business analytics implementation relies on four basic stages: information/data, technology, intelligence and implementation/communication (Chen, Chiang & Storey, 2012).

With regards to information/data, telecommunication industries need to regard information as a strategic resource. These industries should use their analyses of market strengths and weakness by systematically thinking about how this information, combined with their strategies can give them a competitive advantage. Because of strong competency in the telecommunication industry, telecom enterprises need to ensure that it has staff with both strategic and information knowledge represented at the top management level (Schroeck et al., 2012). Correlating an overview from fragments of customer data is a complex endeavour that consumes significant time and effort. The quality of data is suspect and needs significant readjustment, reduplication and cleansing before it can be leveraged effectively. Innovation is a continuous process, so telecommunication enterprises must evaluate their strategies based on publicly and privately open data and business analytics infrastructure. To continue their momentum, telecommunication enterprises need to capture, centralize and distribute customer interaction data. Therefore, some publicly open data including industry-specific market reports should be embedded into the data infrastructure and relevant analytics applications should be built on that (Wiig, 2015).

Traditional data analytics applications could be very expensive with regards to cost-benefit aspect for some service-oriented organizations and often very difficult to operate (Olszak & Ziemba, 2012). Especially in the telecommunications sector, combinations of a wide variety of devices, smartphones and tablets have great influence on the industry. These devices have brought wealthy information sources about their customers into the knowledge bases of the companies. With its volume, velocity and variety, telecommunications industry sees big data as an opportunity to transform them into value. Important information and knowledge companies such as Oracle, Google, etc., open potential opportunities in the ocean of big data. With its SQL and NoSQL based system, Oracle built its structure around Hadoop technology. Oracle's analytical solution is structured around Enterprise architecture via data mining, text mining and graph analysis capabilities. From the perspective of cloud-based big data solutions: Google Analytics brought an important perspective into the market. The data from the company and customer sides are collected by Google Services and distributed to the clients to satisfy analytics requirements through a huge cloud-based infrastructure. Rather than implementing solutions of other companies, tailored solutions specific to the telecom market solutions has an important influence on the market. AT&T implemented big data solution for gaining revenues from the marketing of customer data to the third-party marketing companies. SingTel implements its big data for improving their advertising platform. Cable, IP-TV and satellite-TV companies implement big data solutions for acquiring valuable information from the client devices such as tablets, cell phones and computers.

Tab. 1: Traditional Analytics versus Big Data Analytics. Source: Banerjee, 2013.

Dimensions	Traditional	Big Data Analytics
Memory demand	High	Low
Analytics type	Offline	Real-time
Analytics speed	Low	High
Analytics span	Slow	Fast
Knowledge extraction	Low	High
Data range	Structured	Unstructured/Semistructured/Structured
Volume	Low	High
Velocity	Batch	Real-time
Data compression time	Long	Short
Support cost	High	Low
Response time	Long	Short

Large companies can also face problems regarding focus on the required data and eliminate the unwanted data. Storing all the data that come from online sources means making a lot of investment for data storage systems. Also, it will increase the time needed for data processing and data administration. For not causing these problems, the business analytics functions need an expert tool to separate essential and redundant data to extract strategic information.

Nowadays, telecommunication industries use business analytics function to deliver information to strategy function by some kind of feedback processes like corporate performance management (CPM) and score carding. This is an ongoing cycle where the enterprise as its starting point has defined a strategy to be implemented in the various departments that make up the business. Coordination is performed by critical success factors, which are the elements that are essential to whether the strategy is successful and making sure that they are coordinated. This is typically done via internal meeting across functions and structured data from internal reports. But in this way, key performance indicators such as customer loyalty or market growth are rarely hit accurately, but rather a bit over or under. Therefore, a decision support system platform could be of potential use in analytics solutions tailored to the telecommunications companies.

4. The Proposed Conceptual Model

The proposed conceptual model is centralized in between big data, cloud computing, business analytics and telecommunications. The data created by the telecommunications industry (by service providers and users) turns into the big data with its 4Vs (Volume, Velocity, Variety and Value) dimensions. In order to maximize the value aspect of big data in the context of analytics, the implementation of a cloud computing solution is required that will remove the infrastructure and technical details from the responsibility of the decision maker while enabling them to focus on the real content of the data and knowledge Business analytics technologies are then what is required to turning data value into business meaning (Hiziroglu & Cebeci, 2012). This requires the trans-disciplinary synergy of computer, software and telecommunication engineering fields, information and knowledge sciences and decision support, and business intelligence concepts.

The conceptual design of the model is based on a complete data life cycle that starts with the acquisition of raw/semi-structured/structured input data and ends with business analytics reporting on one hand and re-usable open data on the other as shown in Fig. 2. The functionality and capabilities of the main components are detailed in the followings:

Data Pre-processing: Data pre-processing is a service that includes data acquisition, adaptation and data integration. It is the stage of our model that deals with the structured/semi-structured and structured data that come from telecommunication companies and open public data. This open public data will be gathered from governmental, EU and other international organizations (e.g., state budgets, growth indices, sector-specific data, procurement statistics) as well as unstructured documents such as whitepapers and reports by means of a vocabulary approach combined with statistical processing and data/text mining techniques. The collected data will have different structures, scripts and forms. For dealing with these manners, the data pre-processing service will use semantic mining techniques, ETL and standardization tools to transform these different data types into one common data type and will facilitate further knowledge extraction strategies.

Data Warehouse: This is the function where all the pre-processed data will be stored before conveying them to the engines. The integrated data that come from the former stage will be stored on the required database management systems (i.e., SQL, RDFS, OWL).

Analytics Engine: The business analytics will be done in this engine. Data mining techniques such as web and text mining, business optimization and modeling approaches will be employed at this stage. The aim of this component is to gain information about the customers, market, sales and campaign success. For this purpose, sales and market analytics, customer analytics and campaign analytics will be used in the analytics engine.

Reporting Engine: The telecommunications sector has some strategic questions that should be answered. For example, should they enter this market? Should they change their pricing model? Which segmentation model is the most appropriate? How can they reach specific target segments? What are the values of their customers? What service offers can they provide to specific segments? and so on. The reporting engine uses visualization tools, rules, pattern, documents and reports to show the answers of these questions. The analytics results that are retrieved from analytics engine will be delivered by these reporting tools.

Fig. 2: System Architecture and Data Flow of the Proposed Model (Hiziroğlu & Cebeci, 2013).

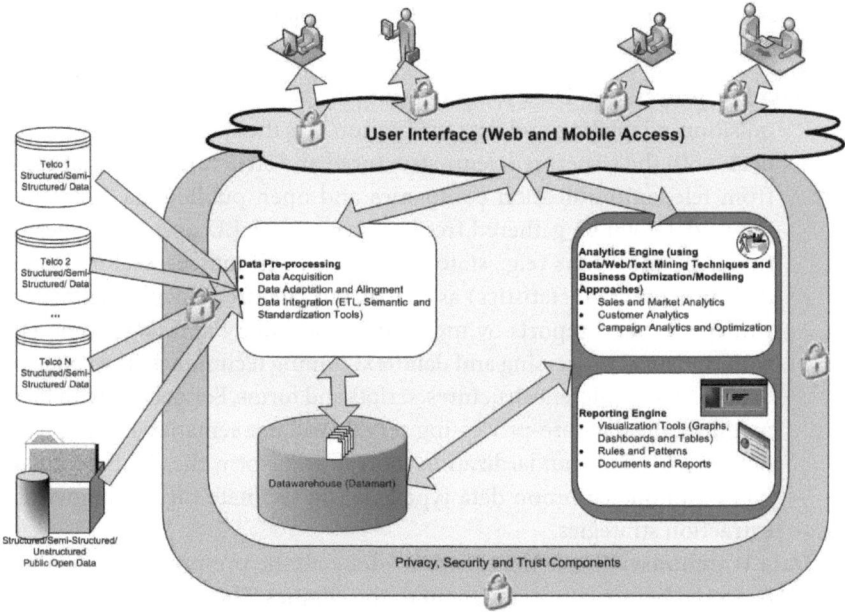

5. Conclusions and Discussions

The proposed model aims to provide telecommunications industry with a cloud-based open platform for big data analytics by promoting growth, increased efficiency and profitability across the entire telecommunication value chain. The proposed model gives a complete and transparent view of customers and enabling a new way of capturing and retaining them. The model provides the tools and environment to support business strategies toward market development, support-ed by various languages and across different countries. The proposed model will allow companies to extend the platform with additional innovative multilingual data products and services, by hooking up to the core environment, making use of the various tools, or integrate additional tools that will support the transforma-tion of big data.

The proposed model will focus to deliver a successful Big Data Analytics para-digm for the telecommunication sector. However, and as emphasized previously, the approach is generic and thus expandable to many other demanding appli-cation domains. This will actually be the scenario of success for the proposed model: build a comprehensive cloud-based platform that, on the one hand is able to meet the emerging telecommunication industry needs, and on the other hand be extendable and applicable to several other industry verticals in the future. The model will therefore address the analytics-related problems of the telecommuni-cations industry by moving the business analytics infrastructure into the cloud in order to facilitate a flexible way to use environment for telecommunications industry and to assist them in taking business decisions via an IT enabled busi-ness analytics process.

References

Acker, O., Blockus, A., & Pötscher, F. (2013). Benefiting from Big Data: A New Approach for the Telecom Industry. *Strategy and Analysis Report*. Booz & Company, 1–12, Retrieved from https://www.strategyand.pwc.com/media/file/ Strategyand_Benefiting-from-Big-Data_A-New-Approach-for-the-Telecom-Industry.pdf.

Banerjee, A. (2013). Big Data and Advanced Analytics in Telecom: A Multi-Billion-Dollar Revenue Opportunity. Technical Report, *Heavy Reading*, New York.

Blumberg, R., & S.Atre (2003). The Problem with Unstructured Data, *DM Review*, 42–46. Retrieved from http://soquelgroup.com/wp-content/uploads/2010/01/ dmreview_0203_problem.pdf

Bose, R. (2009). Advanced Analytics: Opportunities and Challenges. *Industrial Management & Data Systems*, 109(2), 155–172.

Brown, B., Chui, M., & Manyika, J. (2011). Are You Ready for the Era of "Big Data". *McKinsey Quarterly*, 4(1), 24–35.

Chen, H., Chiang, R. H. L., & Storey, V. D. (2012). Business Intelligence and Analyt-ics: From Big Data to Big Impact. *MIS Quarterly*, 36(4), 1165–1188.

Gandomi, A., & Haiza, M. (2015). Beyond the Hype: Big Data Concepts, Methods, and Analytics. *International Journal of Information Management*, 35(2), 137–144. https://doi.org/10.1016/j.ijinfomgt.2014.10.007

Hiziroglu, A., & Cebeci, H. İ. (2013). A Conceptual Framework of a Cloud-Based Customer Analytics Tool for Retail SMEs. *Periodicals of Engineering and Natural Sciences (PEN)*, 1(2), 44–50.

Olszak, C. M., & Ziemba, E. (2012). Critical Success Factors for Implementing Business Intelligence Systems in Small and Medium Enterprises on the Example of Upper Silesia, Poland. *Interdisciplinary Journal of Information, Knowledge, and Management*, 7(2), 129–150.

Raisinghani, M. (2003). *Business Intelligence in the Digital Economy: Opportunities, Limitations and Risks*. Idea Group Publishing, Hershey, PA, 126–140.

Turban, E., Sharda, R., Aronson, J. E., & King, D. (2008). *Business Intelligence: A Managerial Approach* Pearson Prentice Hall, New Jersey.

Schroeck, M., Shockley, R., Smart, J., Romero-Morales, D., & Tufano, P. (2012). Analytics: The Real-World Use of Big Data. *IBM Global Business Services*, 12, 1–20.

Vercellis, C. (2011). *Business Intelligence: Data Mining and Optimization for Decision Making*. John Wiley & Sons, United Kingdom.

Vesset, D., McDonough, B., Wardley, M., & Schubmehl, D. (2012). Worldwide Business Analytics Software 2012–2016 Forecast and 2011 Vendor Shares. *IDC Market Analysis*, Massachusetts.

Van Den Dam, R. (2014). Big Data a Sure Thing for Telecommunications: Telecom's Future in Big Data. 2013 International Conference on Cyber-Enabled Distributed Computing and Knowledge Discovery (CYBERC), Beijing China, 148–154.

Wiig, A. (2015). IBM's Smart City As Techno-Utopian Policy Mobility. *City*, 19 (2–3), 258–273.

Zhang, Q., Cheng, L., & Boutaba, R. (2010). Cloud Computing: State-of-the-Art and Research Challenges. *Journal of Internet Services and Applications*, 1(1), 7–18.

Zümrüt Ecevit Satı[*]

Challenges and Opportunities of Logistics 4.0: Reflections from Industry 4.0

1. Introduction

The industry is undergoing a major change. After mechanization, industrialization and automation, it has now reached the threshold of the fourth revolution: digitization and networking. Used today in factories, manufacturing processes will be made more efficient, smarter, more controllable and more transparent in the near future. Industrial processes will be interconnected through the network by using modern communication and information technologies to create continuous improvements in the production process. With the help of these new digital production methods, companies will be able to respond more easily to individual customer needs and consolidate their competitive edge in global markets.

Today, about 10% of companies in the manufacturing industry have passed the Industry 4.0 components. As an Industry 4.0 supplier, this ratio doubles in the area of machinery and plant engineering. Currently, 5.6% of machinery and plant engineering companies are at the advanced level of application of these technologies. A fewer than 18% of businesses are working to understand Industry 4.0 and are making preparations to put it into practice (IW Consult/FIR, 2015). In addition, there are comparable findings of Industry 4.0 processes and technologies in their work. Both large companies and SMEs have a great desire to use intelligent services, such as assessing big data generated during production processes or networked production processes (IW Consult/FIR, 2015; Techconsult, 2015).

Being able to respond to customer requests with increasingly fluid, complex, uncertain and personal characteristics is the main goal of logistics management. Customers are demanding immediate response to changing needs now more than ever. In order to cope with this challenge, logistics management needs to closely monitor innovations in information and communication technologies, realize real-time information management and integrate decision support systems with all internal and external processes.

---

[*] Corresponding author: Zumrut Ecevit Sati, zsati@istanbul.edu.tr.

According to these developments, the functions that logistics management has undertaken have also changed considerably due to information and communication technologies, innovations in high technology and industrial subjects. Both internal and external logistics systems lead to systems with increased numerical and communication capabilities as they can communicate with each other to exchange information about all the resources of a product and its related components. Logistics 4.0 is used to describe the transformation of logistics management in the new roles that physical elements as well as technological innovations have presented. The Industry 4.0 elements can overcome shortcomings of some areas of logistics such as demand planning, route optimization, warehouse management, real time supply chain monitoring, production management, modeling supply chains, risk analysis, security and innovation opportunities. Logistics has come to a new stage with the development of Industry 4.0 technologies and applications. Large-scale operations in logistics systems require data in order to work efficiently.

2. Method

In logistics, management collaboration and integration with software, hardware and acceptance of other intelligence are crucial for delivering smart warehouse and smart factory solutions and customer specific end-to-end SCM solutions. Smart integration between supply, production, warehouse, pick and pack, unloading, loading, returns and other business processes executed seamless and controlled by intelligent predictive self-learning software algorithms. Everything packed in to standard solutions that are cost-efficient to install, scalable and provide measurable real value to all stakeholders. Intelligent software algorithms combined with smart machines, robotics and others will transform and automate logistics processes today and in the future.

The Industry 4.0 revolution with its characteristic elements of cloud platform, big data, block chain technology, the Internet of Things (IoT), automation and robotic, augmented reality (AR) and 3D Printing has caused a paradigm shift within the logistics industry. This impact is called as Logistics 4.0. Logistics 4.0 can describe the digital transformation from physical logistics to these new technologic innovations-based logistics. The optimization and visibility of logistics systems depend on information accuracy and availability.

This study deals with the features of Industry 4.0, the investigation of the theoretical areas of logistics 4.0, the evaluation of the opportunities and problems of logistics 4.0, the examination of the reflection of Industry 4.0 technologies on Logistics 4.0. Industry 4.0 technologies are being explored on the basis of cloud platform, big data, block chain technology, IoT, augmented reality and 3D printing. The applications of these technologies in the logistics sector are also examined.

3. Findings

3.1 The Characteristics of the Industry 4.0

The development of "Intelligent Manufacturing Systems" has created a paradigm shift in the industry with the digitalization of every stage of the production chain, the provision of machine-human-infrastructure interaction. Steam powered mechanical systems have evolved into a system in which the cyber physical systems take place within about 300 years. The new process, referred to as Industry 4.0, is a structure that will change. Production systems that adapt instantaneously to the changing needs of the consumer in one place and continuous communication and coordination with each other in the other automation systems in the form of a new era entering into the characteristic structure is depicted. The "Smart Factories" of Industry 4.0 recognize the business need with sensors, communicate with other remote production tools via the Internet and produce the products they need, that is intelligent machines that capture knowledge of the big data in cloud systems.

Hermann, Pentek & Otto (2016) state that in the fourth revolution, the physical and the virtual worlds are combined (or fused) due to the use of cyber-physical systems, which are able to connect with other systems via machine-to-machine communication through the Internet of things. These cyber-physical systems are steered via big data that is obtained through sensors or other data gathering devices (Hermann, Pentek & Otto, 2016).

According to Smit, Kreutzer, Moeller and Carlberg (2016), there are features that stand out in Industry 4.0. One of these features of Industry 4.0 is interoperability. The CPS allows people and smart factories to connect and communicate with each other via IoT and IoS. Another feature that stands out in Industry 4.0 is virtualization. It refers to the ability to connect physical systems with virtual models and simulations. Substantive management implies that decision-making and management of individual subsystems are carried out independently and in parallel. Another important feature is the ability to move in real time. Adaptation to these real-time needs is a key requirement for all kinds of communication, decision-making and management of real-world systems. The service orientation feature emphasizes modularity. Accordingly, Industrial Systems 4.0 systems should be maximum modular and should be able to make rapid changes based on automatic detection of a real situation.

According to Smit, Kreutzer, Moeller and Carlberg (2016) the main features of Industry 4.0 are interoperability (CPS allow humans and smart factories to connect and communicate with each other through IoT and IoS); virtualization (capability to connect physical systems with virtual models and simulations); decentralization (decision making and management are performed independently and parallel in separate subsystems); real-time capability (adjustment with requirements in the real time is the key requirement for any kind of communication, decision making and management of systems in the real world); service orientation; modularity (systems of the Industry 4.0 should be maximally modular and capable for quick changes on the basis of automatic detection of a real situation).

Tab. 1 shows the basic features of smart manufacturing based on Industry 4.0.

Tab. 1: Main Features of Smart Manufacturing. Source: De Felice, Petrillo & Zomparelli, 2016.

Features	Description
Automation	Automated, integrated, monitored and continuously evaluated
Connection	Connection the smart factory, smart machines, robots and advanced sensors via standards
Supply chain	Providing ubiquitous use of mined information throughout product value chain
Communication	Communicating between people, equipment and enterprise and operations management applications
Digital Data	Leveraging auto-identified components with tags (e.g., RFID) and smart products with product configuration

Ultimately, these innovations observed in the business environment, and the vertical integration of new business models based on these developments is transformed. In a sense this means that all business functions are directed toward the digital ecosystem. This orientation expands with cloud, big data, Internet of Things, 3D printing, augmented reality, automation, robotics and other digital technologies. The emergence of such new business models, the digitization of products and services together, the digitization and integration of every link in the company's value chain are reflected in every size of business and operations.

3.2 The Emerging of Logistics 4.0

The technology and applications increasing with the Industry 4.0 paradigm have increasingly influenced all business functions and relationships. From this exchange, the logistics function has also taken its share. The development of global communication and travel options significantly improves logistics and logistics services all over the world. To provide large manufactory and international trade opportunities, it is necessary to significantly improve logistics and their relevant processes. Some major logistics systems are goods, assets or other objects delivered to the client. That is why there is a need for logistics to use accurate identification and real-time tracking with location determination. All components combined guarantee effective delivery time.

Logistics activities are one of the main actors of the manufacturing sector and function as a basic support function for many sectors as well. The increasing importance of logistics activities is increasingly driven by factors such as the growing importance of emerging markets, globalization of supply chains, increasing regulatory efforts in environmental management, demands for personalization, changing product range and value-driven approaches (Kückelhaus & Terhoven, 2013). The term "logistics business" characterizes processes that determine the management system and coordinate all material flow from the factory to the end users (Coyle & Bardi, 2003).

The concept of logistics 4.0 was addressed by Jeschke (2016) in two approaches. From short-term perspective, Logistics 4.0 is defined as the management of robust and mutually related processes through the use of large amounts of data generated through the relationships and transactions between independent members. In the medium-term approach, Logistics 4.0 is defined as autonomous, self-organization systems within other systems (Szymanska, Adamczak & Cyplik, 2017). Timm and Lorig (2015) evaluated Logistics 4.0 as the management of logistics systems that are composed of independent subsystems but can be integrated. The behavior of subsystems in logistics systems depends on other surrounding subsystems. The concept of Logistics 4.0 provides support for Industry 4.0 and at the same time it means process automatization and co-organization (Hompel & Kerner, 2015; Szymanska, Adamczak & Cyplik, 2017). Logistics 4.0 definitions seem to combine both procedural (supply chain processes and Logistics 4.0 actions) and technical view (tools and technologies that support internal processes in the supply chain).

In the logistics industry, where 3PLs manage millions of daily shipments, large amounts of data are obtained via orders, transcripts, smart low battery consuming sensors, GPS, RFID tags, weather forecasts or even social media (Neaga et al., 2015). These sensors measure real-world conditions and convert these into digital representation. If logistics is viewed from an activities point of view, the term can be described as a process which grants access to an object in the required quantity, in the required time and for the necessary clients.

An integral part of the logistics process is organization. Theory often deals with material flow, without paying so much attention to the organization. Logistics activities are important for organizing all of the organization's benefits and services. The goal of organizing the logistics process is to achieve that all parts meet customer expectations. For example, in logistics every service provides an important operation. Each service provides information for operations, which helps to organize the overall supply chain. This level of organization allows setting up a system that meets customers' needs better than the competitors, for example, the noticeable benefit for an organization might be a quicker delivery system than others (Rushton, Croucher & Baker, 2010). Such a level of logistics in a company is a significant tool for business and competition.

The European Union Commission, which internalizes the basic development in Germany Industrial 4.0 paradigms, is mainly shaped in three dimensions (European Commission, 2015): The first one is horizontal integration between value creation networks. It integration is cross-company and cross-company intelligent connections and digitalization of value creation. The other one is connection and vertical integration in manufacturing systems. Vertical integration includes manufacturing lines, factories and intelligent connection at the level of creating different accumulation and hierarchical values in the production modules and digitalization. The last one is end-to-end engineering in the product life cycle. It is a smart cross-linking at all stages of the product life cycle and digitalization (raw material procurement, manufacturing, product use and product life end). The differences between Industry 3.0 and Industry 4.0 can be seen in Tab. 2 to assess the impact on the logistics process.

Tab. 2: Comparison: Industry 3.0 and Industry 4.0 with Logistics Process. Source: Pesti & Nick, 2017.

Industry 3.0	Logistic Process	Industry 4.0
Cost and risk reduction	Procurement	Sustainable strategic allowances
Logistic total cost control	Distribution	Web-based resource optimization
Excel based calculations	Inventory management	Web-based and connected databases and real-time big data management
RFID and similar technologies	Warehousing	Intelligent vehicles and storing facilities
Route and time optimization	Transportation	Intelligent vehicles and routes
ERP systems next to each other, link is just the order itself	Order management	Connected, automatized information and material flow
Focus on automation	Their ICT background	Focus on network

The smart factory must adapt to ever-changing market demands in real time to be able to respond to customer needs in the most accurate and timely manner (Azevedo & Almeida, 2011). This requires machines to communicate with each other and to keep processes under control. However, in a smart factory it is important to optimize external processes and relationships, and to build relationships that promote information sharing with suppliers and customers (Shrouf, Ordieres & Miragliotta, 2014).

3.3 The Challenges and Opportunities of Logistics 4.0

In recent years, the industrial sector has undergone a transformation toward full digitalization and intelligentization of production processes to reduce costs, increase responsiveness and ensure high productivity (Erol et al., 2016). Logistics has become ever more important, as demands for delivery times to be more accurate and more flexible increase. With Logistics 4.0 the intralogistics industry, too, will be focusing on the topic of digitalization in future. The demand for individualized products and services is steadily increasing as a result of these developments. At this point, due to the complexity of the increasing diversity, it is critical that supply chain processes (incoming logistics and outgoing logistics) adapt to the changing environment. Because this structure has features that cannot be covered by ordinary planning and control applications (Galindo, 2016; Premm & Kirn, 2015). Logistics 4.0 technologies is the use of cyber-physical systems with feedback loops that monitor and control physical processes, affecting the computation of physical processes. This CPS uses RFID technology to send data to a computer that uses data and can collect and analyze data to identify, detect and find the item. These systems can be shared in real time by communicating with other systems or people so that processes can be used as a communication tool in a coordinated way (Galindo, 2016; Herman, Pentek & Otto, 2016).

Commercial and industrial sector considers logistics service as a factor of cost and competitiveness at the same time. There is an increasing trend for outsourcing and contract logistics to more effectively manage this factor, to benefit from the efficiency of the scale economy and synergies. In today's conditions, customers of logistics service providers are demanding individualized logistics services with flexible and broad service delivery, individual logistics processes, value-added services, transparency of costs and higher performance. These requests increase the following problems, which become apparent for many logistics service providers (Daniluk & Holtkamp, 2015):

- Generally, there is an inability to provide sufficient investment funds (or willingness to invest) for Information Technology (IT) expansion that will be needed to accomplish these transformations.
- There is sufficient IT expertise and human capacity to operate the IT infrastructure.
- It takes time to develop new IT components and integrate them into the existing IT field.

The Center for Retail Compliance reports more companies are turning to enhanced operations to reduce transport costs and inefficiencies wherever possible, which include following (Robinson, 2017): Better management systems to reduce overhead, collecting data and identifying unsustainable operations; load planning reduced redundancies; load utilization eliminated wasted, "empty" space in freight; route planning reduced demand on fleet vehicles and fuel; lane sharing and pooling defined freight consolidation, giving rise to the power of collaboration within the supply chain; grants enabled major logistics providers to upgrade fleets.

Because of these challenges IT has a critical role in logistics as a driver of change, digitalization and innovation. This can be observed in many action fields in logistics, especially in the fast-growing e-commerce sector. For this reason, one of the aims of new technologies in Logistics 4.0 is to reduce human labor in transport and to gain transportation time. In the supply chain as well as in the field of production management, the cycle is determined to meet the needs of today. Optimized planning, transport management and visibility are required throughout the process to ensure agility. In both B2B and B2C models of e-commerce, supply chains are intermingling and overlapping at some point. The different requirements of businesses on the supply chain increase the complexity of shipping operations; different modes of transport and different services need to be optimized.

Information and communication technologies provide significant contributions to logistics through data sharing. Among these contributions are planning for network design and effective use of platforms. Monitoring and visibility from the factory to the last customer is another contributor. It simplifies inventory management by lowering the required inventory, optimizing the warehouse area, integrating with warehouse equipment such as high-bay shelf systems, by working with tools that enable quick placement, selection and processing. However, mobile systems should be exploited to deal with new customers, new roots, new territories, new partnerships and heirs and mergers and acquisitions. To this end, flexible installation and configuration should be presented quickly and easily, standards and interoperability should be built (Kewill, 2013).

3.4 Logistics 4.0 Technologies in Frame of the Industry 4.0

Evaluation of the effects of Industry 4.0 technologies in the management of logistics processes, opportunities and problem areas should also be examined. For this purpose, these technologies such as cloud platforms, big data, Blockchain technology, IoT, automation and robotic, augmented reality (AR) and 3D printing are being discussed.

Cloud Platforms: Mell and Grance (2009) describes the cloud computing platform as a model that provides an appropriate and arbitrary network access to a rapidly deployable, shared, configurable resource pool (e.g., networks, servers, storage, applications and services). This platform is released with minimal management effort and/or service provider interaction. This model also increases usability. And it consists of five basic features, three service models and four distribution models (De Felice, Petrillo & Zomparelli, 2016). The basic characteristics are on demand self-service for consumers, broad network access and resource pooling to serve multiple consumers using a multi-tenant model, rapid elasticity of resources and metering capabilities for service provision. Cloud-based approaches can be customized to suit the logistics industry, transportation, shipping instructions and warehouse specific import-export regulations. Customization in this way will lead to better satisfaction of customer requirements. At the same time, it will reduce resource use and improve processes. Cloud-based logistics management and the ability to manage data in real time makes order processing, storage, inventory management, transportation and general pricing more scalable, thus enabling better cost control.

Big Data: Via analysis of large data sets, organizations are able to obtain patterns, trends or insights for their business activities (Long, 2017; Swaminathan, 2012). The use of this data leads to many benefits. The level of information transparency improves as well as process quality and performance. Predictive analytics are performed on the data sets to calculate resource demands and provide better planning. New or disruptive scenarios are being tested with simulation practices. And, cyber physical systems as autonomous vehicles or robots are improved due to data obtained from sensors, which map their direct environment (Jeske & Grüner 2013; Witkowski, 2017). The analytical data evaluation will provide logistics companies with the possibility to visualize particular logistic processes within the supply chain in order to significantly speed up the decision making.

Blockchain Technology: Because data is increasing in amount and possible functionalities, it is becoming a valuable asset for logistic companies (Zyskind, Nathan & Pentland, 2015). A new way of open data sharing is needed, and is found in "Blockchain technology". Block chain technology was first used for the cyber currency "Bitcoin". Through the new technology, two people were able to transact money (data) over the Internet, without the need of a bank, acting as an intermediary third party. The technology relies on a system in which all transactions ever made by all members within the system are documented, so that everybody knows how the sharing party has obtained his information. The data system itself therefore provides the trust. The data in the system is stored in a decentral database, which is controlled by everybody in the system, therefore power inequalities do not exist (Baker, 2015).

Internet of Things (IoT): Smit, Kreutzer, Moeller and Carlberg (2016) define IoT as "information technology (IT) systems connected to all sub-systems, processes, internal and external objects, supplier and customer networks that communicate and cooperate with each other and with humans". The same authors regard Internet Services (IOS) as a platform for providing internal and cross-organizational services driven by large data and cloud computing to participants in the value chain. Industry 4.0 is said to be compatible with both IoT and IOS. In other words, it can be said that this factor represents the applications in production and service sector as well as linking with smart technologies. It is expected that tools, machines, systems, products and means of transport will establish a permanent link between virtual and physical world, negotiating within a virtual marketplace (Galindo, 2016). Therefore, software-based systems and service platforms will play a major role in tomorrow's manufacturing, since they are the only way to bring connectivity, including data analysis, to machines and work pieces in production (Bosch, 2016). Smart shelves and pallets will come from vehicles with critical prescription in inventory management. With regard to the movement of goods, monitoring may become faster, more precise, predictable and secure. Analysis of the development of the "connected fleet" can help predict failure conditions and can automatically plan actions to improve the supply chain (Witkowski, 2017).

Automation and Robotics: Technology and automation provide technologies that make life easier in the logistics sector as well as every field. Specially developed software and intelligent applications facilitate many inland logistics activities such as planning goods movements, shipment, delivery, inventory of goods, return goods, route planning and vehicle tracking. With automation technologies, operating costs are reduced, customer satisfaction is increased, competition is increased and time and labor loss is prevented. Delivering the service to the right point at low cost and just in time reveals the importance of the automation system. Surveys show that 80 % of the existing warehouses are operated manually by human hands without a supporting automation. These warehouses are faced with the problems of efficient warehouse design, warehouse-production-transport communication, management of mobile material handling equipment and IT workers' productivity improvement and productivity increase by supporting the continuously improved IT functioning (DHL, 2016).

One of the biggest challenges facing the logistics industry today is a labor market. Industrial robots have already undertaken some tasks in manufacturing. Robots are expected to revolutionize the logistics industry. However, they are not intelligent enough to help humans in distribution centers and warehouses. Current technological advances in robotics and automation will change this. Robotics is different from mechanical or automotive, provides a flexible and reconfigurable solution and can seamlessly integrate with human operators in existing infrastructures.

Augmented Reality (AR): Augmented reality (AR) is the representation of a live direct or indirect view of a new perception environment created by computer-generated sensory inputs of data such as sound, video, graphics or GPS. This technology combines animated items with a physical, real-world environment. Inputs that will appeal to the human senses with increased reality and move their senses will be modified and enriched by the computer and presented to the perception of the new reality being created. Enrichment occurs in real time and interacts with the surrounding objects. Increased reality can interact with information that creates the user's reality environment and other issues. Artificial information and articles about the surrounding environment are compatible with the real world.

In this context, information can be obtained from many different environments. The providers of this environment can be any virtual object or content, including text, graphics, video, audio, touch feedback, GPS data and even smell. However, AR is not just a simple imaging technology. It also represents a new type of real-time natural user interface for human interaction with objects and digital devices (Carmignani & Furht, 2011). The definition of augmented reality applies to all activities, the main goal of which is to augment the real world environment with virtual information that enriches human senses and abilities. AR is able to combine virtual information with the real world (Azuma, 1997). The AR concept actually alters the real environment with virtual imagery. This technology is commonly used in real time and semantic context with environmental elements (Bimber & Reskar, 2005). AR technology has been used for supporting manual manufacturing processes, warehouse picking tasks of logistics area, maintenance scenarios for reducing complexity as well as training (Büttner et al., 2017). Besides, AR in logistics is to give consumers the chance to see other settings without ever leaving their homes and it is already being used to review products prior to purchase in greater detail than before (Robinson, 2017). With it AR has so far shown most promise for logistics in warehousing operations (DHL, 2014). These operations are estimated to account for about 20 % of all logistics costs, and the task of picking accounts for 55 % to 65 % of the total cost of warehousing operations (De Koster, Le-Duc & Roodbergen, 2007). This indicates that AR has the potential to significantly reduce cost by improving the picking process. It can also help with the training of new and temporary warehouse staff, and with warehouse planning (DHL, 2014).

> **3D Printing**: 3D printing thus enables the printing of a product, on the basis of a 3D blueprint. This way of production leads to several benefits, including a reduced time-to-market, reduced waste, improved maintenance processes and costs, reduced inventory and reduced costs of spare parts production (Thewihsen et al., 2016). Within the framework of Logistics 4.0, where machine-to-machine communication enables cyber-physical components within integrated supply chains, 3D printing could play a viable role in facilitating increased product to market processes. Linked ERP systems could autonomously demand the printing of a product, which will be quickly made available for the customer. Currently, four scenarios are thinkable in which these cyber-physical 3DP devices could be applied. In traditional manufacturing, companies tended to outsource their production to countries with low labor costs. Due to 3D printing, these organizations are able to produce near their consumer market, which is often in the high-wage countries. This "near shoring" results in the fact that long-distance shipping of goods will be reduced and local 3DP production sites will rise. In order to realize fast response times arise, logistics shifts toward "last mile" shipping, and end-of-runway services, where logistic companies install special 3DP warehouses next to important airport hubs (DHL, 2016).

4. Conclusion

Processes can be controlled automatically by means of computerized systems used in logistics and transportation, and the transport of TIRs, unmanned vessels, unmanned ships, unmanned aircraft, drone deliveries and cloud technology processes to digital platforms is rapidly transforming the sector. All these developments necessitate the development of Logistics 4.0, integration with industry and trade. Logistics 4.0 is at the center of new digital transformation. It is aiming to adapt and to develop solutions for Industry 4.0 technologies. This level of integration forces players to plan collaboration over time using a single set of numbers to run scenarios and estimate potential balances between variables such as capacity, cost, margin, delivery performance and fill rate.

Within these developments, Logistics 4.0 is going to change on all levels in the context of a fourth industrial revolution. The primary areas of this interaction are interconnection of everything with everything in real-time, context and user sensitive system using semantic technologies, distributed artificial intelligent systems of systems, and automated systems and in addition to the development and adoption of new technologies, organizational and social change. These results indicate a challenge, as Industry 4.0 technologies are important for the logistics sector. With these technologies, logistics management can gain significant benefits by shortening the cycle of logistics service processes, making information and document sharing safer, enabling real-time forecasting, increasing logistics service innovations and optimizing costs. Researches show that very soon big data applications will make significant changes in global and local transport chain management and organization. Cloud logistics, AR, IoT and robots are also expected to become a big trend in the coming five years. 3D printing and predictive analytics will enable accurate forecasts, for example, shipping of goods to a nearby distribution center after analyzing customer behavior to reduce delivery time.

The workflow can then be modeled ultimately to integrate all collaboration processes and fast and reliable information about the delivery time of the final products to the customer can be provided. These Logistics 4.0 technologies not only offers new opportunities but also presents new challenges like reduced predictability and transparency of the systems behavior through decentralized decision and action; anticipation of decisions and actions of highly automated systems; trusting all actors (human and machine) is a key requirement for distributed responsibility. These are also the important points that need to be taken into account in terms of the lack of standards and the cost of large investments needed. To meet these challenges a paradigm shift and learning process involving all stakeholders is required. However, software systems and new technologies are increasing the need for data-based and information technology-based employees.

References

Azevedo, A., & Almeida, A. (2011). Factory Templates for Digital Factories Framework. *Robotics and Computer- Integrated Manufacturing*, 27(4), 755–771.

Azuma, R. (1997). A Survey of Augmented Reality. *Teleoperators and Virtual* Environments, 6(4), 355–385.

Baker, J. (2015). *Opening Supply Chain Data on the Blockchain: Jessi Baker—ODI Summit 2015*. YouTube. https://www.youtube.com/watch?v=lWP8Y6NQIrkz. Accessed on May 14, 2017.

Bimber, O., & Raskar, R. (2005). *Spatial Augmented Reality*. Massachusetts: A K Peters.

Bosch. (2016). Connected Manufacturing, I. 4. Retrieved May 12, 2017 from https://www.bosch-connected-industry.com/en/connected-manufacturing/

Büttner, S., Mucha, H., Funk, M., Kosch, T., Aehnelt, M., Robert, S., & Röcker, C. (2017). The Design Space of Augmented and Virtual Reality Applications for Assistive Environments in Manufacturing: A Visual Approach. *PETRA '17*, Island of Rhodes, Greece. June 21–23, 2017, ACM New York, NY, USA ©2017ISBN: 978-1-4503-5227-7, doi>10.1145/3, 056540.3076193.

Carmignani, J., & Furht, B. (2011). Augmented Reality: An Overview. In Furht, B. (Ed.), *Handbook of Augmented Reality* (pp. 3–46). Heidelberg/Dortrecht/London/New York: Springer Verlag.

Coyle, J., & Bardi, E. (2003). *The Management of Business Logistics*. South – Western/Thomson Learning. Mason, OH.

Daniluk, D., & Holtkamp, B. (2015). *Cloud Computing for Logistics*, Hompel M.; Rehof, J.; Wolf, O. (eds.), Lecture Notes in Logistics, DOI 10.1007/978-3-319-13404-8_2, Springer International Publishing: Switzerland.

De Felice, F., Petrillo, A., & Zomparelli, F. (2016). Design and Control of Logistic Process in an Italian Company: Opportunities and Challenges Based on Industry 4.0 Principles, *XXI Summer School "Francesco Turco"—Industrial Systems Engineering*, http://www.summerschool-aidi.it/edition-2016/cms/extra/papers/ final_39.pdf

De Koster, R., Le-Duc, T., & Roodbergen, K. J. (2007). Design and Control of Warehouse Order Picking: A Literature Review. *European Journal of Operational Research*, 182(2), 481–501.

DHL. (2014). Augmented Reality in Logistics, Changing the Way We See Logistics—A DHL Perspective. Retrieved from http://www.dhl.com/content/dam/downloads/g0/about_us/logistics_insights/csi_augmented_reality_report_290414.pdf. Accessed on June 15, 2017.

DHL (2016). Robotics in Logistics. Retrieved from http://www.dhl.com/content/dam/downloads/g0/about_us/logistics_ insights/dhl_trendreport_robotics.pdf. Accessed on June 15, 2017.

Erol, S., Jager, A., Hold, P., Karel, O., & Sihn, W. (2016). Tangible Industry 4.0: A Scenario-Based Approach to Llearning for the future of Production. The 6th Conference on Learning Factories. Gjøvik, Norway Procedia CIRP 00 (2016) 000–000, Published by Elsevier B.V.

European Commission. (2015). *Factories of the Future*. Retrieved from http://ec.europa.eu/research/industrial_technologies/factories-of-the-future_en.html. Accessed on January 17, 2016.

Galindo, L. D. (2016). *The Challenges of Logistics 4.0 for the Supply Chain Management and the Information Technology*. Master Thesis. Norwegian University of Science and Technology, Department of Production and Quality Engineering Master Science in Mechanical Engineering.

Hermann, M., Pentek, T., & Otto, B. (2016). Design Principles for Industrie 4.0 Scenarios, Institute of Electrical and Electronics Engineers -IEEE-; IEEE Computer Society: 49th Hawaii International Conference on System Sciences, HICSS 2016. Proceedings: 5–8 January 2016, Kauai, Hawaii Los Alamitos, Calif.: IEEE Computer Society Conference Publishing Services (CPS), 2016, ISBN:978-0-7695-5670-3, pp. 3928–3937.

Hompel ten, M., & Kerner, S. (2015). Logistik 4.0 Die Vision vom Internet der autonomen Dinge, *Informatik Spektrum*, 38(3), 176–182.

Institut der deutschen Wirtschaft Consult (IW Consult)/FIR. (2015). Industrie 4.0-Readiness, Commissioned by the IMPULS-Stiftung of the VDMA, Frankfurt, http://www.impuls-stiftung.de/studien;jsessionid=D371AF8942274C06B67E99721D269ED4. Accessed on October 10, 2015.

Jeschke, S. (2016). Quo Vadis Logistik 4.0, Retrieved from www.ima-zlw-ifu.rwth-aachen.de/fileadmin/user_upload/INSTITUTSCLUSTER/Publikation_Medien/ Vortraege /download/ /Quo_vadis_Logistik4.017March2016.pdf. Accessed on February 18, 2017.

Jeske, M., & Grüner, M. (2013). *Big Data in Logistics—A DHL Perspective on How to Move beyond the Hype*, DHL Customer Solutions and Innovation, December 2013. http://www.delivering-tomorrow.com/wp-content/uploads/2014/02/CSI_Studie_ BIG_DATA_FINAL-ONLINE.pdf. Accessed on April 4, 2017.

Kewill Ltd. (2013). *White Paper: Logistics in 2020.* Retrieved from http://info. kewill.com/. Accessed on June 16, 2017.

Kückelhaus, M., & Terhoeven, M. (2013). *Key Logistics Trends in Life Sciences 2020+. A DHL Perspective on How to Prepare for Future Growth.* Powered by Solutions & Innovation: Trend Research. http://www.dhl.com/content/ dam/downloads/g0/about_us/innovation/whitepaper.pdf. Accessed on April 15, 2016.

Langley, J. & Capgemini (2016) 2017 third-party logistics study – The state of logistics outsourcing results and findings of the 21st annual study, http:// www.3plstudy.com/media/downloads/2016/09/2017-report.pdf, Accessed on 21 June, 2017.

Mell, P. & Grance, T. (2009).The NIST Definition of Cloud Computing. https://nvl pubs.nist.gov/nistpubs/legacy/sp/nistspecialpublication800-145.pdf National Institute of Standards and Technology, U.S. Department of Commerce, Special Publication: 800–145, Computer Security Division Information Technology Laboratory National Institute of Standards and Technology Gaithersburg, MD 20899–8930 September 2011

Neaga I., Liu S., Xu L., Chen H., Hao Y. (2015). Cloud enabled big data business platform for logistics services: A research and development agenda. In: Delibašić, B., Hernández, J.E., Papathanasiou, J., Dargam, F., Zaraté, P., Ribeiro, R., Liu, S., Linden, I. (Eds.), Decision Support Systems V – Big Data Analytics for Decision Making, Lecture Notes in Business Information Processing, First International Conference, ICDSST 2015, Belgrade, Serbia, May 27–29, 2015, Proceedings, Springer.

Pesti, I., & Nick, G. A. (2017). *Industry 4.0 from the Aspect of Logistics Innovation,* Society for Regional Science and Policy, Region Direct, Depend, http://ersa.sk/ Zbornik/files/Pesti_Nick.pdf. Accessed on May 21, 2017.

Premm, M. & Kirn, S. (2015). A Multiagent System Perspective on Industry 4.0 Supply Networks. In J. P.Müller, W.Ketter, G.Kaminka, G.Wargner, &N.Bulling (Eds.), Multiagent Systems Technologies MATES 2015 Revised Selected Papers of the 13th German Conference on Multiagent System Technologies – Volume 9433 Pages 101–118, Cottbus, Germany-September 28-30, 2015, Springer-Verlag New York, Inc. New York, NY, USA ©2015, ISBN: 978-3-319-27342-6 doi>10.1007/978-3-319-27343-3_6

Robinson, A. (2017). *Internet of Things.* Retrieved from http://cerasis.com/ 2017/01/23/augmented-reality-in-logistics/. Accessed on June 15, 2017.

Rushton, A., Croucher, P., & Baker, P. (2010). The Handbook of Logistics and Distribution Management Handbook of Logistics and Distribution Management, 4th edition, The Chartered Institute of Logistics and Transport, New Delhi, India.

Shrouf, F., Ordieres, J., & Miragliotta, G. (2014). Smart Factories in Industry 4.0: A Review of the Concept and of Energy Management Approached in Production Based on the Internet of Things Paradigm. In International Conference on Industrial Engineering and Engineering Management, 2014 IEEE International Conference on, 9–12 December, 697–701, Malaysia, http://dx.doi.org/10.1109/IEEM.2014.7058 728

Smit, J., Kreutzer, S., Moeller, C., & Carlberg, M. (2016). Industry 4.0, Study for the ITRE Committee, *Policy Department A: Economic and Scientific Policy*, European Parliament, Brussels.

Szymanska, O., Adamczak, M., & Cyplik, P. (2017). Logistics 4.0- A New Paradigm or Set of Known Solutions, *Research in Logistics &Production*, 7(4), 299–310. DOI: 10.21008/j.2083-4950.2017.7.4.2

Swaminathan, S. (2012). The Effects of Big Data on the Logistics Industry. Retrieved 10 April 2016 from http://www.oracle.com/us/corporate/profit/archives/opinion/021512-sswaminathan-1523937.html.

Techconsult. (2015). Business Performance Index BPI, BPI Fertigung, https://www.techconsult.de/business-performance-index/studie-bpi-mittelstand-fertigung. Accessed on 5 May 2017.

Thewihsen, F., Karevska, S., Czok, A., Pateman-Jones, C., & Krauss, D. (2016). If 3D Printing Has Changed the Industries of Tomorrow, How Can Your Organization Get Ready Today? *?, https://www.ey.com/Publication/vwLUAssets/ey-3d-printing-report/$FILE/ey-3d-printing-report.pdf 2016 EYGM Limited. EYG no. 02810-163GBL.*

Timm J., & Lorig, F. (2015). Logistics 4.0—A Challenge for Simulation, Yilmaz, L., Chan, W. K. V., Moon, I., Roeder, T. M. K., Macal, C., & Rossetti, D. (Eds.), *Proceedings of the 2015 Winter Simulation Conference*, IEEE Press Piscataway, New Jersey, pp. 3118–3119.

Witkowski, K. (2017). Internet of Things, Big Data, Industry 4.0—Innovative Solutions in Logistics and Supply Chains Management, *Procedia Engineering*, 182, 763–769. www.sciencedirect.com

Zyskind, G., Nathan, O., & Pentland, A. (2015). Decentralizing Privacy: Using Blockchain to Protect Personal Data. In 2015 IEEE Security and Privacy Workshops, May 18–20 2015, 180–84, Son Jose, CA.

Index